Springer Biographies

The books published in the Springer Biographies tell of the life and work of scholars, innovators, and pioneers in all fields of learning and throughout the ages. Prominent scientists and philosophers will feature, but so too will lesser known personalities whose significant contributions deserve greater recognition and whose remarkable life stories will stir and motivate readers. Authored by historians and other academic writers, the volumes describe and analyse the main achievements of their subjects in manner accessible to nonspecialists, interweaving these with salient aspects of the protagonists' personal lives. Autobiographies and memoirs also fall into the scope of the series.

More information about this series at
http://www.springer.com/series/13617

William Dean Brewer ·
Alfredo Tiomno Tolmasquim

Jayme Tiomno

A Life for Science, a Life for Brazil

 Springer

William Dean Brewer
Fachbereich Physik, FU Berlin
Berlin, Germany

Alfredo Tiomno Tolmasquim
Museo de Amanhã
Rio de Janeiro, Brazil

ISSN 2365-0613 ISSN 2365-0621 (electronic)
Springer Biographies
ISBN 978-3-030-41013-1 ISBN 978-3-030-41011-7 (eBook)
https://doi.org/10.1007/978-3-030-41011-7

This Springer imprint is published by the registered company Springer Nature Switzerland AG
The registered company address is: Gewerbestrasse 11, 6330 Cham, Switzerland

Preface

This book tells the story of the life and work of Jayme Tiomno, one of the most eminent Brazilian physicists of the twentieth century. But in addition, it is the story of a long and continuing effort to establish the institutions and conditions necessary to foster excellent scientific research and education in Brazil, in spite of the vicissitudes of politics and economics, and of long-established habits and prejudices.

Jayme Tiomno—together with his physicist wife, Elisa Frota-Pessôa—was an important contributor to that effort during much of his life, and he carried it on with patience and persistence, with modesty and dignity and respect for others. He remained devoted to the country of his birth, even after his forced exile in the 1970s, during the military regime; and on returning to Brazil, he again took up the banner of science there. His was truly a life for science—in his case, theoretical physics—but also, and equally, a life for Brazil.

The title of this book reflects the two great passions of Jayme Tiomno: To perform relevant and productive research in physics, and to guide and train younger physicists so that they could carry forward the work of research and teaching—both goals for the advancement of science and society in his native country. Many difficulties were laid in Tiomno's path by authoritarian political regimes, who had no respect for his ideals and accomplishments; and he had numerous opportunities to leave Brazil and establish himself in some 'easier' environment abroad, where he might have gained more personal acclaim and success—but he remained dedicated to Brazil and to his ideals, which he shared with Elisa, his wife of many years. His long life was indeed lived for science and for his country.

There are many accounts available in Brazil of Jayme Tiomno's life and work, mostly in the form of interviews and articles, but they are in Portuguese and not readily accessible to an international audience, and they do not tell the complete story of his life and career, nor his science. We hope to fill this gap with the present book, part of Springer's Scientific Biography series.

In an obituary published shortly after Jayme Tiomno's death in a popular Brazilian science magazine,[1] a 'Physicists' Triad' was proposed, analogous to the famous 'Tiomno triangle' which Tiomno drew to clarify the concept of a Universal Fermi Interaction (Chap. 6). The vertices of this triad are the three distinguished twentieth-century Brazilian physicists *César Lattes, José Leite Lopes*, and *Jayme Tiomno*, of whom Tiomno was the last to pass away, in early 2011. The article closes with the remark that it is particularly sad that many younger Brazilian physicists today no longer know who these predecessors were and what they accomplished. We hope that this book will also make at least a small contribution toward remedying that situation.

At the end of the book, we have included an Appendix giving a brief history of Brazil, as a guide to various references made in the text. A second Appendix presents a chronology of Jayme Tiomno's life and career, and gives a list of his scientific works and publications.

Backnotes with references to sources and the scientific literature, some offering supplementary information, are indicated by sequential numbers in square brackets [..], usually as raised indices, sometimes on line in the text as source references. The numbered backnotes are given in order at the end of each chapter, to facilitate finding specific references. Their numbering begins from '1' in each chapter. *Source archives* are indicated by abbreviations, e.g. [JT] refers to the archive of Tiomno's papers at the *Museum of Astronomy and Related Sciences* (MAST) in Rio de Janeiro. A summary of these abbreviations and their sources can be found at the beginning of the book. References to items in the Literature List are indicated by the authors' (or the editors') name(s) and the year of publication (in parentheses). In addition, an Index is provided at the end of the book.

The scientific *articles and contributions* of Jayme Tiomno are collected in the second part of Appendix B, and denoted as [JT5], [JT12], etc. When

---

[1]Cássio Leite Vieira, *'Dia triste para a História da Ciência'*, in: *Ciência Hoje*, January 2011. This is an obituary for Jayme Tiomno, in a well-known science magazine in Brazil, published shortly after his death in early 2011. The correspondence between César Lattes and Jayme Tiomno mentioned in it is to be found in the *César Lattes Archives* at Siarq, University of Campinas. Vieira speaks of a 'Lattes-Leite-Tiomno triad', analogous to the 'Tiomno triangle', whose vertices are those three distinguished twentieth-century Brazilian physicists and founders of the CBPF. See Vieira (2011).

referred to in the text or in the source references, they are cited as '[JT*N*]' without a date, so as to avoid confusion with the articles and books from the Literature List as cited in the text or in the backnotes.

Longer *quotations* from original sources are inset; and our supplementary remarks within them are set in italics and enclosed in square brackets.

We would also like to clarify our use of the names (*Jayme*) *Tiomno* and *Elisa* (*Frota-Pessôa*). In contrast to Tiomno and most other scientists, who are generally referred to by their last names in formal writing, Elisa Frota-Pessôa was known in Brazilian academic circles simply as *Elisa*. This might be due to the fact that her first husband, Oswaldo Frota-Pessôa, was also a well-known scientist and was known publicly as *Frota-Pessôa*; or it might be in recognition of Elisa's salient characteristic of being very communicative and willing to approach people in a familiar way, breaking the usual barriers of formality; or, finally, to her status as a pioneering Woman in Science, emphasized by using her (clearly feminine) first name. In any case, we have opted for the convention usual in Brazil during her lifetime, and which she used herself, of referring to Elisa Frota-Pessôa simply by her first name, Elisa (to close friends, she was known as 'Lili').

Preparing this book has been an interesting journey through time and space, bringing both of us back to some familiar places. We have attempted to explain the science at a level that will be accessible to most interested readers, including those who are not themselves scientists. Chapter 4 introduces and summarizes many of the scientific ideas and terminology that appear in later chapters. We hope that reading the book will be as absorbing and interesting to a curious public as writing it has been for us.

We wish to thank all of those who have aided in the writing and production of this book, in various ways; in particular *Antônio Augusto Passos Videira, Cassio Leite Vieira,* and *Suely Braga.* This includes also those persons who consented to be interviewed and provided some materials: *Nicim Zagury, Mario Novello* and *Henrique Lins de Barros,* as well as those responsible for the archives at the *Museu de Astronomia e Ciências Afins,* the CPDOC/*Fundação Getúlio Vargas,* Casa de Oswaldo Cruz/*Fiocruz,* Public Archive of the State of Rio de Janeiro (APERJ), Library of the American Philosophical Society and at the *Centro Brasileiro de Pesquisas Físicas,* the Central Archive of *AC/Siarq* (archives at Unicamp), the library of the Institute of Physics at UFRJ, and the *Mario Peixoto* Archive.

We thank most especially the members of Jayme Tiomno's family: his sister, *Silvia Tiomno Tolmasquim,* and his stepchildren, *Sonia Frota-Pessôa* and *Roberto Frota-Pessôa,* for providing information, interviews and photos, for transcribing correspondence, and for critical readings which have made this book possible. We also thank the Springer-Verlag, in particular *Angela Lahee*, for supporting and encouraging the idea and the realization of the book.

Berlin, Germany   William Dean Brewer
Rio de Janeiro, Brazil   Alfredo Tiomno Tolmasquim
September 2019

Contents

Abbreviations

APhS	American Philosophical Society
APP	Archive of Political Police (DOPS—Department of Political and Social Order), at the *Archivo Publico do Estado do Rio de Janeiro* (APERJ)
APS	Archives of the American Physical Society
CR	Archive of Joaquim da Costa-Ribeiro's papers and correspondence, at MAST, Rio de Janeiro
EFP	Archive of Elisa Frota-Pessôa, at MAST, Rio de Janeiro; other items are in the personal records of her family *Frota-Pessôa, Sonia (2018/19)* (the latter have however mostly been transferred to [EFP])
GB	Archive of Guido Beck, at the CBPF, Rio de Janeiro
HM	Archive of Haity Moussatché, COC/Fiocruz, Rio de Janeiro
IF/USP	Archive of the *Instituto de Física, Universidade de São Paulo* (USP)
JLL	Archive of José Leite Lopes' papers and correspondence, at the *Centro de Pesquisa e Documentação Contemporânea* (CPDOC) of the *Fundação Getulio Vargas* (FGV), Rio de Janeiro
JT	Archive of Jayme Tiomno's papers and correspondence, at MAST, Rio de Janeiro
JWA	John Wheeler Archive, Library of the American Philosophical Society (APhS)
STT	Private archive of the Tiomno and Aizen families, with Silvia Tiomno Tolmasquim

1

Prologue

In order to do justice to Jayme Tiomno's life story, we need to take a step back and examine the history and situation of his country, at the time of his birth and beforehand, as well as throughout his own lifetime (see *Appendix A*). His personal and scientific development and his life's decisions are inseparably bound up with the modern history of Brazil, and with his goals for science there. A chronological summary of the important events in Tiomno's life and career, as well as a list of his scientific publications and contributions, are collected in *Appendix B* at the end of this book. That list spans the 7 decades of Tiomno's scientific productivity and includes 124 entries.

Tiomno was born in Rio de Janeiro, the son of a family of Russian Jewish immigrants recently arrived in Brazil. He grew up in the mainly rural state of *Minas Gerais*, in the small towns of *São Sebastião do Paraíso* and *Muzambinho*, returning to Rio at age 14 to complete his high school and complementary education there, leaving a distinguished record as a talented pupil. He then entered the *Universidade do Brasil*, initially intending to study medicine. After a year, he discovered his strong interest in physics, and later changed his major. Following graduation and military service, having started some research in experimental physics, he moved to the *Universidade de São Paulo* (USP), at that time the foremost (in fact, the *only*) research university in Brazil, and began working on gravitational theory and particle physics with *Mario Schenberg*'s guidance. He then went on a fellowship to Princeton University in the USA, completing his Masters and Ph.D. there under the supervision of the renowned physicists *John A. Wheeler* and *Eugene P. Wigner*, and collaborating with other contemporaries including *David Bohm* and *Chen*

© Springer Nature Switzerland AG 2020
W. D. Brewer and A. T. Tolmasquim, *Jayme Tiomno*, Springer Biographies,
https://doi.org/10.1007/978-3-030-41011-7_1

Ning Yang. With the latter, he introduced the term 'Universal Fermi Interaction' (UFI). He made important contributions to the early understanding of muon physics and weak interactions, which later played a role in the construction of the Standard Model of Particle Physics, currently still the best available (if not the ultimate) description of the sub-microscopic world.

While at Princeton, he participated in the conception and planning of the *Centro Brasíleiro de Pesquisas Físicas* (CBPF), the Brazilian Center for Physics Research in Rio de Janeiro, which was intended to provide an environment for excellent research in physics in Brazil that was not available elsewhere at the time, in particular since the Brazilian universities, with the exception of USP, were not research-oriented and lacked the necessary facilities and organization. The CBPF rapidly advanced to become a highly successful research and teaching institution of international renown in the decade of the 1950s. Towards the end of that decade, it began to suffer problems due to limited funding and support, and undertook some (however unsuccessful) efforts to join a university as an adjunct research institution.

In the 1960s, an ambitious plan was launched to create a new university in the recently-constructed capital city of *Brasília*, the *Universidade de Brasília* (UnB). It would, like the new capital itself, be directed towards the future, and was to have a new plan of organization and curricula, conceived by the eminent Brazilian educator *Anísio Teixeira*, who had also founded a university (the UDF) in 1935 in Rio de Janeiro—it had unfortunately been short-lived. The Brazilian physicist *Roberto A. Salmeron* returned from his work at the European Nuclear Research Center (CERN) in Geneva to coordinate setting up the Central Institute of Sciences at UnB, and he was joined by Jayme Tiomno and his physicist wife, *Elisa Frota-Pessôa*, who were responsible for establishing the new Physics Institute.

Interference by the military regime, which had taken power in a *coup d'état* in 1964, led to the mass resignation of most of the faculty and the collapse of the new university. Tiomno and Elisa went back to São Paulo, where he entered—and won—a competition for a professorship in Advanced Physics. He established a core group there, which later became the nucleus of the Department of Mathematical Physics at USP.

But Jayme and Elisa were put on the blacklist of the dictatorial government, and were banned in 1969 from employment as teachers or researchers in Brazil. They went to Princeton in 1971 at Wheeler's (and Freeman Dyson's) invitations, spending one and a half years there. After optimistically returning to Brazil in mid-1972, Tiomno was able—with the aid of a special dispensation from Pope Paul VI—to obtain a professorship at the Catholic University (PUC) in Rio de Janeiro. He spent 8 years there, not without suffering the professional and psychological effects of his persecution, finally returning in 1980 to the CBPF that he had helped to found. He continued working there, both in research and in support of education and international scientific cooperations, even after his retirement as *researcher emeritus* in 1992, and was active until his death in 2011, receiving many honors late in life. Elisa survived him by nearly 8 years. With her passing, nearly the last of the 'Founders' generation' of Brazilian physicists has departed.

Yet their heritage survives, in the form of their many students, who have themselves begun numerous research activities and schools of physics, so that it is very much alive in Brazil today. And the lives and careers of Jayme Tiomno and Elisa Frota-Pessôa stand as examples of how difficulties can be overcome, and even when the problems and hurdles seem insuperable and overwhelming, one can still plant the seeds of knowledge and enlightenment, which often blossom after some time in unexpected ways. As Tiomno himself said, rather late in his life, "*We had a conviction that we were going to get there*".

2

Growing Up

Rio de Janeiro

The month of April, 1920 is passing by in Rio de Janeiro, the Brazilian capital. Autumn is in full swing, but the trees remain green, as they often do during this season, and even in the middle of winter. Nearly every day, the sun shines down from a clear, azure sky, and the heat is relieved only by occasional rainy days.

In a modest single house, situated in a small street, the *Rua Marquês de Pombal*, near the square *Praça Onze*—that street, and the square itself (Fig. 2.1), no longer exist—the Jewish immigrant Mauricio Tiomno is living with his family, recently arrived from *Salvador*, in the State of *Bahia* in the northeast of Brazil.[1] Besides Mauricio himself, the family includes his wife Annita, their two-year-old son Benjamin, and Annita's father, the widower Salomão. Although they are not at all well off, they frequently offer shelter in their house to other recent immigrants to Brazil, for the most part compatriots. Their guests are lodged in a room below the entrance stairs to the house; they sleep on mattresses there, and are given food and clothing.

The region is located near the old city center, where there are many public buildings, various shops and services, the central railway station, and some factories (Fig. 2.2). Hundreds of poor families live there, mainly freed slaves and recent immigrants—many of whom are Jewish, from Eastern Europe, but also including Portuguese, Italian, Spanish and Arabian immigrants. It was on the *Praça Onze* that, under the influence of former slaves, the samba was born, and the famous Rio Carnival celebrations evolved. A Jewish community

© Springer Nature Switzerland AG 2020
W. D. Brewer and A. T. Tolmasquim, *Jayme Tiomno*, Springer Biographies,
https://doi.org/10.1007/978-3-030-41011-7_2

Fig. 2.1 Praça Onze, early twentieth century. Courtesy of the *Archivo da Cidade*, Rio de Janeiro (public domain)

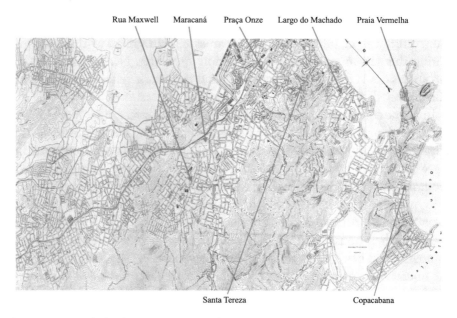

Fig. 2.2 Map of Rio de Janeiro, around 1928, showing locations where the Tiomno family lived or worked. Courtesy of the University of Texas Libraries, Austin, Perry-Castañeda Library Map Collection (public domain).

had begun to grow, as evidenced by synagogues, libraries, newspapers and social and beneficial institutions.

On the 16th of April, 1920, Annita, who is pregnant, enters into labor. In contrast to the usual practice of the day, she is accompanied not by a midwife, but by a medical doctor. The doctor was apprehensive since, in a prenatal examination, he had been able to hear only a single heartbeat, although he could distinctly feel the presence of two heads. Mauricio was kept informed about the potential problem, but Annita was spared from worrying about it. Finally, the news that relieves Mauricio's fears arrives: A pair of twins has just been delivered. On the positive side, both are healthy; on the other hand, the layette is far from sufficient for two babies. As soon as Annita has recuperated from the birth, she sits down in front of her ancient pedal-driven sewing machine, bought second hand, and begins to hem a collection of diapers.

Once the first practical problems have been overcome, the newborns receive their names: the girl is Mariam (later called *Mira*), the boy Jayme— *Jayme Tiomno*.

From Eastern Europe to the Tropics of South America

The Tiomno family had its immediate origins in the small city of *Sokorone*,[2] located since 1991 in the Ukrainian province of *Chernivtsi Oblast*, near the border between the Ukraine and Moldova (Fig. 2.3). The province today includes two main regions, the *Bukovina* in the west, with its principal city Chernovitz (Ukrainian *Chernivtsi*), famous for its mixture of Slavic, German and Jewish cultures. It was part of the Austro-Hungarian Empire throughout the nineteenth century.—And to the east lies the northern tip of *Bessarabia*, previously belonging to the Russian Empire. Sokorone is in the easternmost corner of this region, just above the border with Moldova. (The major portion of the original province of Bessarabia lies across the border to the south, in Moldova.)

The clan had lived in the area since at least the middle of the nineteenth century, a time when Sokorone belonged to the Russian province of Bessarabia. It had previously been part of Moldova and the Ottoman Empire. In the course of a century filled with much pain and suffering, the region again passed to Moldova, then to Romania, to the Soviet Union, and is now, finally, a part of the smallest province of the Ukraine.

Mauricio, whose name was originally Meir, lived with his parents Ersch Tiomne (the family name was changed to Tiomno after his arrival in Brazil[3])

Sokorone   Birzula   Dnipropetrovsk

Fig. 2.3 Modern map of the southern Ukraine, showing the locations of places mentioned in the text. *Source* UN map of Ukraine, 2005, cropped (public domain)

and Eva, and his 5 siblings—3 sisters and 2 brothers. They lived in the *shtetl*—a small, poor Jewish settlement separated from the other parts of the town by custom and decree. The members of the Tiomne family were religious, and, although they knew Russian and were educated using that language, they spoke Yiddish within the family; it was their mother tongue.

Eva, Mauricio's mother, died tragically: She was standing in the entrance door of the house, protecting her children during a *pogrom*, when a soldier ran his sword through her abdomen, killing her instantly. Later, the widower Ersch remarried. His new wife, Ruchl, bore him no children, but she helped care for the children from his first marriage. Ersch earned his living as a grain merchant, arranging sales of cereals and flour at the town market. His two older sons, Mauricio and Francisco, leased a plot of land and planted tobacco with the help of hired hands; they sold the leaves still green, uncured.

In 1910, Mauricio, still single and barely 20 years old, decided to cross the Atlantic in search of a better and more secure life. He arrived in Rio de Janeiro in 1911, and thus avoided the devastation and upheavals wrought in his previous homeland by the First World War a few years later.

Brazil was a former Portuguese colony.[4] The Brazilian economy had in previous centuries been based mainly on agricultural products—monocultures of sugar cane, later also coffee and rubber latex—and on mining, mostly carried out with the labor of slaves imported from Africa between the 16th and the nineteenth centuries. Sugar from sugar cane, planted near the coasts, was the dominant export product in earlier times. In the late seventeenth century, gold was discovered in the province of *Minas Gerais*, in the south-central region of the country, causing a rapid development and population growth there. In addition, in the early eighteenth century, coffee was introduced to the province of *São Paulo*, and it became the principal export product of the colony (remaining today one of the major economic factors). Somewhat later, the extraction of rubber latex in the Amazon region led to considerable economic and population growth in that remote area, which however by 1920 was winding down due to competition from rubber plantations in Southeast Asia.

An unusual episode in the history of empires occurred in 1807: Napoleon invaded Portugal, and the entire Portuguese court, led by King João VI, fled to the colony with the aid of the British navy. His entourage, including around 15,000 persons carried by 17 ships, arrived in Brazil in 1808. For the first time, a colony became the seat of the royal court of its mother country. Various measures were taken to support the establishment of the court in Rio de Janeiro, including the building of a library to house the 60,000 volumes brought from Lisbon, as well as the founding of the Bank of Brazil, of the Royal Printing Establishment, of the Royal Military Academy, and of medical schools, among other things. In addition, industrial facilities were set up—they had previously not been permitted—such as iron works and gunpowder factories. The ports were opened to direct commerce with other countries, i.e. those held to be 'friendly nations', in particular Great Britain. The Brazilian state was promoted from the status of a colony to that of a kingdom, becoming part of the United Kingdom of Portugal and the Algarve.

In 1821, following the restoration of the European monarchies by the Congress of Vienna and the final defeat of Napoleon in 1816, the court began its return to Portugal. Soon after, in the following year, Pedro I, the son of João VI, who had remained in Brazil as Viceregent, declared independence from Portugal, establishing the Empire of Brazil. After a brief and unsuccessful attempt at resistance on the part of forces loyal to Portugal, independence was achieved.

Little by little, an intellectual class was emerging, and some scientific institutions were established, for example the National Observatory, the Imperial Botanical Garden, and the National Museum. There were also technical schools, medical schools and law schools, specialized institutions intended mainly to provide services and to train new professionals.

Slavery was formally abolished in Brazil only in 1888, under pressure from other countries and from the local intellectual class. Brazil was one of the last countries to abolish the slavery of imported Africans. During roughly 400 years, between 4 and 4.8 million enslaved Africans were brought to Brazil. After the abolition of slavery, there were few efforts to make it possible for the freed slaves to find work and lead independent lives. They gravitated to the cities, without employment or places to live. Already in the following year, a group of military officers, supported by major rural landowners who had lost their slaves without any compensation, and by an intellectual elite who were striving for a new vision of progress, forced the abdication of the emperor Dom Pedro II (the son of Pedro I), transforming Brazil into a republic. In order to replace the workforce which had been provided by the slaves in the fields, and to provide workers for the developing industries, as well as with the intention of increasing the proportion of 'whites' in the overall population, Brazil opened its doors to immigration. At the end of the 19th and the beginning of the twentieth centuries, several million immigrants arrived, coming in the main from Portugal, Spain, Italy, Germany and Japan, but also including Syrian-Lebanese and Eastern Europeans.

Mauricio Tiomno was one of those immigrants! On arriving in 1911, he encountered a country which was preeminently agricultural, its economy based on the production and export of coffee, with the beginnings of industry, some public institutions, higher education distributed over individual faculties or schools, and some few research institutes dedicated to providing services. The population was 24 million, the majority illiterate, including immigrants from various parts of the globe, former slaves and their descendants.

Rio de Janeiro (often abbreviated simply as 'Rio'), then the capital of the country, had 920,000 inhabitants in 1911 and had recently experienced a period of great renovation, based on the example of Paris: A hill in the center of the city, the *Morro do Castelo*, was excavated—and numerous buildings had been demolished, making space for grand avenues. A large theater was erected, the *Teatro Municipal*, inspired by the Paris Opera; and even pigeons were imported to give the city a more European air. At some distance from the remodeled center, the first *favelas* had begun to appear, built by people who had been forced out of their houses and had constructed huts illegally on public lands.

Mauricio Tiomno began selling pots and pans from door to door, with the aid of some compatriots who had arrived in the country earlier. However, a short time later, he moved to *Salvador*, the capital of the state of Bahia, in the Brazilian northeast, considerably smaller than Rio de Janeiro. He managed to set up a sawmill—and it was there to which various relatives found their way, emigrating from Bessarabia, among them his two brothers, Francisco and Arnaldo.

Mauricio returned to Rio in 1915 for the wedding of a cousin, and there he met Annita, the sister of the bride. She was also of Jewish origin, having emigrated from Russia with her parents, Salomão and Mariam Aizen, and her 4 siblings (3 girls and a boy), arriving in Brazil in 1911, the same year as Mauricio, when she was 14 years old.

In Russia, Salomão, Mariam and their children had lived in a large city—*Ekaterinoslav*, the capital of the province of the same name (*Ekaterinoslav Oblast*). Today, the city is called *Dnipro*, and is the capital of the Ukrainian province of *Dnipropetrovsk Oblast.* The municipality, well provided with transportation by rail and river barges, was established on the banks of the Dnieper river, which flows into the Black Sea to the northwest of the Crimea. It was economically important for its steel works and other industries, as well as for service providers such as flour milling, treatment of agricultural products and commerce in wood products.

In spite of many regulations at the time which limited the activities and residence of Jewish citizens in Ekaterinoslav, Annita's father had a special permission to live and work there as a grain merchant—an occupation that was considered to be important for the provincial economy. Their life in the urban environment had distanced the family from religious practices, and their mother tongue was Russian, and not Yiddish. Nevertheless, given their situation, only a few members of the clan had been allowed to learn to read and write; typically, only one of the children from each Jewish family was permitted to go to school. Most of the children had to help support the family by rolling cigarettes or performing other simple tasks.

Having decided to go in search of new opportunities for a better life, the family emigrated from Russia to South America, first moving to Buenos Aires, the capital of Argentina, where two of Salomão's brothers already lived. However, a short time later they moved again to São Paulo and finally to Rio de Janeiro, searching for a better location to engage in commerce.

Annita Aizen and Maurício Tiomno were married in Rio de Janeiro in 1916, and returned to Salvador, where Mauricio was living. Shortly thereafter, Salomão and Mariam, Annita's parents, also went to live in Salvador.

In November 1917, the young couple experienced the joy of the birth of their first son, Benjamin. His birth was overshadowed shortly afterwards by the sad loss of Mariam, a victim of the Spanish Influenza, which caused a worldwide epidemic that year and had arrived in South America. The widower Salomão went to live with Annita's family. In 1919, they all moved back to Rio de Janeiro, hoping to find better conditions for their work there, as well as being nearer to Annita's siblings.

Initially neutral in the 1st World War, Brazil declared war on the Central Powers, the German and Austro-Hungarian Empires, only in April of 1917, following attacks on two Brazilian ships near the French coast. However, its participation in the war was limited to sending medicines and medical teams to aid the Allied forces, and helping to patrol the Atlantic ocean. International commerce, on which Brazil was very dependent, had been drastically curtailed by the war, reducing coffee exports on the one hand, but giving a strong impulse for building up local industries on the other. The war was not felt very strongly by the general population, excepting the immigrants, who were unable to maintain contact to their families who had remained in Europe, and could no longer travel themselves.

Travels in the Interior of Brazil

With the birth of the twins Mariam and Jayme Tiomno in April 1920, the modest little house on *Praça Onze*, already occupied by Mauricio, Annita, their son Benjamin and father-in-law Salomão, became too small to shelter the family with three children (Fig. 2.4). They moved to a house in the *Rua Maxwell*, at that time home mainly to middle-class citizens. But their financial situation continued to be difficult, leading the family to move again to the south of the state of *Minas Gerais*.

Minas Gerais (for short, simply 'Minas') had enjoyed a period of rapid growth and prosperity throughout the eighteenth century, after the discovery of gold there. Following the depletion of the gold veins and the decline of the mining industry, the economy of the region shifted gradually to cattle raising and agriculture, and near the end of the nineteenth century, the southern part of the state had become a region dominated by coffee production.

Since Annita was once again expecting a child, they waited for the birth of the new baby, a girl, which took place four days before the beginning of the year 1924. The fourth co-heir of the couple was given the name *Feiga Rebeca* (later, her nickname was *Riva*).

Mauricio and Salomão left for Minas before the others, in order to set up their commercial enterprise and prepare accommodations for the family. A few months later, Annita traveled alone with the four children—Benjamin,

Fig. 2.4 The Tiomno family, around 1922. Grandfather Salomão is standing behind, and seated in front are Annita and Mauricio. The twins are standing between their parents (Jayme is on the left), and Benjamin is at right front. *Source* Private collection, STT

7 years old, Jayme and Mariam, 4, and the baby Feiga Rebeca—by steam train from Rio to Minas Gerais.

Annita indeed had need of considerable courage and resolve to make the trip—in those days difficult at best—with 4 young children, including a practically newborn baby, bringing all the necessary baggage with her, to an uncertain future in a provincial, undeveloped region. They settled first in the small town of *Cambuquira*, where they were installed in a modest hotel; then they spent a brief time in the city of *Varginha*, and finally moved to *Guaranésia*, where they opened a store. Interestingly, Jayme Tiomno returned to Cambuquira nearly 60 years later under very different circumstances.[5]

After living in Guaranésia for some time, they moved again, this time to *São Sebastião do Paraíso*, since the coffee planters with whom Mauricio did business were located there. In the new town, Mauricio Tiomno set up a larger business, a general store selling "a little bit of everything"—including textiles and shoes. It was initially called the *Casa Russa* (House of Russia), but its name was later changed to *Loja Fluminense* ('Rio Shop').

Jayme's earliest memory was from that unsettled time when the family was still looking for a home, in Cambuquira or Varginha, when he was 4 years old. He liked to swing on the entrance gate to the (rented or borrowed) house where they were living temporarily. Their mother warned him that if he broke the gate, the owner would come with the police. One day, the gate indeed broke while he was swinging on it, and he ran away, in fear of the (presumably) approaching police… a memory which stayed with him for life.

Located on the main street—a wide avenue leading from the rail station—the store in São Sebastião occupied the same building as the family residence: It was at the front, while their private living quarters were in the back. There was no bathroom inside the house; baths were taken using a basin. An outdoor toilet was in the back yard, a rudimentary fixture called the "*casinha*" (little house), which consisted of a cubicle with wooden walls and a deep hole surmounted by a wooden toilet seat. When the hole was nearly full, it was covered with earth and a new one was dug nearby.

Only three Jewish families lived in São Sebastião do Paraíso. When one of them moved away, the Tiomno family took over the house where they had lived, in front of the city gardens, with music playing on Sundays in the band shell. This was the center of social life, where the townspeople promenaded, children played and dances were held during Carnival. The store was also moved to the new location, again at the front of the building, and in back, there was a large yard where one could find the "casinha".

There, in 1929, the third daughter of the family was born—Silvia. Jayme, now 9 years old, attended the *Escola Municipal* de São Sebastião do Paraíso (elementary school), while his mother, 33 years old and with five children, worked hard, with the help of a private teacher, to learn to read and write in Portuguese, which she had not had an opportunity to do in her mother tongue.

Muzambinho

After some years in São Sebastião, the family moved once more, this time to *Muzambinho*, about 65 km to the southeast and very near the state border between Minas Gerais and São Paulo. Even smaller than São Sebastião, Muzambinho was a center of coffee production and cattle ranching, but it did have a good secondary school, the *Ginásio de Muzambinho*. After a brief stay in a house similar to those in São Sebastião, they were able to find a much nicer site, on the main street near the church (Fig. 2.5), and most importantly, with an indoor bathroom! It seemed like a dream. It also had large gardens for vegetables, fruit trees and herbs; Jayme was the 'proprietor' of the herb garden. His father again opened a general store at the front of the house.

The years in Muzambinho were not entirely tranquil for the Tiomno family. Since the establishment of the Republic in 1889, Brazil had been a union of states with great autonomy and a weak central government, directed by the president. The presidential candidates were chosen by a tacit agreement from among the 20 state governors, and generally selected in rotation: a politician from Minas Gerais, and the next from São Paulo, and so on, in what was

Fig. 2.5 Muzambinho around 1927, showing the church and the main street. Reproduced from an old postcard (public domain)

called the "milk and coffee alliance" (later termed the Old Republic). How-ever, for the new election in March 1930, the governing president Washing-ton Luis (from São Paulo), broke with this tradition and, with the support of some of the governors, nominated Julio Prestes, then governor of São Paulo, to be the next candidate for the presidency.

The unrest and instability among the states and the federal government, which had been building up during the decade of the 1920s, thus finally 'boiled over' in 1930, when three states, *Rio Grande do Sul, Paraiba*, and *Minas Gerais*, together with the Democratic Party of São Paulo, rebelled against the previous government and its chosen president, Júlio Prestes. In October, 1930, an armed conflict had broken out between the formerly allied states of Minas Gerais and São Paulo, endangering Muzambinho, just across the border in Minas Gerais. Many families feared an armed invasion or even bombing, and fled to the surrounding countryside.

This was of course difficult for the Tiomno family with their 5 children; the oldest, Benjamin, was just 13, while the youngest, Silvia, was still a baby, born the year before. They were able to find shelter in the surrounding area. There was in fact some fighting in the neighborhood of Muzambinho, and some fatalities as well as politically-motivated repression, and also tragedies among those who had fled their homes and wandered unprotected in the countryside. The Tiomno family escaped unharmed, but shaken.

The São Paulo forces were defeated, and on November 3rd, 1930 *Getúlio Vargas*, from Rio Grande do Sul, was installed as the new president. But after two years, another uprising occurred, initiated by forces in São Paulo who were dissatisfied with the dictatorial tendencies of the Vargas government (it was called the 'Constitutional revolution' or 'Paulista war'). This time, Annita refused to leave their house, although preparations were made for a possible last-minute flight (or fight).[1] Troops from Minas and from the north were quartered in the high school in Muzambinho, using the town as a base for incursions into São Paulo; this prevented a repetition of the invasion by the Paulistas as had occurred in 1930. Still another revolt against São Paulo was prepared briefly in the years 1933/34, and led to the formation of a local vigilante group in Muzambinho, in which Jayme's brother Benjamin Tiomno participated; but it ended without serious conflicts.

Jayme, who was in his early teens during this period, was spared from mil-itary actions, but his schooling was complemented by various other activities, including working in the family store, and, around 1932, by his preparations for the *Bar Mitzvah*, the traditional Jewish rite of passage for boys coming of age. He spent some weeks in Rio, living with his uncle Francisco (Chico) and making his preparations. Chico was a very religious man and Jayme accompanied him on the long walk on Saturday mornings (*Shabat*) to pray in the synagogue; travel by bus or tram was not an option. However, Jayme had to return to Muzambinho before his 13th birthday, owing to the beginning of the school year; otherwise he would have lost the whole year's school credit— and thus, he never completed the actual ceremony.

A testimonial to Tiomno's achievements as a talented pupil (and, later, as a successful scientist) can be seen on a website devoted to the history of Muzambinho.[6] Jayme Tiomno left a good impression among his fellow pupils and his teachers at the *ginásio*. When his class graduated, the principal of the school directed great praise to Jayme, who had left in the middle of the course and didn't graduate with the rest of his class. Those words of praise for Jayme were warmly applauded by the assembled pupils and teachers[7] in his former home.

Back to the Capital

The time in Muzambinho was on the whole a happy one for the Tiomno family. Their business flourished, they were well integrated in the local society and the older children were all in school; Jayme and Benjamin even began violin lessons. But their parents wished to send all of their children to college, an opportunity which they had not been offered themselves. The local high school, the *Ginásio de Muzambinho*, however did not provide higher-level college preparation courses, so that they would have to move again to a larger city if their dream of higher education for the children were to be fulfilled. In 1934, Benjamin was nearly 17, and it was time to make the move if he were not to lose too much momentum in his studies. Furthermore, they were faced with the delicate situation of Benjamin's budding love for his teacher. He was sent to Rio immediately, and Mauricio went there to arrange housing for the rest of the family, who were installed in a hotel while waiting to make the move.

By mid-year, an old but suitable apartment in the *Rua Dona Zulmira* in the *Maracanã* district had been found, and the family moved there. It was not far from the Rua Maxwell, where they had lived before going to Minas. Some 15 years later, when some members of the family were still living in that district but on a different street, they witnessed the construction of the famous soccer stadium of the same name, and experienced the games there from close up.

Benjamin had already completed high school in Muzambinho, and began his complementary course, required for university admission, after arriving in Rio. Silvia was only 5, and not yet in school. But the other three siblings were allowed to transfer to the well-known *Colégio Pedro II* near the center of the city. It was supported by the federal government and was free of charge. It had been created during the imperial period, in 1837, and it was one of few options as a non-religious secondary school. The intervention of a congressman from Muzambinho, Licurgo Leite, smoothed their transfer, a practice common at the time. Jayme was 14 years old, already in his fourth year, and graduated the following year. Much later, he was lauded by the *Colégio Pedro II* as an eminent pupil due to his successes as a student and a scientist.

Opening a new store in Rio would have been much more expensive than it had been in a small town in Minas, and Jayme's father's financial reserves were not sufficient for that. In addition, he was beginning to have lung problems, perhaps related to his work in the tobacco fields in Bessarabia. Annita, who had been a housewife in Minas Gerais (an ample occupation with 5 children!), now set up a business, designing and confecting clothing and accessories, which augmented the family's income and permitted the children to have a university education.

In Rio de Janeiro, Jayme's immediate family, who generally did not observe traditional Jewish religious practices, gradually began to be re-assimilated into their extended family and into the local Jewish community. Thus, at the beginning of the school year in 1935, little Silvia started first grade in a Jewish primary school, and the family began to participate in Jewish life and attended the synagogue on religious holidays.

They also joined some Jewish groups, such as the *Cabiras* Club, a recreational organization set up in 1939 by five Jewish friends and intended to unite young people from the Jewish community. Jayme, along with the older two of his sisters, Mariam and Feiga, started to go to dances, excursions, picnics, lectures and other activities. After a few years, Jayme was elected president of the *Cabiras*, and he promoted the *Campanha dos mil sócios*, a campaign to attract a thousand members, which more than fulfilled its goal. Mariam and Feiga met their future husbands in the group, and married. Jayme also had several girlfriends from the group: Ruth Fridman, Rachel Kaufman, and a girl called Geny. However, in contrast to his sisters, he did not become serious in any of these relationships, and remained single.

Jayme had strong ties to his family, playing dominoes with his grandfather Salomão and accompanying him on doctors' visits, staying by his side when he had to undergo surgery and even when he was on his deathbed. He helped his mother, copying patterns for clothing and going to pick up Silvia when she was visiting friends, and he gave injections to their father, whose health was declining. These functions led to his several nicknames, as the "designer of the family", the "nurse of the family", or the "electrician of the family" (Fig. 2.6).

Their return to the capital had also brought the Tiomnos into closer contact with the political movements that were affecting the life of the country. They observed with considerable uneasiness the rise of the *Ação Integralista Brasileira*, a nationalist, extreme right-wing movement based on fascist models, whose motto was "God, the Fatherland and Family", which agitated against non-Christian religions. The members wore green uniforms embroidered with the symbol Σ and organized street demonstrations, foreshadowing the difficult times to come.

Fig. 2.6 Jayme Tiomno (lower left) with his family in 1941, celebrating their parents'. 25th anniversary. (Behind, standing): their parents Annita and Mauricio; (seated below): his sisters Feiga Rebeca (Riva), Silvia, and Mariam (Mira). *Source* Private, STT

Notes

[1] Many of the biographical details of the Tiomno and Aizen families given here are recounted in the book '*Histórias de invernos e verões*' ('Stories of Winters and Summers'), by Silvia Tiomno Tolmasquim, the youngest sister of Jayme Tiomno and mother of one of the present authors (ATT); see Tolmasquim (2014) (in Portuguese).

[2] The town is also called *Sokyriany* (Ukrainian), *Secureni* (Romanian), Сокиряны (*Sokiryany*) (Russian), and *Sekurian* (Yiddish).

[3] The origin of the name *Tiomno* is quite interesting: Mauricio's paternal grandfather was named *Shloime Leizer*. In his later years, his vision became very poor, so that he could barely distinguish light and shadow. He was given the Russian nickname тёмный (dark, dim), which became his surname and was passed on to his children and grandchildren! Its transliteration in Roman letters is *Tiomnii* or *Tiomne*, which was changed by the immigration official who registered Mauricio in Brazil to 'Tiomno', using the argument that Brazilian surnames must end in 'a' or 'o'; cf. [1].

[4] A compact history of Brazil is given in Appendix A.

[5] In 1980, Tiomno attended a special session of the annual meeting of the Brazilian Physical Society—of which he was a co-founder—in Cambuquira/MG, in honor of his 60th birthday. See Chap. 15.

[6] See the Muzambinho historical website on Jayme Tiomno, at https://historiademuzambinho.blogspot.com/p/um-aluno-ilustre-o-grande-fisico.html. Many years later, Jayme Tiomno returned to Muzambinho to be celebrated as a '*Cidadão Honorário*', an honorary citizen, by the Municipal Chamber of Muzambinho. The visit and ceremony were held in the year 2000, when he was 80 years old, and he was accompanied by his wife and siblings; cf. [1] and Chap. 17.

[7] Irene? to Jayme Tiomno, August, 6th, 1934; [JT].

3

Student Days in Rio de Janeiro: Medicine Becomes Physics

The Faculty of Medicine

Jayme Tiomno completed his secondary schooling at the *Colégio Pedro II* in Rio in 1935, and then had to attend two years of complementary courses (*'curso complementar'*), required for matriculation as a student at a college-level faculty. At the age of 15, he was faced with choosing the major subject for his university studies. He was in doubt as to whether he should choose medicine or engineering. His parents favored engineering, since his older brother, Benjamin, was already studying medicine, and they thought that Jayme should enter a different field. But on the other hand, engineers were less well paid than physicians. As a result, their opinions didn't help much, so in the end Jayme decided to register for the pre-medical complementary course. At its completion, after two years, he entered the Faculty of Medicine as a student.

The Faculty of Medicine (Fig. 3.1), together with the Engineering and Law Faculties, made up the University of Rio de Janeiro, founded in 1920. This university combined professional schools with a strongly positivist tradition, based on the idea that nearly everything important was already known and only a few new principles remained to be discovered through research. Therefore, the instruction was based in the main on textbooks without any reference to research laboratories or scientific journals. The principal function of knowledge was considered to be for increasing the progress of the country and the physical well-being of its citizens.

© Springer Nature Switzerland AG 2020
W. D. Brewer and A. T. Tolmasquim, *Jayme Tiomno*, Springer Biographies,
https://doi.org/10.1007/978-3-030-41011-7_3

Fig. 3.1 The old building of the Faculty of Medicine, UB, in *Praia Vermelha*. Behind and and to the left is the *Morro da Urca*. This building no longer exists. Image courtesy of the *Acervo Casa de Oswaldo Cruz/Fiocruz*

However, in July of 1937, while Jayme was completing his complementary course and preparing for his entrance exams to the university, the Federal government combined the University of Rio de Janeiro together with various previously independent faculties, creating the *Universidade do Brasil* (UB; later renamed the *Universidade Federal do Rio de Janeiro*, UFRJ, which remains its modern designation). The main purpose of the various faculties within the new university was the education of professionals (i.e. it was not intended to be a research university).

In addition, in November 1937, President Getúlio Vargas carried out an internal *coup d'état* within his own government, instituting a totalitarian regime (called the '*Estado Novo*'). It was inspired to a great extent by the Italian and German regimes of that era. Vargas had considerable sympathy for the Fascist and Nazi movements in Europe, mirroring the attitudes of Mussolini and Hitler to strengthen the power of the Brazilian central government and weaken that of the states. The coup was claimed to be a response to the so-called Communist Revolt led by Luis Carlos Prestes in 1935, which had been rapidly suppressed. The government responded two years later with a tightening of authoritarian controls, becoming essentially a dictatorship.

Neither the academic nor the political instability could however prevent Tiomno from taking the entrance exams and entering the *Faculdade Nacional de Medicina* at the beginning of 1938. (The academic year began in March, following the summer vacation period in January and February). The Faculty of Medicine was still located in a beautiful building at the *Praia Vermelha*, south of the city (the first Urca campus). During his studies, he met the biophysicist Carlos Chagas Filho, and at the end of his first year, he became Chagas Filho's student assistant (a sort of unpaid recognition for the best students).[1,2] Chagas Filho was one of the few professors in the Faculty who were involved in research. At that time, he was studying the process of accumulation of electric charge by the *poraquê*, an electric eel found in the rivers of the Amazon basin. Working with Chagas Filho aroused an interest in research in Jayme Tiomno, accompanied by the desire to attend the Faculty of Sciences at the University of the Federal District (UDF).

Established in 1935, the UDF was the result of a process initiated in the early years of the twentieth century by Brazilian scientists and intellectuals who were interested in setting up a university that would value research and would encourage critical thought, favoring the creation of a new class of intellectuals. Those scientists and intellectuals wanted to establish a free space for research and reflection. Their concept of a research university was intended to have a faculty of sciences at its core, based on research, which would nurture the other faculties with new knowledge. It would also have the task of educating new secondary teachers, who could then carry out a reform in the schools. As another result of this process, the University of São Paulo was established in that city in 1934; its core was its *Faculdade de Filosofia, Ciências e Letras* (FFCL – Faculty of Philosophy, Sciences and Letters). And the following year, the *Universidade do Distrito Federal* opened in Rio de Janeiro, having as its nucleus the *Escola de Ciências* (Faculty of Sciences). In contrast to the University of Rio de Janeiro (later called the *Universidade do Brasil*, now UFRJ), which was linked to the Federal government, both USP and the UDF were established by their respective local and state governments.

The founding of the UDF was mainly due to the efforts of the educator *Anísio Teixeira*, who regarded education as a powerful instrument for forming aware and participative citizens. For him, Brazil at the time was a country of 'honorary' university degrees whose main function was to prepare professionals for practical purposes. The new university, in contrast, would have the task of educating groups of intellectuals, providers of critical ideas and bearers of culture; professors, writers, journalists, artists and politicians, who up to that time in Brazil were mainly self taught.[3]

The UDF was regarded with considerable suspicion by the Federal government under Getúlio Vargas. There was a potential rivalry between the Municipal government (the government of the Federal District in which Rio was located as national capital) and the Federal government, following Vargas' original coup in 1930. After the attempt to overthrow the central government in 1935, the Vargas regime decided to intervene in several institutions. This affected even many people who were not involved in the communist coup attempt, but rather simply did not accept all of the ideas of the Vargas government. The mayor of Rio de Janeiro, Pedro Ernesto, was dismissed from the city hall, Anisio Teixeira was forcibly retired as Director of the Department of Education, as was Afranio Peixoto, the Rector of the UDF.

Nevertheless, the UDF project continued, and its new Rector, Afonso Penna Junior, was able to attract the best scientists as professors; they were gathered from various research institutions, as well as among independent intellectuals and younger teachers who were sympathetic to the principles of the new university. Furthermore, professors were invited from France, mainly in the humanities, including Émile Brehier (philosophy); Eugène Albertini, Henri Hauser and Henri Tronchon (history); Gaston Leduc (linguistics); Pierre Deffontaines (geography); and Robert Garric (literature). The Dean of the Faculty of Sciences was the engineer *Roberto Marinho de Azevedo*, a member of the Brazilian Academy of Sciences (ABC) and an important voice within the former School of Engineering in favor of research and opposing positivistic dogmas.

The Faculty comprised four divisions: Natural History, Physics, Chemistry, and Mathematics. It was a center of innovative teaching in Brazil, with many talented and well-known professors, but few resources. It had been installed in the old building of a public school (Fig. 3.2) in the *Largo do Machado* (to the south of the city center; see map in Chap. 2). It had no laboratories, and practical instruction was given in other institutions, usually the workplaces of the principal teachers, such as the National Institute of Technology, the Laboratory of Minerals Production, and the *Oswaldo Cruz* Institute, among others.

In 1937, a law was promulgated prohibiting the accumulation of positions in public institutions, which affected many of the teachers who, aside from their work at other institutions, also taught at the UDF (with only a small bonus salary). Examples were Bernhard Gross, Plinio Süssekind Rocha and Roberto Marinho de Azevedo. They gradually began to leave the University and to work exclusively at their home institutions. As a consequence, the University also lost many students.

Fig. 3.2 The teaching faculty in Physics at the FNFi, 1942. They are standing in front of the old site of the *Colégio Estadual Amaro Cavalcanti* on the *Largo do Machado*, used by the FNFi at the time. *From left*: Paulo Alcântara Gomes, Elisa Frota-Pessôa, Jayme Tiomno, Joaquim da Costa Ribeiro, Luigi Sobrero, Leopoldo Nachbin, José Leite Lopes, and Mauricio Matos Peixoto. *Source* Private collection of Jayme Tiomno (public domain)

In order to increase the number of students and to try to survive the bottleneck imposed by the central government, the UDF started to facilitate the entry of students who had matriculated in other university courses or had already completed the complementary course, even in a different major subject. In 1938/39, Jayme Tiomno was spending his summer vacation in *São Lourenço*, a therapeutic spa in the state of *Minas Gerais*. His brother Benjamin, recalling Jayme's earlier desire to attend the *Escola de Ciências* of UDF and to become a secondary-school teacher, registered him for Natural History, which didn't require additional examinations. Tiomno was grateful for his registration in the Faculty of Sciences, but he was by then sure that he wanted to study physics, and not natural history.

Since he was already enrolled in a university course, Jayme was able to obtain a special permission from the Dean of the Faculty, dispensing him from the usual entrance examinations. He was, however, required to pass an internal examination in mathematics, in order to demonstrate his qualification to study the new major subject. At the time, the Dean of the Faculty of Sciences at the UDF was *Luis Freire*, who had succeeded Roberto Marinho de Azevedo. Freire was a very enthusiastic professor of physics from Recife, in the northeast of Brazil, and was well known for his teaching and for his pioneering work in identifying promising students from his home region (the state of *Pernambuco*) and encouraging them to go to Rio de Janeiro or to São Paulo, the main scientific centers in Brazil, to continue their studies.

Jayme Tiomno was allowed just 15 days' time to prepare for the internal qualifying exam, to be administered by Lélio Gama, head of the Department of Physics and Mathematics. In that brief time, he was able to study only part of the material to be covered. He answered just one of three questions on the written exam, and again in the oral exam. He told Prof. Gama that he had not been able to study all of the topics. Gama asked him what he had prepared, and concentrated on those areas in the rest of the examination. At the end, he remarked, "All right, you meet the necessary standards in those areas that you have prepared. I am going to let you enter the course." Due to the fortunate conjunction of Freire and Gama, Tiomno was able to matriculate in physics. He later credited those two professors with opening his way to a career in physics.[4]

Had he waited longer, that might not have been possible; a few days after he was accepted, in January 1939, the UDF was shut down by a presidential decree, and its courses, teachers, students and building transferred to the *Universidade do Brasil*. This act had the support of the Rector of the University of Brazil, Leitão da Cunha, who was interested in strengthening his own institution. The courses at the UDF were transferred to similar courses at the UB, with the exception of those of the Faculty of Sciences, which remained initially under the responsibility of its own Rector. In April, the *Faculdade Nacional de Filosofia* (FNFi) was created as an entity of the UB, and the courses, teachers and students of the former Faculty of Sciences of the UDF were allocated to it. Thus, Tiomno, who had applied to the UDF, found himself again studying at the *Universidade do Brasil*. In the end, he was studying two majors at the same university: medicine at the *Faculdade Nacional de Medicina*, in *Praia Vermelha*, and physics at the *Faculdade Nacional de Filosofia*, in the *Largo do Machado*.

Two Majors, at One University

After the new Faculty was created, it was necessary to structure it and to rebuild its cadre of teachers, since many had left. Once again, it was decided to search abroad, particularly in Europe, for faculty members to set up departments and give the new courses. This search followed the same pattern used by the corresponding Faculty of Philosophy, Sciences and Letters (FFCL) at the *Universidade de São Paulo*.

The *Universidade de São Paulo* (USP) had been established in 1934 by the State government, as we have seen. It was composed of a number of professional schools in addition to the FFCL, intended to carry out the education of future academics, and also to be a research institution. The strategy adopted for its development was to search for highly-qualified scientists in Europe, who would be able to establish a research environment in Brazil. Given the increasingly fascist tendencies in many European countries at the time, this was quite realistic, since many academics had been forced into retirement or were eager to move to a more democratic environment. Emphasis was on recruiting excellent scientists in France, Germany, and Italy. France contributed in the main social scientists and humanists, Germany naturalists and biologists, and Italy provided professors of the exact and mathematical sciences. Italy was a particularly rich source of talent at that time; one can recall that both Enrico Fermi and Emilio Segré emigrated from Italy to the USA in 1938, Fermi in the end to Chicago and Segré to Berkeley. Both of them became Nobel prize winners, Fermi in 1938 and Segré in 1959.

Above and beyond the criterion of searching for faculty members from countries with a longer tradition of research in each of those areas, the USP project had a strongly ideological character. The local intellectual elite was intent on defending the cause of liberalism and was opposed to totalitarian regimes, in particular to the centralist government of president Getúlio Vargas, as evidenced by the 'Paulista war' of 1932. Thus, in the fields of the humanities and social sciences, professors were recruited in the main from France, where there was no fascist regime in the 1930s, in the hope that they would have a positive ideological influence on their future students. In more technical areas, in which ideological considerations were held to be secondary, faculty members were acquired from Italy and from Germany.

On the other hand, many European nations were looking to increase their zones of influence in peripheral countries, and transferring academics to Brazil was considered to be an effective method for advancing that goal. In Italy, the Brazilian representative of USP was even received by Mussolini himself! The Italian and French governments offered a salary supplement, to be paid through the Brazilian host university, as an enticement to academics to spend some time teaching abroad. Apart from that, they would enjoy a sort of diplomatic status in Brazil.

For the new Faculty of Philosophy in Rio, the same model of bringing high-level scientists from abroad was adopted. Even a similar criterion for the distribution of subjects by nationality was applied: Italians were recruited to teach physics and mathematics, Germans for the natural sciences, and French for the humanities and the social sciences. This was especially surprising, since unlike the University of São Paulo, the University in Rio was directed by the Federal Government of Getulio Vargas, which leaned towards the Italian and German ideologies at that time.

For general and experimental physics, the Italian *Dalberto Faggioani* was chosen; for theoretical, mathematical and advanced physics, *Luigi Sobrero*, previously professor at '*La Sapienza*' in Rome and assistant to Levi–Civita; and for rational and celestial mechanics, *Benedetto Zanini*. In mathematics, also, Italian professionals were recruited: *Achille Bassi* for geometry and *Gabrielle Mammana* for calculus and advanced analysis. *Joaquim da Costa Ribeiro*, a Brazilian professor who had moved from the UDF to the new FNFi, was nominated as temporary occupant of the Chair of General and Experimental Physics.

The Faculty remained in the same building, with no laboratories, which had been used by the UDF, and had to continue requesting from other institutions the use of their facilities for holding classes and giving laboratory courses. It also lacked a good library with scientific journals. Tiomno and other physics students were accustomed to go to the *Instituto Oswaldo Cruz* to consult the library there. Although it was an institute dedicated to research in public health, it had the best collection of scientific journals in the city, even in the area of physics (Fig. 3.3).

While the establishment of the two most important universities in Brazil at the time, the *Universidade do Brasil* in Rio de Janeiro and the *Universidade de São Paulo* in that city, was carried out using similar strategies for both, the details and history of their development, and its results, were nevertheless very different, due to the differences in environment and personnel in the two cities.[5] In São Paulo, the establishment of the physics, chemistry and mathematics departments of the new university was guided by the mathematician *Teodoro Ramos*, a Brazilian who had spent time in various European countries and knew their universities from the inside. The professors were chosen for their professional excellence and were given a free hand to set up their respective departments. As a result, USP was able to 'jump start' as a research

Fig. 3.3 The *Instituto Oswaldo Cruz*, where Jayme Tiomno and his fellow students made use of the library during their studies at the FNFi. *Photo* Peter Ilicciev, Courtesy of *Acervo Casa de Oswaldo Cruz/Fiocruz*

university, quickly becoming the foremost university in Brazil, a position that it still holds in many fields even today, more than 80 years later.

In Rio, under Getúlio Vargas' Minister of Education, Gustavo Capanema, the departmental structure was first meticulously mapped out over a five-year period, following the model of fascist Italy, and even its architecture was planned to imitate that in Rome at the time (this latter project was however fortunately never carried out!). The professors who were invited from Italy were chosen by the (Mussolini) regime, or else they took the opportunity to flee from the country. The results, as experienced by Jayme Tiomno, were for the most part only moderately successful, at best:

> …in Rio, the choice of the professors from abroad for the *Faculdade de Filosofia* was an imitation of what was done in São Paulo, but not carried out with the same criteria as in Sao Paulo. There, *Amoroso Costa* [*in fact, Teodoro Ramos*][6] went personally to Europe and chose mathematicians, physicists and chemists. They provided a very secure foundation; while in Rio, the Minister of Education of the Vargas [*regime*] asked the Ministry of Education of the fascist Italian government to send professors… thus, candidates appeared who were mainly people wishing to flee the regime, or, in other cases, people who came here to make propaganda for the regime. Sobrero, for example, was one of

those who were fleeing the fascist regime, and he was the best of all who came to Rio. Sobrero had an impact, an influence on the development of physics in Rio which was very great.[4]

The physics course was the least popular of those offered by the FNFi.[7] In the year 1939, 29 students entered for Natural History, 20 for Chemistry, 17 for Mathematics, but only 5 students for Physics, Tiomno being one of them.[3] In the following year, 1940, only four students entered Physics. Two of them, José Leite Lopes and Hervásio de Carvalho, had already graduated in engineering in Recife, and went to the FNFi to study physics at the suggestion of Luis Freire. That same year, a young woman named Elisa Frota-Pessôa (just over 1.50 m tall, and already married, despite being only 19 years old), very determined and self-confident, was one of the remaining two entering students.

Tiomno, Leite Lopes and Elisa became good friends, and they called themselves 'the Three Musketeers'. Later on, Lepoldo Nachbin joined the group. He also came from Recife, and was studying engineering at UB, but he was interested in research and began to attend classes at the FNFi. He became the 'd'Artangnan' of the group! They, along with other students, including Mauricio Matos Peixoto, formed a seminar group, called the 'Seminário dos Alunos' (today, it would be termed a 'journal club'). Compare Fig. 3.2.

Already in his second year as a physics major, Tiomno began working in 1940 as a student assistant to Joaquim da Costa Ribeiro. The latter relied heavily on Tiomno's abilities, so that he determined only the subject to be studied and gave Tiomno complete freedom to conduct its study as he wished, both theoretically and experimentally. Costa Ribeiro worked experimentally in the field of thermodielectric phenomena, and had discovered an effect which bears his name. He was investigating the influence of the presence of radioactive impurities on the behavior of dielectric solids, in particular in Carnaúba wax.

In the course of the experiments, he observed an increase in electric current associated with a change of the physical state (melting, solidification), which he called the *thermodielectric effect*. It was exhibited by both Carnaúba wax and by napthalene in paraffine. With the help of his experimental work, he was able to establish the laws governing this phenomenon, relating the rate of the phase transition to the current strength and the electric potential associated with the variation of mass in each phase when passing from one phase (solid, liquid) of the system to another. Furthermore, he established the value of a new constant K, known as the thermodielectric constant. Later, Elisa Frota Pessôa was invited to work as student assistant to Costa Ribeiro

in his research on the thermodielectric effect. This was the topic of the thesis of Costa Ribeiro which he submitted to the competition for the Chair
of General and Experimental Physics at the FNFi. In his thesis, he thanks
among others Jayme Tiomno "for lucid discussions of various theoretical and
experimental aspects of the topic at hand" and Elisa Frota Pessôa "for efficient
assistance in carrying out the measurements".[8]

Thus, for about two years, Jayme Tiomno was majoring both in medicine,
at the *Faculdade Nacional de Medicina*, and in physics, at the *Faculdade
Nacional de Filosofia*, both now at the *Universidade do Brasil*, as well as working as a teaching assistant for general and experimental physics. However, in
1940, a new rule was established by the University, prohibiting students from
majoring in two subjects at the same time. For the second time, Tiomno was
required to choose between medicine and the physical sciences. In contrast to
the first time, and to the disappointment of his mother, he opted for the latter. She tried to convince him to complete the course in medicine, for which
he needed only one more year, but in vain. He had decided to dedicate himself to physics. Throughout the rest of her life, even after Jayme had received
considerable international recognition as a physicist, his mother would now
and then lament, "But he could at least have finished his medical studies!".[9]

Beginning Research and Graduate Work

Tiomno carried out his first research in those years, in the laboratory of Costa
Ribeiro. However, his contacts with Luigi Sobrero had stimulated his interest
in theoretical physics, and his first publications, in 1942, were theoretical
([JT1, JT2]; cf. Appendix B).

Tiomno's first research papers (published in the *Revista da FNFi*, an in-
house journal of the FNFi faculty at the UB) were guided by Luigi Sobrero
and by Costa Ribeiro, respectively, and appeared in 1942. The work guided
by Sobrero was purely theoretical, '*On a problem in the theory of elasticity*',[10]
while that guided by Costa Ribeiro described a theoretical analysis of the
'*Theorem of uniqueness of the charge distribution in conductors*' [11]; it related
to Costa Ribeiro's work on the thermodielectric effect. Early work on a
(frequency) analyzer based on mechanical harmonic oscillators has also been
mentioned.[12,13] These were basic exercises in formulating research results.

At that time, Tiomno's research was essentially self-guided, since he was not
given any formal orientation in research procedures by his immediate mentor,

Costa-Ribeiro.[1,4,13] This experience may have left him with a life-long aversion to that sort of do-it-yourself introduction to research[1] and stimulated his interest in mentoring of physics education and graduate courses.

It was very likely the influence of Sobrero, a brilliant mathematician and an excellent teacher, that had made Tiomno aware of the beauty of mathematical physics and guided him towards his vocation as a theoretician. In 1941, after nearly 4 years at the university and three years studying physics at the FNFi, Tiomno received his Bachelor's degree in physics. And in the following year, he began studies for his *licenciatura* (teaching certificate) in physics, which would enable him to fulfill teaching duties in schools and at universities.[13]

Tiomno and his colleagues in physics also began to attend the lectures and courses given at the Brazilian Academy of Sciences (the '*Academia Brasileira de Ciências*', ABC), which had become a forum for presenting research work in progress. In August of 1941, the *Academia* hosted a 5-day symposium on cosmic rays, taking advantage of the presence of a scientific commission led by Arthur H. Compton, from the University of Chicago, which was investigating methods for the measurement of cosmic radiation as part of the "Ryerson Physical Laboratory project".[14] Besides the North American researchers, various researchers from the *Faculdade de Filosofia, Ciências e Letras* at USP gave talks, including Gleb Wataghin, Giuseppe Occhialini, Mario Schenberg, Marcelo Damy de Souza Santos and Yolande Monteux, as well as some from other institutions in São Paulo, Rio de Janeiro, and even from Paris.[15] It was beginning to become clear to Jayme Tiomno and his colleagues that the most active theoretical physics and the highest-quality research were to be found in São Paulo, and that they should make it their goal to go there. Many years later, Tiomno recalled this period:

> I well remember the excitement with which I, a student of physics at the *Faculdade Nacional de Filosofia*, heard a seminar by Professor Gleb Wataghin, the Father of modern physics in Brazil, in early 1940. He spoke of the recent progress in physics of which he had become aware during his latest trip to Europe. One item was that of the experiments of [*Otto*] Hahn, and the results of [*Lise*] Meitner, whose most satisfactory interpretation raised the possibility of obtaining great amounts of energy from atomic nuclei. It was thus no surprise for me, and for the majority of physicists, in spite of the extreme secrecy of the continuing research on that topic, to learn of the explosion of the first nuclear bomb in 1945, and its terrible consequences....[2]

However, in August of 1942, after an attack on five Brazilian ships in the Atlantic and under pressure from the United States – which had first remained neutral but entered the 2nd World War in December, 1941 following the Japanese attack on its Pearl Harbor naval base and the declaration of war by Germany – Brazil also declared war on the Axis powers (Nazi Germany, Fascist Italy). Tiomno, who at 18 had already served one year of compulsory military service (essentially basic training, called *Tiro de Guerra*, 'military marksmanship'), was inducted into the army. Despite this, he succeeded in finishing his *licenciatura* degree, and he started to teach at a secondary school in order to augment his salary. In addition, he continued to give lectures at the *Universidade do Brasil* as assistant to the Chair of General and Experimental Physics, occupied by Joaquim da Costa Ribeiro. Serving in the military, he was exempted from going to university, but he didn't want to leave Costa Ribeiro alone with his duties, and continued to go to the faculty and attend to his teaching, returning to his quarters in the evenings. He often gave lectures in uniform, having gone directly from the barracks to the university.

In the army, he worked in the area of radio communications, and after two years as a soldier, in 1944 he entered reserve-officer training school (CPOR). He had no intention of pursuing a military career, but it was the only way to avoid remaining a low-ranking soldier (*soldado raso*). Besides patrolling the Atlantic, Brazil sent a regiment to fight in Italy. Tiomno however remained in Brazil.

At the same time, the Italian professors, including Luigi Sobrero, had to leave Brazil and return to Italy, being classed as 'enemy aliens'. In a letter to Tiomno, his colleague Leite Lopes clearly expressed the disappointment caused by Sobrero's departure[16]:

> Unfortunately, [*the news*] is not good. I am sad to report that our Prof. Sobrero is planning to leave. Together with him, the other Italian professors will also be going. We here are trying to do everything possible to prevent the departure of Sobrero, in particular. But, unhappily, our good intentions and the goodwill that he holds for us are unable to overcome the difficulties which belong in a different sphere, absolutely inaccessible to us. The situation is extremely delicate, and the only thing that we can do is to express to Sobrero and to the others our sadness at their departure and our deep gratitude for what they have done to support scientific culture in Brazil.

Tiomno lost his mentor and the chance to publish his first article in an Italian journal, where it would have been guided by Sobrero. But the seeds of theoretical physics had already been sown in his mind.

Another consequence of the suspension of relations with Italy was that the *Faculdade Nacional de Filosofia*, which had previously occupied an old school building on the *Largo do Machado*, was transferred to the *Casa d'Itália*, the former seat of the Italian embassy, which had been confiscated by the Brazilian government. There, they had much more space to set up laboratories.

During this period, José Leite Lopes, one of the 'Musketeers', who had not been inducted into the military, finished his Bachelor's in physics in late 1942 and went to São Paulo, where he became teaching assistant for the courses given by Gleb Wataghin, and carried out research work with Mario Schenberg. At the end of 1943, he was awarded a fellowship by the U.S. Department of State (an assistance granted in return for the Brazilian engagement in the war), and went to pursue his doctoral degree at Princeton University. There, he initially worked with *Josef-Maria Jauch*, and later with *Wolfgang Pauli*, on the recommendation of Schenberg. Both were European physicists who were in exile in the United States. Princeton was nearly emptied of younger American physicists at that time, since they were mostly involved in war work elsewhere, but Leite Lopes could interact with some of the older, famous-name physicists such as Albert Einstein, Wolfgang Pauli, and John von Neumann, all refugees from the fascist governments in Europe. He finished his Ph.D. in early 1946 and returned to Rio de Janeiro. There, he was invited to occupy the Chair of Theoretical and Advanced Physics temporarily; it had been vacant since Sobrero's departure in 1943.

With the surrender of Germany in May, 1945 and the conclusion of his officer's training at the CPOR, Tiomno was able to return to his earlier path, pursuing a career in theoretical physics, which was his highest goal. It was clear to him that his next step should be to go to São Paulo.

Notes

[1] The period 1942–1950 in Jayme Tiomno's scientific career, when he was beginning independent research and was in graduate school in Rio, São Paulo, and Princeton (USA), is described in detail in an article by Karin S. F. Fornazier and Antonio A. P. Videira; see Fornazier and Videira (2018) (in Portuguese).

[2] Jayme Tiomno, interview given in Rio de Janeiro in 1977; see Tiomno (1977).

[3] Some of the history of the University of the Federal District (UDF) is given by Paim (1981). The parent institution of the *Universidade do Brasil*, the *Universidade do Rio de Janeiro*, had been founded on Sept. 7th, 1920, just 5 months after Jayme Tiomno's birth, on the national holiday (Independence Day) that year.

[4] Tiomno (1977).

[5] The parallel developments at UB and USP were described for example by R.A. Salmeron in the lecture he gave upon receiving an honorary doctorate from the *Universidade de Brasília* (UnB) in 2005; cf. the reprint in Salmeron (2013).

[6] *sic.* In fact, this was Teodoro Ramos.

[7] The course of study leading to the Bachelor's degree in Physics took three years, and comprised the following individual courses: (First year): Calculus (I), Analytic and Projective Geometry, General and Experimental Physics (I); (Second year): Calculus (II), Descriptive Geometry and Related Topics, Rational (Classical) Mechanics, General and Experimental Physics (II); (Third year): Advanced Calculus, Advanced Physics, Mathematical Physics, Theoretical Physics. Silva Filho (2013, p. 84).

[8] Costa Ribeiro (1945).

[9] Tolmasquim (2014, p. 155).

[10] [JT2]; cf. Appendix B.

[11] [JT1].

[12] Listed in Tiomno's resumé submitted to the FFCL, USP, in 1966 as part of a *concurso* (competition) for an open position there; [JT]. See Tiomno (1966a), and Chap. 12.

[13] cf. the article on '*Wheeler, Tiomno, and Brazilian Physics*': Bassalo and Freire (2003).

[14] The commission led by Compton consisted of William P. Jesse, of the University of Chicago; Norman Hilberry and Ann Hepburn Hilberry, of New York University; Ernest O. Wollan and Donald Hughes, of the University of Chicago; and Paulus A. Pompéia, of the *Universidade de São Paulo*. Source: Silva Filho (2013, pp.128–129).

[15] The following additional scientists gave lectures: J.A. Ribeiro Saboya, of the *Escola Politécnica* in São Paulo; Pater F. X. Roser, SJ, of the *Colégio Anchieta*; Adalberto Menezes de Oliveira, of the *Escola Naval*; Bernhard Gross, from the *Instituto Nacional de Tecnologia*; René Wurmser, from the *Collège de France*, and Joaquim da Costa Ribeiro, of the *Faculdade Nacional de Filosofia, Universidade do Brasil*. Source: Silva Filho (2013, pp. 128–129).

[16] José Leite Lopes to Jayme Tiomno, February 8th, 1942; [JT].

4

A New World of Subatomic Particles: A Summary of the Physics

The end of the 19th and the beginning of the 20th centuries represent a period of rapid discoveries of atomic and subatomic particles and the forces acting on them. Physicists in Europe and later in the USA and elsewhere were excited at these discoveries and were hard at work trying to understand this newly-revealed microscopic world. Thus, in order to appreciate the work on physics that was being carried out at USP by the mid-20th century, as well as Jayme Tiomno's interests and research, both in São Paulo and later on in Princeton, it is helpful to have an overview of the developments that led up to the forefront of physical research in his time in the late 1940s.

For readers who may not be familiar with the history of these topics, we describe in this chapter their development, a natural evolution of the general topic of *nuclear physics*, beginning in the early 20th century. It led to what is now called 'High-Energy Physics' (HEP), and to the study of the fundamental building blocks and interactions in the natural world (*particle physics*), culminating in the late 1970s in the Standard Model of Elementary Particles (SMEP). It is considered by most physicists to be a highly successful, although not the final step in the process of understanding the microscopic world.

Those readers who are already informed about these developments, or who are less interested in the scientific-historical aspects of Jayme Tiomno's life and career, may skip over this chapter without forfeiting the continuity of our story.

By 1900, the science of motion and of the mechanical properties of matter ('*Classical Mechanics*') was highly developed and held by many to be the final and exact description of those phenomena. Great progress had also been made

© Springer Nature Switzerland AG 2020
W. D. Brewer and A. T. Tolmasquim, *Jayme Tiomno*, Springer Biographies,
https://doi.org/10.1007/978-3-030-41011-7_4

in the description of processes involving heat ('*Thermodynamics*') and electromagnetism (the *Maxwell Theory*, which predicted electromagnetic waves, initially identified with light, ultraviolet and infrared radiations, and radio waves). The question of the existence of *smallest particles* (atoms, molecules) as building blocks for matter remained open, and many physicists (but fewer chemists) were still skeptical, although the success of *Statistical Mechanics*, based on the atomic hypothesis, was convincing evidence to some scientists of the existence of atoms and molecules.

How Nuclear Physics Began

Near the end of the 19th century, several important developments had opened up whole new fields of study in fundamental physics. Among them were Wilhelm Conrad Röntgen's identification in 1895 of what are now called *X-rays* (an energetic form of electromagnetic waves); Henri Becquerel's discovery of *radioactivity* (1896); the discovery of the *electron* by J.J. Thomson and others in 1897[1]; and the theoretical description of *thermal radiation* by Max Planck in 1900. These discoveries, along with the problem of the emission and absorption of electromagnetic waves by *moving bodies*, revealed chinks and gaps in the structure of classical physics, and made it clear that much was yet to be understood.

Nuclear physics grew out of Becquerel's discovery of radioactive decay. While he did not pursue the topic much further himself, younger researchers, including Marie and Pierre Curie in France, Ernest Rutherford and Frederick Soddy in England, Otto Hahn and Lise Meitner in Germany, and Stefan Meyer in Austria, among others, continued investigating the modes and properties of radioactive decays, opening the door to a whole new world of microscopic phenomena.

The decays were soon found to obey an exponential law, their rate decreasing exponentially with time, leading to the definition of a characteristic 'decay constant' which could range over a vast domain of times from the smallest fractions of a second to millions of years. The radioactively-decaying elements could thus be "short-lived", decaying so fast that they would never be obtained in more than microscopic quantities; or "long-lived", so that considerable amounts of them have survived over the 4.5 thousand million years since the formation of the Solar System, and are still found in nature. These naturally-occurring, long-lived radioactive elements, such as uranium and thorium (and the long-lived isotope of potassium, ^{40}K), were the first to be extracted and studied.

Their modes of decay were classified as 'alpha', 'beta', and 'gamma', according to the penetrating power of their accompanying radiations. *Alpha* decays, with the least penetrating radiations, were found by Rutherford's group to emit microscopic particles, the 'alpha particles', which were soon identified as the main constituent of the rare gas *helium* (helium nuclei). *Beta* decays likewise emit particles, the 'beta particles', found to be simply electrons of high energy. And *gamma* decays were determined to be accompanied by energetic electromagnetic waves ('gamma rays'), similar to X-rays, but of even higher energies.

In 1911, Rutherford (with his assistants Hans Geiger and Ernest Marsden) also discovered the *atomic nucleus*, by allowing energetic alpha particles to pass through a thin metal foil and studying their scattering by the material of the foil. If the matter in the foil were continuously distributed, with a positively-charged, massive "jelly" providing the background framework for pointlike, negatively-charged electrons ('Jellium Model'), as was then thought, then the alpha particles would pass through the foil almost without deflection. Instead, the experiments showed that while most of them passed through the foils with only minimal scattering, some few were deflected to large angles, up to 180°, their number decreasing sharply with increasing scattering angle (Fig. 4.1).

Rutherford was able to show that this was exactly what would be expected from scattering by pointlike, fixed positive charges (Coulomb scattering). Those 'point charges' must contain all of the positive charge and most of the mass of the atoms in the foil. By varying the material of the foils (gold, silver,

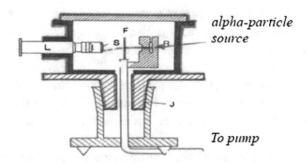

Fig. 4.1 The apparatus used by Rutherford, Geiger and Marsden. L is a small microscope used to observe the fluorescent screen S, which produces flashes when struck by the alpha particles. The particles leave the source R (for 'radium'), collimated by a thin hole in the lead block, and pass through the foil F where some are scattered. The upper part can be rotated around the glass joint J to vary the angle of observation. The chamber is evacuated to avoid scattering by the air. *Source* Private collection, WDB

copper, etc.), he could show that the electric charge on these pointlike '*atomic nuclei*' corresponded simply to the order number of the corresponding element in the Periodic Table. And by examining the small deviations from pure Coulomb scattering observed at large angles, he could estimate the size of the nuclei (found to be around 100,000 times smaller than the atoms themselves), and could infer that besides the electric repulsive force (Coulomb force), there must be an even stronger force acting between the alpha particles and the nuclei, of very short range (the *nuclear binding force*, now attributed to the *strong interaction*). These simple experiments thus revealed a whole new microscopic world within ordinary matter, and led to the new field of nuclear physics, as well as to the development of Bohr's atomic model (and later to modern quantum mechanics).

More and More Particles ...

Initially, two so-called *elementary particles* were known: the light, negatively-charged *electron*, and the much heavier *proton*, the nucleus of the simplest atom, hydrogen; its positive charge is precisely equal in magnitude (but opposite in sign) to the charge of the electron. (The proton is today no longer considered to be an *elementary* particle, but instead is a composite particle, containing other, still smaller particles, the *quarks* (*fermions*, basic matter particles) and the *gluons* (*bosons*, the messenger particles of the strong interaction). Just what is meant by these terms will be explained further on).

The nuclei of heavier atoms were thus initially thought to contain a number A of protons (A is the 'nuclear mass number', the number of massive particles within the nucleus), and a number $A–Z$ of electrons (Z is the number of elementary charges on the nucleus, equal to the order number of the element). The mass of the nucleus, proportional to A, would thus be essentially due to the protons, and its net positive electric charge would be due to the excess of protons over electrons in the nucleus. That a radioactive nucleus might undergo beta decay and emit an energetic electron could thus be readily understood.

However, this simple model was contradicted by quantum mechanics in the 1920s. It was found to be impossible to confine electrons in such a small volume as that of an atomic nucleus. Heisenberg's uncertainty relations require that the position and the momentum (velocity times mass) of a quantum particle cannot be simultaneously fixed: A finite uncertainty remains, and the product of the uncertainties in the position and the momentum cannot be smaller than the characteristic constant of quantum physics, the Planck constant h (a very small number in everyday terms).

Confining an electron within a nucleus would fix its position to such a precise extent that the uncertainty in its momentum (and thus its average momentum) must become quite large, in order to satisfy the uncertainty relation. And that would mean, due to the small mass of electrons, that our electron must have an enormous velocity and could not be confined within the nucleus—a logical contradiction!

[This phenomenon—an increase in kinetic energy when a particle is confined to a small region of space—is a general quantum-mechanical effect, known as 'delocalization energy'. It plays a role in the binding of the electrons in atoms (determining the size of the smallest stable 'orbit'), and in the bonding of atoms to form a metal (delocalized electrons; quasi-free electron gas). It also plays a role in determining the observed mass of complex composite particles like protons, which is determined by the balance between the delocalization energy and the separation energy of their constituent matter particles, the *quarks*; cf. p. 147].

Thus, the nuclei of heavier elements must contain some other particle—as heavy as the proton, but without electric charge. These so-called *neutrons* were discovered in nuclear reactions by Frédéric and Irène Joliot-Curie in Paris, and by James Chadwick in England in 1932. Chadwick's conclusive experiments led to his 1935 physics Nobel prize for the discovery of the neutron.

In the meantime, measurements of *beta decays* had exposed another puzzle. While the radiations from alpha and gamma decays exhibit 'line spectra'—that is, the emitted radiations have discrete, well-defined energies, which can be associated with the difference in energy between the parent and the product nuclei, confirming energy conservation in the decay—in contrast, the energy spectrum of the emitted electrons from beta decays is *continuous*: thus, a small number of electrons is emitted at very low energies, increasing to a maximum in the middle of the spectrum, then decreasing again, falling to zero at the "endpoint energy" of the spectrum, which is a constant for a particular beta decay, and indeed does correspond to the energy difference between parent and product nucleus in the decay. But why do most of the emitted electrons have less energy? Must one doubt the fundamental principle of energy conservation for the special process of beta decay? Even Bohr asked himself this question.

The Beta Decays

A precise measurement of the momenta of all the bodies involved in the decay (parent nucleus, product nucleus, emitted electron) can give the answer, which was anticipated by the theoretician Wolfgang Pauli in 1930: There is a *fourth particle* involved in beta decay, called the *neutrino*, of very low mass (originally thought to be zero), and without electric charge. It interacts with matter only very weakly, and is therefore extremely difficult to detect (requiring enormous detectors with tons of 'absorber' material). Beta decay is thus a *four-body process*, and its prototype is the decay of the free neutron[2]:

$$n \rightarrow p^{(+)} + e^{(-)} + \bar{\nu}_e. \tag{1}$$

Here, n represents a neutron, p a proton (with its positive charge as exponent in parentheses), e a (negative) electron, and $\bar{\nu}_e$ an (anti-)neutrino. Electric charge as well as mass and energy are conserved in this process; the excess energy (small mass difference!) between the neutron and the proton is shared statistically between the electron and the neutrino (both termed 'light particles' or *leptons*), giving rise to the statistical, continuous spectrum shape observed for the energy spectrum of the electrons alone. The neutrino is not directly observable in most experiments (Fig. 4.2).

When bound within a nucleus, in the "potential well" created by the strong nuclear binding force, the neutron is stable for indefinitely long times. Outside a nucleus, as a free particle, it undergoes the above decay with a mean

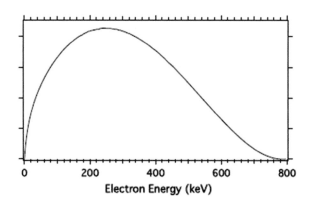

Fig. 4.2 The beta spectrum from the free neutron decay: The number of 'beta particles' (electrons) emitted, plotted against their energies. The 'end-point energy' Q is 782 keV. *Source* Nico (2009). Used with permission of the author and the publisher (J. Phys. G, IOP Press)

decay time of 882 s (just under 15 min.). If a nucleus has an excess of neutrons, it can reduce its total energy by undergoing nuclear beta decay, converting one of its neutrons into a proton (and thus increasing its nuclear charge Z by one unit, so that the product nucleus is one element higher in the Periodic Table). Such a nucleus will then be unstable towards beta decay (i.e. it is *beta radioactive*), and it will decay with a time constant depending on how much excess energy it possesses.

Some other important discoveries were made by the early 1930's. Paul Adrien Maurice Dirac's theory of relativistic electrons (1928) yields an equation with four components in its solution. Initially, their interpretation was unclear, but it soon became apparent that they represent two pairs of solutions, each pair referring to the two possible orientations ('up' and 'down') of the 'spin' or intrinsic angular momentum of the electrons. Subatomic particles behave like tiny gyroscopes, as though they spin around their own axes, maintaining an angular momentum like that of a spinning top. However, unlike a top, their angular momentum is 'quantized', i.e. it can take on only certain values, multiples of the constant $h/2\pi$ (h is again Planck's constant). The particles with half-integer multiples of $h/2\pi$ ("spin 1/2, 3/2, …") are referred to as '*fermions*' for historical reasons; the particles with integer multiples ("spin 0, 1, …") are called '*bosons*'. Electrons are fermions of spin ½. This is associated with a magnetic moment, so the particles also behave like miniature bar magnets.

These properties had been noticed in various experimental results (Zeeman effect, Einstein-de Haas effect, Stern-Gerlach experiment) between 1900 and 1922, but were not understood until 1925, when G.E. Uhlenbeck and Samuel Goudsmit proposed the idea of *spin*, later treated phenomenologically by Pauli and found directly from his theory by Dirac. The 'nucleons' or heavy particles, the protons and neutrons, are also fermions, and their angular momenta are coupled together within an atomic nucleus ('vector coupling'), resulting in an overall 'nuclear spin' I which can be measured (e.g. in high-resolution atomic spectra, by atomic-beam experiments (Rabi method), or by nuclear magnetic resonance methods).

The other pair of solutions of the Dirac equation has negative energy and can be interpreted as corresponding to the *anti-electron*, or positron. Antiparticles have the same masses, but opposite electric charges, as the corresponding particles. When a particle and its antiparticle meet, they 'annihilate' each other, releasing energy (the energy of their rest masses) as gamma rays. Particle-antiparticle pairs can also be produced by gamma rays of sufficient energy, passing through a strong Coulomb field (e.g. near to an atomic nucleus). The anti-electron, the *positron*, was detected in cloud-chamber tracks of cosmic ray products in 1932 by C.D. Anderson. A positron can be emitted from an energetic nucleus in an analogous way to the usual beta decay:

$$p^{(+)} \rightarrow n + e^{(+)} + \nu_e. \tag{2}$$

Such '*positron decays*' can occur only if the parent nucleus (proton-rich nucleus) has sufficient excess energy. Such nuclei do not occur in the natural decay chains (U, Th), but can be produced by nuclear reactions, for example in a cyclotron or by cosmic-ray reactions. Since the neutron's mass is larger than the proton's by about 2.6 electron masses, an excess energy equivalent to nearly 4 electron masses must be available on the left side of reaction (2) to permit it to occur.

A third type of 'beta decay' can take place as the reverse of reaction (1). The inner electrons in medium to heavy atoms spend considerable time passing through the atomic nucleus. There, they can interact with the protons present in the nucleus, according to the following reaction scheme:

$$p^{(+)} + e^{(-)} \rightarrow n + \nu_e. \tag{3}$$

This '*electron capture*' decay requires less additional energy (the equivalent of about 2 electron masses) than positron decay, since an electron is *added* to the nucleus rather than being created. If the excess energy of the parent nucleus is sufficient, both (2) and (3) can occur with different probabilities from the same parent nuclei. The long-lived, naturally-occurring isotope ^{40}K is an example of another possible combination, exhibiting both decay modes (1) and (3).

A 'Nuclear Binding Force'

By around 1935, there were considerable data on atomic nuclei (masses, charges; decay modes and lifetimes of radioactive nuclei; nuclear spins and magnetic moments; gamma-ray spectra), but the nature of the '*nuclear binding force*' that holds the nucleons within the nucleus was still not clear. Its high strength and short range could be estimated from scattering experiments, but it had not been possible to describe it by a potential, as can be done for the Coulomb force that binds the electrons within atoms. And there was no coherent description of the energy levels that could be occupied by the nucleons within a nucleus (and are observable in their gamma-ray spectra).

In analogy to the Electromagnetic interaction, presumed to be mediated by exchange of 'messenger particles', the *photons* (quanta or particles associated with electromagnetic radiation), it was postulated by the Japanese theoretician *Hideki Yukawa* in 1935 that the nucleons were bound in an effective potential due to exchange of a hypothetical particle, the 'π *meson*' (now often simply called the *pion*). Since the *nuclear binding force* acts between particles of the same or different charge (p-p, n-n, n-p), the π mesons could be expected to exist as either charged or neutral particles. The short range of the nuclear force gives an estimate for their mass. (The photon, by contrast, is neutral and massless, since the Electromagnetic interaction has an infinite range and connects only charged particles). And the π meson, like the photon, should be a boson (of spin 0 or 1).

The energies attainable by cyclotrons in the 1930s were not sufficiently great to produce the postulated π mesons artificially. Therefore, they were sought as products of cosmic rays, whose primary radiations (mainly protons) strike the earth's upper atmosphere with very high energies and can interact there with nuclei of the atoms and molecules in the air. Detectors (film emulsions or Geiger counters) were carried to high altitudes on mountaintops or by balloons (and later by rockets). Magnetic fields and absorber plates were used to distinguish the different types of particles which had left their tracks in the emulsions. In 1935, Anderson and Neddermeyer detected negatively-charged particles of intermediate mass from cosmic-ray tracks in cloud chamber experiments, calling them '*mesotrons*'. They were initially thought to be Yukawa's π mesons, but they proved not to interact strongly with nuclei, as the pion must. This prompted I. I. Rabi's famous remark, "Who ordered *that?*"

These particles were later renamed 'μ *mesons*' and are now called '*muons*' (μ^-). Their corresponding antiparticle has the same mass but a positive charge (μ^+). They are spin-½ particles of mass about 207 times that of the electron ('heavy electrons'), and are now classed as *leptons*. They decay with lifetimes of 2.2 μs (2.2 millionths of a second) into electrons and neutrinos: $\mu^\pm \rightarrow e^\pm + \mu_0 + \nu$, where ν is an ordinary (electron) neutrino or antineutrino (now termed ν_e), and μ_0 is a '*muon neutrino*' (or antineutrino), as was later discovered (now denoted by ν_μ). Muons can be captured in atoms, replacing an electron on an inner orbit, very near the nucleus (because of their large mass). They can also be captured *into* the nucleus, analogously to electron capture.

The *pions* were finally detected in high-altitude cosmic-ray observations, analyzed in 1947 in Britain by Cecil Powell, the Brazilian César Lattes, and the Italian Giuseppe Occhialini, and verified by Lattes at the Berkeley

184" cyclotron, where they were produced artificially and detected in 1948. They are now classified as *mesons*. Many other mesons have in the meantime been discovered. They are not elementary particles, as originally thought, but instead are composite particles like the nucleons, having two 'valence quarks' (quark-antiquark pairs). (The heavier *nucleons* have three valence quarks). In the following years, the pion decay modes were studied intensively. There are, in addition to negative pions π^- and their positive antiparticles π^+, also *neutral* pions, π^0, as predicted by Yukawa. All of them have much shorter decay lifetimes than the muons, which are the main products of the decays of $\pi^\pm$:

$$\pi^\pm \rightarrow \mu^\pm + \nu_\mu, \tag{4}$$

with a decay lifetime of only 26 ns (26 thousand-millionths of a second). A direct decay yielding electrons, $\pi^\pm \rightarrow e^\pm + \nu_e$, is also possible, but is extremely rare, occurring in only about 0.01% of pion decays, as was found in emulsion observations by Elisa Frota-Pessôa and Neusa Margem[3a] and later precisely measured at CERN[3b]. The charged pions can also decay into neutral pions, electrons, and electron neutrinos. The neutral pion itself decays still faster, emitting two gamma rays (or, more rarely, gamma rays and electron-positron pairs, or only electron-positron pairs). The decay lifetime for the most common mode is only 84 as (84 million-million-millionths of a second).

In 1948, none of these details were known, but the main decay mode of the charged pions, reaction (4), was already clear. Jayme Tiomno's interest in that decay and the subsequent muon decays was aroused by those discoveries, and it was shared by his first mentor at Princeton, John A. Wheeler.

The *Weak interaction*, which mediates nuclear beta decays, as well as the neutron decay and muon and charged pion decays, was first described theoretically by Enrico Fermi in 1933/34. His description was intended at the time for 'normal' beta decay (reaction (1) above). He formulated a 'four fermion interaction' as a contact interaction between vector currents (nucleon and lepton currents). The generalization to other operator ranks and as a *universal interaction* for all weak processes was a major task carried out in the 1950's, in which Jayme Tiomno would play an important (although often unrecognized) role.

Thus, the four fundamental interactions known today are the *Electromagnetic* interaction EI (Maxwell theory; quantum electrodynamics); the *Strong* interaction SI (nuclear binding force; quantum chromodynamics); the *Weak* interaction WI (nuclear beta decays, etc.); and the *Gravitational* interaction (General Relativity, GR).

The Quantum World

Several puzzling experimental results, unexplained by classical theory, had emerged by the late 19th century. One of these was the spectrum of the light emitted by hot, glowing objects: thermal or 'black-body' radiation. It had been studied carefully, especially since the invention of incandescent lamps, with their considerable economic and industrial importance. Thermal radiation is emitted as a continuous spectrum; at its low-frequency ('red') end, its intensity increases as the square of the light frequency, then passes through a maximum (in the spectrum of the Sun, the maximum is in the green, where the human eye also has its greatest sensitivity), finally decreasing exponentially towards higher frequencies (the 'blue' end of the spectrum). The behavior at low frequencies (Rayleigh-Jeans law) and at high frequencies (Wien's law) can be understood classically, but the shape of the curve in between could not be explained. Max Planck succeeded in deriving it in 1900, at first by a careful interpolation procedure, then from a complicated derivation based on entropy. For the latter, he had to assume that radiation could be emitted only in small 'energy packets', which he called '*quanta*', each with an energy given by $E = h\nu$, where ν is the frequency of the radiation (light), and h is a constant of nature, now called *Planck's constant*. He considered these 'quanta' to be simply a mathematical device, not reflecting a true physical phenomenon.

A second unexplained experiment was the *photoelectric effect*, detected by Heinrich Hertz in 1886 during his experiments on electromagnetic waves. In this effect, light falling on a solid object (e.g. a metal plate) releases electric charges (electrons) from the object. It was studied in some detail by Hertz's assistant Wilhelm Hallwachs, who found that the *number* of electrons released is indeed proportional to the intensity of the light, as one might expect, but their *energy* is determined only by the light's frequency, and is given by the simple formula $E_e = h\nu - \Phi$, where ν is again the frequency of the radiation (light) and Φ, called the 'work function', depends on the material of the object. This relation was explained by Albert Einstein in 1905 with his *quantum hypothesis*, which emphasized the reality of Planck's light quanta. The light acts as 'particles' of energy $h\nu$, each of which can release just one electron from the object. Einstein received the Nobel prize for physics in 1921 for this work.

The principle of *quantization* was later applied by Niels Bohr in his successful model of the hydrogen atom, based on Rutherford's nuclear model of the atom. In the 1920s, a more quantitative theory, now called *quantum mechanics*, was developed by Werner Heisenberg, Erwin Schrödinger,

Max Born and others. It was later generalized to the relativistic case by P. A. M. Dirac, and applied to electromagnetic fields (quantum electrodynamics, QED). A major unsolved problem of fundamental physics is the combination of quantum theory with the gravitational theory introduced by Einstein in his General Relativity (1915); this *unified theory* is often referred to as 'quantum gravity'. Einstein himself worked unsuccessfully for the last 30 years of his life attempting to solve the problem of *unification*, and it became a generally active field in the latter half of the 20th century, still continuing today.

The *Weak interaction*, which mediates nuclear beta decays, as well as the neutron decay and muon and charged pion decays, was first described theoretically by Enrico Fermi in 1933/34, as we have seen above. He formulated his 'four fermion interaction' as a *contact* interaction (occurring at one spacetime point) between *vector* currents (nucleon and lepton currents), in analogy to the Electromagnetic interaction between an electric-charge vector current of charged particles and the vector potential of the electromagnetic field.

The Fermi Four-Particle Current-Current Interaction

Fermi modeled his description of beta decay on the quantum electrodynamic (QED) description of the Electromagnetic interaction. Electromagnetic fields (and thus photons) are created by accelerated electric charges (i.e. charges that are changing their speeds or directions), as had been known since the classical theory of Maxwell (1865). In the QED description, the operator for the energy of the interaction is the product of two vectors, the charge-current vector and the vector potential of the electromagnetic field. They contain operators in four-dimensional spacetime (Dirac operators), which can be expressed in terms of the four Dirac matrices $\gamma_1 \ldots \gamma_4$ (the indices are often denoted instead as 0 … 3 in more modern notation).

The diagrams below represent the corresponding interactions in the cases of Electromagnetism and of the Weak interaction (here the neutron decay). In such diagrams, the *time* increases vertically from below to above. A vertex, where the timelines of different quantum objects meet, represents the spacetime point where the interaction occurs. In the Electromagnetic interaction, at left, a charged particle (here a proton) is accelerated at the vertex and emits a photon (γ). In the Weak interaction, at right, a neutron decays at the vertex, becoming a proton and emitting an electron and an (electron) antineutrino at the same time (Fig. 4.3).

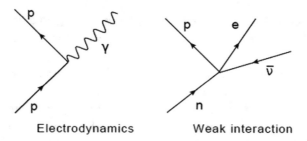

Electrodynamics   Weak interaction

Fig. 4.3 Diagrams representing the Electromagnetic interaction (left: emission of a light quantum (a photon) by an accelerated proton); and the Weak interaction (right: neutron decay, where the neutron is transformed into a proton, emitting an electron and an electron antineutrino). *Source* Rajasekaran (2014). Used with permission of the author

This latter process is also described as the product of two vector operators, one representing the nucleon current (more generally, a *hadron* current, where 'hadron' is the generic name for particles that participate in the *Strong interaction*; or, in modern terms (Standard Model, SMEP), a *quark* current). The other current operator represents the lepton current ('*leptons*', or light particles, include the electron and its neutrino, the muon and its neutrino, and in the SMEP, also the tau particle and its neutrino, as well as their antiparticles). In the case of the neutron decay, the hadron current connects the neutron and the proton, while the lepton current connects the electron and an electron-antineutrino. For more details, see for example [4], or the essays in [5].

The current operators can assume one of five possible forms, or ranks, classified by their *symmetry properties*, and still be compatible with Lorentz invariance (i.e. obey Special Relativity). Fermi assumed them to be *vector* operators (V), in analogy with the Electromagnetic interaction. Vectors represent quantities which have both a *magnitude* and a *direction* in space, such as a velocity or an electric field. Vector quantities change their signs when going from a right-handed to a left-handed coordinate system[6] (spatial inversion). A corresponding quantity which does *not* change its sign on spatial inversion is called an *axial vector* (A). Examples are spin (i.e. angular momentum), or a magnetic field. In addition, in general also *scalar* operators (S) can occur; they represent quantities which have only a *magnitude*, but no *direction* (such as mass or electric charge), and they do not change under spatial inversion. The corresponding quantities which *do* change their signs on spatial inversion are called *pseudoscalars* (P) (e.g. a magnetic charge or monopole, whose existence in nature is uncertain). Finally, *tensor* operators (T) with a more complicated symmetry are also allowed by Lorentz invariance. Thus, in a general formulation of the Weak interaction, one would first have to consider all five operator types, S, V, T, A, and P. The theory must then be chosen so as to reproduce all of the known observable properties of the weak processes (decay and reaction rates, spectrum shapes, directional and spin correlations, etc.).

Symmetry operations are changes in space and time, and other background conditions of a system, which leave its physical description unchanged, or *invariant*. For example, <u>translations</u> (linear displacements) or <u>rotations</u> in space should not change the physical description of a system, since space is presumed to be *homogeneous* (translationally invariant) and *isotropic* (rotationally invariant). Physical systems are also said to be 'invariant' under such changes.

Associated with invariances are *conservation laws*, as required by Noether's Theorem, proved by the mathematician Emmy Noether in 1915–18. Translational invariance corresponds to the conservation of *linear momentum*, and rotational invariance to the conservation of *angular momentum*. Note that translation (linear displacements) and rotation are *continuous* symmetries—they can be carried out continuously, in a series of infinitesimally small steps. Some symmetry operations are however *discrete*—they are a yes-no proposition. For example, a coordinate system can be left-handed or right-handed, but nothing in between. Time can proceed 'forward' or 'backward', with no other possibilities. A group of quantum objects can be particles or antiparticles, but not anything in between. Some important 'discrete symmetries' are the invariance under mirror imaging of space (*spatial inversion*), associated with 'parity conservation', P[6]; under time reversal (*time-reversal invariance*, T); and under particle-antiparticle exchange ('*charge conjugation invariance*', C).

Those three symmetry operations, P, T, and C, obey a very general theorem, known as the 'CPT Theorem', which was proved by Wolfgang Pauli (1955) and independently by Gerhart Lüders (1954). It is based on the very general principle of Lorentz invariance, i.e. that physical systems should obey Special Relativity, as well as locality and causality, and states that all physical laws and processes are invariant under the combined operation C, P, T (that is, all particles are converted to their antiparticles (and vice versa); space is inverted (right-handed becomes left-handed and vice versa); and the direction of time (of all processes involving time) is reversed). Individual symmetries may be violated, but the combination of all three should be obeyed by all interactions and physical laws. For example, the Weak interactions

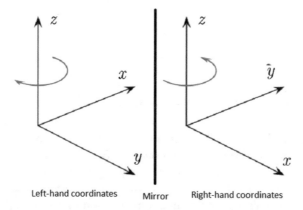

Fig. 4.4 Reflection by a mirror of a left-handed coordinate system, yielding a right-handed coordinate system. This is equivalent to spatial inversion, and the behavior of a physical object under such a transformation determines its *parity*. *Source* Wikimedia. Used under Creative Commons License. [Online at: https://commons.wikimedia. org/wiki/File:Cartesian_coordinate_system_handedness.svg (file modified)]

were found to violate parity conservation, P. But they also violate charge-conservation invariance, C, so that the two violations taken together cancel and overall, CPT invariance is maintained.

[Note that a capital 'P' is used for two different quantities in Weak-interactions theory—on the one hand, it refers to a 'pseudoscalar' operator in the current-current interaction, one of the five ranks of operators which are allowed there. On the other hand, it also represents the 'parity operation', i.e. the symmetry of a physical quantity or interaction under *spatial inversion*, one of the 'discrete' symmetries. These two meanings should not be confused; they are generally distinguishable from context].

The parity transformation is closely related to *helicity*. Many objects in the world exhibit a 'handedness', for example left-handed and right-handed screw threads, circularly-polarized light ('left-' or 'right-hand'-), or left-rotating and right-rotating molecules (sugars, amino acids). Helicity is an intrinsic property, and a relativistic invariant, of massless particles such as photons[7]. A related property, which also applies to elementary particles with mass, is their *chirality* (from the Greek word for 'hand'). Jayme Tiomno found a quantum-mechanical operator for chirality (calling it *mass reversal*) in his thesis work[8]. Chiral symmetry is a possible starting point for the discussion of the appropriate description of the Weak interactions[5] and is a fundamental property of nature.

Notes

[1] These three discoveries are related in a bizarre way. *Röntgen* discovered X-rays (still called '*Röntgenstrahlung*' in German) using a *Crookes tube*, an evacuated glass tube with two metal electrodes inside. Applying a voltage between the electrodes causes 'cathode rays' to be emitted from the negative electrode (cathode) and to be accelerated towards the positive electrode (anode). If they strike the glass walls of the tube, a greenish flourescence light is emitted. The French mathematician and scientist *Henri Poincaré* theorized that this fluorescence was the origin of the X-rays, and began a series of experiments with fluorescent materials to verify that. His colleague *Becquerel* knew that certain minerals (containing uranium or thorium) were also fluorescent, and carried out his own experiments with pieces of such minerals attached to a photographic plate wrapped in black paper (to keep it from being exposed by normal light). He however confused *fluorescence* with *phosphorescence*, in which light energy (e.g. from sunlight) is 'stored' in a material and given off slowly as phosphorescence light. He believed that the minerals would have to be exposed to sunlight in order to produce the rays. His first experiment gave a positive result (a shadow on the photographic plate after developing it), and he tried to repeat the result—but the weather had become cloudy, there was no sunlight; so he put the mineral-photoplate packet into a drawer. Some days later, he developed the plate, 'just for fun', and found a much darker shadow from the mineral (which had been in contact with the plate for a longer time). He thus concluded that the 'radiation' was intrinsic to the minerals, and called the phenomenon 'radioactivity'. The English physicist J. J. Thompson meanwhile identified the 'cathode rays' as a stream of particles that carried a negative electric charge, and called them 'electrons'. They were later found to be a component of radioactive radiations, the 'beta rays'. Thus, new and fundamental discoveries were made by three scientists in three different countries within two years, via an incorrect theory (of Poincaré's). This story is told by Abraham Pais in an oral history interview (see Pais (1974)), where he attributes his interest in the History of Science to hearing it.

[2] Nico (2009).

[3a] Frota-Pessôa and Margem (1950). Note that Neusa Margem was later known under her married name, Neusa Amato; cf. Chap. 7, Backnote [11].

[3b] Fazzini et al. (1958).

[4] Rajasekaran (2014); see also Wu (1964).

[5] See for example the essay by Sudarshan and Marshak, '*Chirality Invariance and the Universal V−A Theory of Weak Interactions*', in MacDowell et al. (1991), and other chapters in that book.

[6] A 'right-handed coordinate system' can be represented by the first two fingers and the thumb of the right hand, extended mutually perpendicular to

each other. The index finger, pointing vertically upwards, is the 'z axis'; the second finger, pointing horizontally towards the observer, is the 'x axis'; and the thumb, pointing horizontally to the right, is the 'y axis'. If the left hand is held in the same way, its thumb points horizontally to the left, and it represents a 'left-handed coordinate system'. These two systems can be converted into each other by spatial inversion: all three axes are reversed (x $\rightarrow$ $-$x, y $\rightarrow$ $-$y; z $\rightarrow$ $-$z), and then the whole system is rotated by 180° around the y axis. Left and right hands are, as is well known, mirror images of each other; mirroring a coordinate system in a plane is thus equivalent to spatial inversion. How physical systems or processes change under such a coordinate inversion determines their 'parity'.

It was long believed that all natural laws should be invariant under *parity transformations*; in other words, nature should not distinguish between 'left-handed' and 'right-handed' in its fundamental laws (Fig. 4.4).

[7] Strictly speaking, particles of *light* (photons) have a 'handedness' described by their *helicity*, which is relativistically invariant. This is because the photon is massless, and thus travels at the speed of light c. No observer can travel faster, so the helicity of a photon (the projection of its spin onto its momentum) is the same for all observers.

In the case of particles with *mass* (e.g. electrons, or also neutrinos, as is now known), which can never attain the velocity c, it is theoretically possible for an observer (an 'inertial frame') to travel *faster* than the particle, thus reversing its momentum direction as seen from that observer's frame, and changing the sign of its apparent helicity—the corresponding intrinsic quantity is then the particle's 'chirality'. See also Chaps. 6 and 9, as well as Mehra (1994) and Wu (1964).

[8] Tiomno's γ_5 operator—see Chap. 6—determines the chirality of a particle's wavefunction. He originally called it the 'mass-reversal operation', and noted its connection to time reversal. The name 'chirality' was suggested later by Abdus Salam for the corresponding property of a massive particle; see also Chaps. 9 and 15.

5

The Young Researcher

Graduate Studies—Theoretical Physics in São Paulo

Physics began at the *Universidade de São Paulo* (USP) with the work of *Gleb Vassielievich Wataghin*, who was recruited from Italy in 1934. Wataghin was born in 1899 in the Russian town of *Birzula*, later known as *Kotov* in honor of a Soviet military hero (and now called *Podilsk*). Today, it is in the northwest of the Ukrainian province of Odessa, as can be seen on the map in Chap. 2. His father was descended from a Russian noble family and was a design engineer for the Russian Imperial Railways.

The young Wataghin, together with his whole family, emigrated to Italy in 1919 when he was 19, after the end of the 1st World War and during the chaos of the ensuing Russian Civil War. They fled via the Crimea and Greece in an adventurous manner. In Italy, he subsequently studied physics and mathematics at the University of Turin, where he finished his undergraduate work and continued his graduate education, obtaining his *libera docenza* (university teaching qualification) in 1929. He began research on cosmic rays in 1931. In spite of his youth, Wataghin had by 1933 already met with Ernest Rutherford and P. A. M. Dirac in Cambridge, and with Niels Bohr in Copenhagen.

Wataghin was suggested to Teodoro Ramos as a professor for USP by Enrico Fermi. At first, he rejected the idea of going to Brazil, fearing that he would be scientifically isolated there. However, he was convinced by Fermi with the argument that, as a foreigner, he would have difficulty obtaining a professorship in Italy. Apart from that, the salary offered by USP was good

© Springer Nature Switzerland AG 2020
W. D. Brewer and A. T. Tolmasquim, *Jayme Tiomno*, Springer Biographies,
https://doi.org/10.1007/978-3-030-41011-7_5

compared to those in Europe, which was still suffering from the effects of the 1st World War and the Great Depression. He initially agreed to go to Brazil for only a few months, but when he returned to Italy at Christmas, 1934, he was met by an increasingly rigid fascism, and chose to stay in Brazil for a longer period.

At the *Faculdade de Filosofia, Ciências e Letras* (FFCL) in São Paulo, Wataghin founded a very innovative and productive physics department, both in experimental and in theoretical physics (his own specialty); and in particular he inspired and educated many younger physicists. His rapid success in establishing a functioning research institute is still considered unique by many scientists who know the story.

In 1938, Wataghin was entered in a competition in Italy by his former mentor in Turin, Elizio Teruca. His publications were given to Teruca by Wataghin's brother, who was still in Italy. To his surprise, Wataghin received notice in early 1939 that he had won! He was invited by several Italian universities to apply for a position, accepting only the invitation of the University of *Sássari*, in *Sardegna* [Sardinia], since he could be officially on the list of faculty there and still remain abroad. By 1939, the conditions in fascist Italy were not conducive to his returning there permanently.

Some of Wataghin's early work while he was still in Turin dealt with the quantization of the gravitational field, and the concept of a 'minimal length'. It is now considered to be among the forerunners of modern theories of quantum gravity,[1] although it met with considerable skepticism at the time.

In 1937, Wataghin was also able to bring Giuseppe ('Beppo') Occhialini to São Paulo. Occhialini was also already a recognized physicist, mainly for his confirmation, together with P.M.S. Blackett in Cambridge in 1933, of the existence of the *positron* (the anti-electron, described in Chap. 4). It had been proposed by P. A. M. Dirac and first detected by C. D. Anderson at Caltech (Nobel prize for physics, 1936. Anderson was also the co-discoverer of the *muon*). Occhialini was openly antifascist and had to emigrate from Italy. Together with Wataghin, he helped to establish the Physics Department at USP.

Wataghin began to attract talented young students with an interest in physics, using the strategy of giving them a basic education and then sending them to study abroad. He was responsible for the education of numerous Brazilian physicists (he is considered by many to be the "Father of modern Physics" in Brazil).[2] His many students formed the basis for physics research and teaching of an entire generation. As he himself said later, "I educated them with the help of great physicists from all over Europe, from Germany, England and Italy".[3]

One of those young physicists was *Mario Schenberg*, considered by Wataghin himself to have been his best student. Schenberg was born in 1914 in Recife (state of *Pernambuco*), where he completed his schooling and began his studies at the School of Engineering in 1931. There, he soon came to the attention of Luiz Freire, who recognized the young man's unusual abilities and convinced him in his third year (1934) to move to São Paulo to attend the *Escola Politécnica*. His obvious abilities in mathematics attracted Wataghin's attention, and owing to political questions – Schenberg was Jewish and a Communist – Wataghin himself decided to attend to his progress. As soon as he concluded his time at the *Politechnica*, he was contracted by Wataghin as assistant for General and Experimental Physics and then as Assistant Professor of Theoretical Physics.

After two years in São Paulo, in 1936, Schenberg was invited to go to England to work with Dirac; but on the way there, together with Wataghin, he visited Enrico Fermi in Rome, and Fermi persuaded Schenberg to stay and work with him on quantum field theory, in particular on the theory of electron showers. Fermi had already been in Brazil in 1934, invited by Wataghin to address a conference on Pauli's proposal of the existence of neutrinos. When Fermi was forced to leave Italy in 1938 by the increasingly aggressive fascist government, Schenberg went to Zurich to work with Wolfgang Pauli. After that, he spent some time at the *Collège de France* in Paris, working together with the Joliot-Curie group, where he delivered a seminar on nuclear physics, based on John Wheeler's work on the energy levels associated with nuclear rotation. As he once mentioned, in all of these places he was given a topic to study and then the opportunity to present a seminar lecture about his results.

When Mario Schenberg arrived back in Brazil, Wataghin later recalled,[3]

> Schenberg returned a different person. He had learned much more than I could have taught him. From then on we collaborated. He did beautiful work on cosmic rays, and then started to work in electrodynamics, under the direction of Dirac. He had learned a lot in Rome, and I decided that he did not have much more to learn from me and should travel again soon.

Later, Wataghin introduced Schenberg to the mathematician *George Gamow*, who was visiting Brazil, and Schenberg went to work with him in 1940 at George Washington University in the USA, on a Guggenheim fellowship. There, he participated in Gamow's studies of the theory of stellar novas and supernovas, which presupposed the existence of neutrinos, and completed one of his most important works (on the 'Urca process', an astrophysical phenomenon, named by Gamow humorously for the Casino that then operated in the Urca district of Rio de Janeiro[4]). Schenberg subsequently became a fellow at the Institute for Advanced Studies at Princeton, where he met John Wheeler, and finally he spent some time at the Yerkes Observatory, working with Subrahmanyan Chandrasekhar, who was later to became a Nobel laureate for his astrophysical research. In 1942, Schenberg again returned to Brazil, obtaining a professorship at USP in 1944 (the Chair of Rational (Classical) and Celestial Mechanics) (Figs. 5.1, 5.2). Another of Wataghin's assistants at this time was *Marcelo Damy de Souza Santos*, an experimentalist. He went abroad the same year as Schenberg, to Cambridge/UK to work with Hugh Carmichael there. We shall meet up with him again in later chapters as an important actor at USP and then at Campinas.

Fig. 5.1 Glab Wataghin in São Paulo in 1936. *Source Acervo IF USP*. Public domain. Reproduced with permission. Online at: http://acervo.if.usp.br/bio01

Fig. 5.2 Mario Schenberg, 1936. *Source* Wiki commons. Public domain. Online at: https://commons.wikimedia.org/wiki/File:M%C3%A1rio_Schenberg_formatura_1936. jpg

Tiomno's Interlude in São Paulo—Encountering Modern Physics

As soon as the 2nd World War ended in Europe in May of 1945, Jayme Tiomno was discharged from the military, and in 1946, he obtained a fellowship for graduate work at USP, under the supervision of Mario Schenberg.[5] At USP, Schenberg gave Tiomno a first assignment to study the textbook *'A treatise on the analytical dynamics of particles and rigid bodies'* by Edmund Whitakker, as well as some articles by Max Born on the quantum-mechanical theory of electrons. Tiomno was very enthusiastic and aimed to finish his reading within three months in order to be able to work on Schenberg's theory and to devote himself as soon as possible to research. He however suffered considerable difficulties at the beginning, since Sobrero had concentrated on classical mechanics and Tiomno had not yet had the opportunity to systematically study quantum physics. In a letter to Costa Ribeiro, he confessed that "I have so much to learn that I don't know where to begin".[5] To Elisa Frota-Pessôa, who remained in Rio and was working as Costa Ribeiro's assistant, he suggested that she should attend the course given by Leite Lopes, who had recently returned from Princeton, and should study hard so that she would not encounter similar difficulties when she also went as planned to São Paulo.[6]

Tiomno's deepest previous contact with quantum mechanics had arisen from the idea that it would be useful to explain the causes of the thermodielectric phenomenon described by Costa Ribeiro. The latter had mentioned in a lecture to the *Academia Brasileira de Ciências*[7] that:

> Another approach to finding the fundamental mechanism of the phenomenon was suggested by J. Tiomno, and consists of considering it from a purely electronic point of view, assuming that, in a given substance, the normal electron densities are different in the solid and liquid phases. In accord with this hypothesis, one would expect an electron transfer across the phase boundary to occur during solidification, which would tend to re-establish the normal value of the electron density in the recently-formed solid layer, and this transfer would constitute the thermodielectric current.

Tiomno continued his studies. In the laboratory, he helped to give practical instruction on electric circuits to the students, and he began constructing amplifiers with coincidence circuits for cosmic-ray counting equipment. He was delighted to see that the students were given research projects to be carried out during the course, contributing to their education and initiating them into experimental research work. This was a model for instruction that stimulated interest in research.

At the same time, Tiomno attended Schenberg's seminars on P.A.M. Dirac's famous textbook, '*The Principles of Quantum Mechanics*'. Schenberg made use of the method applied in Europe, passing out material for the students to study and then having them present it in the seminars. Tiomno was encouraged to give a seminar on the '*Theory of the properties of Metals and Alloys*' by (Sir) Neville Francis Mott.

The FFCL continued its policy of inviting physicists from outside the University and from abroad to give seminar talks. Franco Dino Rassetti visited São Paulo in August of 1946 to give a seminar on nuclear physics. Schenberg also wanted to invite Wheeler, whom he had met in Princeton. The latter, however, had already agreed to give a vacation course at Columbia University, and could go to São Paulo only sometime later.

Tiomno began collaborating with another of Schenberg's assistants, Walter Schützer, on a review of the calculations on the classical theory of pointlike electrons with spin, based on Schenberg's ideas on the theory of electrons which, according to Tiomno, "were labor-intensive".[8] When they had concluded the review, they began working on new calculations which had not previously been carried out, and which were "still more laborious". This work led Tiomno and Schützer to publish the article '*Radiation field derivatives of an oscillating point-like electron with spin*' ([JT3]) in the *Annals of the Brazilian Academy of Sciences.*[9]

About this period in São Paulo, Tiomno later said,[10]

In 1946, I obtained a fellowship for post-graduate studies at the Department of Physics of the *Faculdade de Filosofia, Ciências e Letras* at the University of São Paulo, under the guidance of Professor Mario Schenberg. This was the beginning of my initiation into modern physics, since my previous education had encompassed only classical physics. I filled in the gaps in my education and recuperated from the interruption of my physics work during the war, carrying out my first research work in modern physics with the guidance of Professor Schenberg.[10]

When his fellowship concluded, Tiomno returned to Rio in early 1947 and resumed his activities as assistant professor to the Chair of General and Experimental Physics at the *Faculdade Nacional de Filosofia*. He returned with various ideas for research and with the certainty that he wished to work in theoretical physics. During that time, he met a visitor to the FNFi, the theoretician *Guido Beck*, with whom Tiomno began a collaboration. Beck was Jewish, from Austria, and had left Germany after 1933, where he had been Heisenberg's assistant in Leipzig, to work in various countries (the USA, the Soviet Union, France and Portugal), fleeing the Nazis and searching for a suitable place to continue his work. He arrived in Argentina in 1943 to work at the Observatory in Córdoba. With Beck's guidance, Tiomno began to work on an extension and systematization of the theory of certain types of nuclear collisions. Beck was enthusiastic over the progress of these calculations.[11]

Tiomno also started to interact with his colleagues from the FNFi. Together with the mathematician Leopoldo Nachbin, whom he knew from his undergraduate years, he prepared a paper '*On Sobrero's theorem of hyper-complex algebra*'. However, it was never submitted for publication.[12] And he also started to discuss some ideas with José Leite Lopes, which would lead to joint articles in the near future.

His most notable encounter, though, was with Elisa Frota-Pessôa. She had in the meantime separated from her husband, Oswaldo Frota-Pessôa, and she and Jayme began an intimate relationship. However, in mid-year of 1947, Jayme received an invitation to serve as assistant to the Chair of Theoretical and Mathematical Physics of the FFCL, at USP. He couldn't pass up this opportunity. He decided to return immediately to São Paulo, and they would try to get a fellowship for Elisa also to go to São Paulo in the near future.

Back at USP, he began giving a lecture course on classical mechanics, and prepared a seminar on the resonances in the scattering of neutrons from helium, which was particularly interesting to him because of its connection to the work that he had begun developing with Guido Beck. Tiomno was also studying the publications of Werner Heisenberg and of Hans Bethe, and had started to learn German, which he considered indispensable for being able to read much of the physics literature at the time. [For example, Eugene Wigner's important text on the applications of group theory and symmetry principles in physics was for many years available only in German; likewise Max Born's textbook on quantum mechanics].

Tiomno also started to work on a paper with José Leite Lopes which would become his first article to appear in an international journal (the *Physical Review*, in 1947; [JT4])[13]; it dealt with proton-proton scattering at a (then) accessible energy of 14.5 MeV, and may have been suggested by Schenberg as a theoretical analysis of current experiments. A second paper on an aspect of the same topic, the '*Angular distribution from proton-proton scattering at 14.5 MeV* '; [JT7],[14] was later published by Tiomno alone in the *Annals of the Brazilian Academy of Sciences* (1949); it dealt with a specific feature of the work begun at USP in 1946/47.

At the same time, Schenberg introduced him to General Relativity and gravitational theory. They prepared an article describing an alternative formulation of gravitation, constructed in Minkowski space.[12, 15] In this work, they found a deviation of light [the bending of a light beam by a large mass such as the Sun] which was 25% greater than that predicted by Einsteinian General Relativity. In the solar eclipses observed by a Russian expedition in 1936 and by the Americans in 1947, the data evaluations also pointed to a larger deviation.[16] This was however unknown to Tiomno and Schenberg at the time; they knew only that some observed deviations had been greater than predicted by Einstein.

Schenberg sent the article to the *Astrophysical Journal*, but it was not well received. Based on the evaluation he received from the referee, Schenberg concluded that they "would not publish material on General Relativity". Subrahmanyan Chandrasekar, a member of the editorial board (and future managing editor after 1952) sent the article to England, but it was not well received there, either. The referees had doubts as to whether the work fulfilled the equivalence principle (the foundation of General Relativity). Schenberg maintained that their doubts were unfounded, and planned to submit the article to the *Physical Review* and to the *Annals of the Brazilian Academy of Sciences*.[17] It failed to appear in either, however.[12] Leite Lopes later maintained that no non-Einsteinian interpretations of gravitation were accepted by mainstream journals at that time.[18]

In February 1946, just while Tiomno was in São Paulo for his first stay there, another assistant of Wataghin's, the young experimentalist *César Lattes*, traveled from USP to the University of Bristol in Great Britain to work there with Cecil Powell, who had been developing a technique for observing the tracks of ionizing radiations in photographic emulsions, and with Giuseppe Occhialini, who had gone from Brazil to England in 1944.

[The use of photographic plates to detect radiations was not new—indeed, Becquerel had originally discovered radioactivity in 1896 by exposing a wrapped photographic plate to a radioactive mineral, finding its shadow when he developed the plate. More recently, the technique had been further perfected by *Marietta Blau*, in Vienna, who was the first to observe 'stars' in photographic emulsions due to nuclear disintegration by cosmic-ray impacts. Her contributions have only recently been acknowledged, although it was in fact her work that stimulated Powell's program to improve photographic emulsions as particle detectors; she worked with the British firm Ilford, as did Powell later.][19]

César Lattes had already been working on cosmic-ray detection; cosmic rays had been a favorite research topic of Gleb Wataghin's since the early 1930's. In Bristol, Lattes further improved Powell's nuclear emulsions by having borax added to them, increasing the visibility of the particle tracks by reducing fading. Occhialini tested the new emulsions on the *Pic du Midi*, where the high altitude favored the observation of cosmic rays. They observed a few tracks which might indicate a new particle, but only two presumed *pion* events, not sufficient evidence for a publication. Lattes' youthful energy and enthusiasm doubtless accelerated the discovery considerably—supported by the laboratory in Bristol, he travelled to the *Chacaltaya* peak in Bolivia to expose his plates at a still higher altitude. With the newly-exposed plates, which showed over 30 pion events, he, along with Powell and Occhialini, was able to identify the meson and visualize its decay chain via muons. They quickly published the discovery.[20]

Their observation resulted in the 1949 physics Nobel prize for Hideki Yukawa, who had predicted the particle 14 years earlier. Powell also received the Nobel prize in 1950 for his development of the emulsions used in the pion detection.[21]

A number of physicists began to study this new particle, the π meson or *pion*, and its decays. Among them, Tiomno also became interested in pion-muon physics, stimulated by a seminar given by César Lattes in the second semester of 1947.[22] Lattes and Tiomno became very good friends at the time, and Tiomno was invited to be best man at the wedding of César Lattes to Martha Siqueira Neto, from Pernambuco, in January of 1948.

A Hasty Departure for Princeton

Already foreseeing that Elisa would go to São Paulo, Jayme suggested to her that she should study the works of Heisenberg and of Marcel Schein, since she would probably be working on cosmic rays at USP. In the second semester of 1947, Tiomno's colleague at USP, Walter Schützer, went to Princeton to begin graduate work with John Wheeler and Eugene Wigner. Probably on the recommendation of Schenberg and of José Leite Lopes, who had also done graduate work at Princeton, Tiomno himself applied for a fellowship there, expecting to go to the USA in the middle of 1948. Guido Beck had obtained a scholarship for him to spend the vacation months of January and February in Córdoba, where he could continue with the work that they had jointly started.[12]

During this time, Tiomno continued studying and 'catching up' on developments in modern physics; he read works by Walter Heitler, P. A. M. Dirac, and Wolfgang Pauli, and on the Compton effect, bremsstrahlung, pair production and the absorption of radiation by matter, etc. He also continued studying German. He was now calculating the modification of Coulomb's law for the electric field of a proton near a neutron, due to the mesonic charge during virtual particle formation. Such 'mesonic corrections' are important for exact calculations of the properties of the nucleons, which are composite particles rather than true elementary particles.

Shortly after the confirmation of Elisa's fellowship to spend a year in São Paulo, Tiomno received a positive answer from the U.S. Office of Education (Department of State), promising support for his graduate work at Princeton in 1948. He had expected to go there only in July, at the beginning of the next academic year. This would have allowed him to carry out his plan to go to Córdoba to work with Guido Beck, and would still have left some time to be with Elisa in São Paulo. However, at the beginning of January, only a few days before he was to leave for Argentina, he was informed that he would need to be in Princeton by the 7th of February at the latest, risking the loss of his fellowship there if he arrived later.

He had to make many arrangements in a very short time. Apart from that, the fellowship would not pay Tiomno's passage to the USA, and he didn't have sufficient funds to afford it himself. He attempted to apply for aid from Brazilian sources, such as the *Fundação Getúlio Vargas*, but without success. Arthur Moses, the president of the Brazilian Academy of Sciences (ABC), arranged for him to ride as a 'supernumerary' on a Panair flight (a Brazilian airline which no longer exists); it took a plane back to New York for servicing. One of its engines was overdue for service, so the flight couldn't carry regular passengers. In Tiomno's own words,[10] "Thus, only the crew and some two or three adventurers, psychologically prepared to jump out with a parachute at any moment, went along...". In a much later lecture,[23] he describes their landing in snowy New York, where the plane threatened to slide off the end of the runway, resulting in a last-minute aborted landing by the pilot and a second, successful landing attempt, with firefighters at the ready. At least his military training was finally being put to good use!

Notes

[1] Hagar (2014).
[2] e.g. Frota-Pessôa (1990); Tiomno (1977).
[3] Wataghin (1975).
[4] 'Urca (Rio)' (an acronym for *Urbanização Carioca*) refers to the development of what was then a suburb of Rio in the early 20th century. The process described by Gamow and Schenberg is the cooling of compact stellar objects (e.g. neutron stars) by the emission of neutrinos. Schenberg suggested that the heat must disappear like the money of roulette players at the Casino in the Urca district, being quickly exchanged, and Gamow found that idea to provide an amusing name for the process.
[5] Jayme Tiomno to Joaquim da Costa Ribeiro, April 18th, 1946; [CR].
[6] Private correspondence, property of the family of Elisa Frota-Pessôa; and personal communications by Sonia Frota-Pessôa, 2018/19; cf. [EFP].
[7] Silva Filho (2013, pp. 197–198).
[8] Jayme Tiomno to Joaquim da Costa Ribeiro, June 8th, 1946; [CR].
[9] cf. [JT3], Appendix B.
[10] Tiomno (1977).
[11] Archive of Guido Beck, Report from Guido Beck to the head of the Physics Department at the FNFi/UB, July 8th, 1947; [GB].

[12] Fornazier and Videira (2018).

[13] See [JT4].

[14] See [JT7].

[15] cf. Bassalo and Freire (2003).

[16] Mario Schenberg to Guido Beck, October 8th, 1947; [GB].

[17] Mario Schenberg to Jayme Tiomno, July 30th, 1948; [JT].

[18] Leite Lopes (2004).

[19] See Sime (2012).

[20] Lattes et al. (1947, 1948).

[21] Compare Chap. 17, section on 'The Nobel prize question'.

[22] Tiomno (1984/94).

[23] [JT83].

6

Princeton—In Pursuit of the Universal Fermi Interaction

Princeton, 1948

On February 6, 1948, Jayme Tiomno, the son of Russian Jewish immigrants, raised in small towns in the interior of Brazil, arrived in New York City on his way to continue his postgraduate studies at Princeton University, in neighboring New Jersey. In spite of his harrowing flight on a plane due for servicing, he arrived safely in the USA. His São Paulo colleague Walter Schützer, who had also gone to Princeton somewhat earlier to obtain a Masters degree there, working under Eugene Wigner, was awaiting Tiomno's arrival.

Two things impressed Jayme Tiomno from the outset. The first was discovering that Princeton did not accept women as students, only men. He was encountering an even more 'patriarchal' society than in Brazil! The other was the ritual of dining together with all the graduate students, which took place promptly at 6:30 pm every day: "… for dinner, we go into an enormous and imposing hall (there are 200 students living in the Graduate School) and wait, standing, for a person wearing an academic robe; when he enters, he waits for perfect silence and then says a brief prayer (everyone bows his head and some pray along with him)—then the meal begins" [1]. He also discovered that he was expected to write term papers and prepare for final exams, a new custom at Princeton at the time, which he considered ridiculous (although he admitted that it had the advantage of forcing him to study!). However, unlike his colleague Schützer, who took the exams on arrival in Princeton, Tiomno decided to defer them until the following year. He wanted to begin research immediately, and become integrated into the department and his research group as soon as possible. [1]

© Springer Nature Switzerland AG 2020
W. D. Brewer and A. T. Tolmasquim, *Jayme Tiomno*, Springer Biographies,
https://doi.org/10.1007/978-3-030-41011-7_6

Tiomno's advisor at Princeton was *John A. Wheeler*, who recommended that he take some graduate courses, as is usual for beginning graduate students at American universities. He registered for the courses of Rudolf Ladenburg on nuclear physics and of Eugene Wigner on statistical mechanics,[1],[2] as well as continuing his studies of German ("I wouldn't recommend to anyone to go outside Brazil [*to study*] without knowing some German", he wrote to Elisa). He was also auditing Wigner's course on General Relativity, "to accustom my ears [*to English*]"[1] (Fig. 6.1).

Wheeler, like Tiomno, was at the time thinking about pion-muon physics. However, he had heard from Schenberg of Tiomno's work in São Paulo on gravitational theory, and had begun thinking about that, as well. He wanted to calculate a 'radiation damping' term corresponding to the one introduced by Dirac for a pointlike electron. Thus, Wheeler initially set Tiomno to work on General Relativity problems—this was somewhat before the period when Wheeler was beginning his famous shift from "Everything is Particles" to "Everything is Fields"; he himself dates that shift to May, 1952.[3] Wheeler suggested that Tiomno should study the motions of particles of negligible mass in General Relativity. Tiomno himself was not very enthusiastic about that suggestion. He did not believe in gravitational radiation and felt that

Fig. 6.1 Jayme Tiomno at Princeton, 1948/49, studying. *Source* Private, archive [JT]

Wheeler's proposal was premature, more speculation than theory. In addition, he was still somewhat traumatized by the 200 pages of calculations produced in his work with Schenberg in São Paulo (which had *not* led to a publication). However, he applied himself to the problem to see if he could obtain useful results, since it was important for him to establish a working relationship with Wheeler [4].

After two months, he still had no results and was not satisfied with Wheeler's guidance on the project. It seemed to him that Wheeler was not familiar with the relevant literature. "In fact, it appears that he has never worked in General Relativity", he wrote to his colleague Leite Lopes. Tiomno and Wheeler even had a meeting with Einstein to ask for his guidance; it did not help much, however, although he received some "good criticisms". Tiomno felt that at least he was learning a lot about General Relativity; he gave himself another 15 days to arrive at a satisfactory result. If that failed, he would ask Wheeler to suggest another problem and allow their ideas some time to mature more.[5] After considerable work, his difficulties were overcome and the article was completed, but in the end it was never published. Leopold Infeld and Alfred Schild[6] presented similar results in a more complete formulation just as it was about to be submitted for publication.[7]

Meson Physics with John Wheeler

John Archibald Wheeler (1911–2008) studied physics at the Johns Hopkins University from 1927–1930, then continued with graduate work there to obtain his Ph.D. in 1933, aged 22, with a thesis on the '*Theory of the Dispersion and Absorption of Helium*'. His thesis advisor was Karl Herzfeld. Under a National Research Council fellowship, he spent postdoctoral sojourns with Gregory Breit at NYU (1933/34) and with Niels Bohr in Copenhagen (1934/35). From 1935–1938, he was assistant professor of physics at the University of North Carolina (Chapel Hill), where he was offered a tenured position, which he however turned down in order to move to Princeton.

At Princeton, Wheeler was initially an assistant professor, advancing to Full Professor (during the period of Tiomno's stay), and finally *Joseph Henry Professor* for the last 10 years there before his retirement in 1976. He then worked for 10 years at the University of Texas in Austin, as a Senior Professor, before returning to Princeton, where he spent his final years.[3] Wheeler died in April, 2008, near Princeton.

Among Wheeler's important early contributions were his introduction of the S-matrix in 1937, his work with Edward Teller on the 'liquid-drop' model of nuclei in 1938, and his collaboration with Niels Bohr on nuclear fission in 1939/40. He developed the concept that positrons are electrons traveling backwards in time, which was employed by his graduate student Richard Feynman in his 'Feynman diagrams'.

Wheeler, although he never received the Nobel prize himself, was certainly one of the most influential theoreticians of the twentieth century, and he had many well-known and highly successful students, of whom an unusual number were later themselves Nobel laureates. He introduced numerous important ideas into physics, working in three major fields which he himself termed 'Everything is Particles', 'Everything is Fields', and 'Everything is Information'.[3] Wheeler is known for injecting ideas and catchwords into physics research, and some of them became widely known outside the professional literature. For example, he re-quoted the motto (slightly modified), "Time is what keeps everything from happening at once", originally expressed in a science-fiction story by Ray Cummings in 1922. And he popularized the term 'black holes' for gravitational singularities (or 'collapsed stars') beginning in 1967/68, after an (anonymous) listener at one of his lectures shouted it from the audience; it had however already appeared in print by that time. Both are often incorrectly attributed to Wheeler as originator. A list of his more prominent quotes can be found on the Web.[8]

Wheeler came into contact with Brazilian physicists early on: he certainly met Mario Schenberg while the latter was working at the Institute for Advanced Studies in Princeton in 1941/42, and likewise José Leite Lopes during his doctoral work at Princeton in 1944/46. Wheeler himself was on leave during the War, and worked from January, 1942 until August, 1945 for the Manhattan Project on reactor design and plutonium production, in Chicago (at the 'Metallurgical Laboratory'); in Wilmington, Delaware, where the DuPont company designed the reactors for plutonium production; and in Hanford, Washington, where the plutonium facility was built. He returned to Princeton in 1945.[3],[9]

In early June of 1948, after nearly four months of calculations on General Relativity, Tiomno heard a seminar talk given by Wheeler, presenting a preview of papers to be delivered at the forthcoming Caltech conference on cosmic rays, scheduled for June 21–23, 1948 in Pasadena/CA. Wheeler described calculations of the electron energy spectrum from the 3-body decay of the muon ($\mu \rightarrow \mu_0 + e + \nu$), assuming a Fermi-type interaction, which Tiomno had also already considered while still in São Paulo. It developed that Wheeler had presented a preliminary version of this work at the Pocono meeting in April, 1948, unbeknownst to Tiomno. He had found a coupling constant (describing the strength of the interaction) which was comparable to the Fermi coupling known from nuclear β decays.

This was quite exciting for Tiomno, who had already started working on similar problems and "could have done the first calculation in Brazil".[10] It gave him the opportunity to return to the work that he had interrupted on leaving Brazil, first to make the trip to Princeton and then for his project on gravitational theory. In a letter to Leite Lopes written while he was still in São Paulo,[11] he had speculated at length about the interactions between nucleons, intermediate mesons, and electrons, considering the relative strengths of their possible coupling constants and the depths of the corresponding potential wells. (This touches on what today is called the *Strong interaction*, which is responsible for binding the nucleons within an atomic nucleus). In that letter, Tiomno also quotes an article by Bruno Touschek,[12] who examined data for the excitation of nuclei by scattering of electrons and concluded that the Fermi interaction is too weak to explain the observed effects. Tiomno speculated that an analogous but stronger interaction might be at work. Schenberg however had not accepted that suggestion, preferring to calculate the electrostatic interaction between the charges of the electrons, the protons, and the intermediate mesons.

35 years later, in his paper for a conference celebrating the 50th anniversary of the Weak interactions, Tiomno recalled the initial insights that he had while still in Brazil:[10]

After the 2nd World War, I started to study in Rio de Janeiro and then in São Paulo, cosmic rays and particle physics. I [*would*] like to mention a seminar during the second semester of 1947, given by [*César*] Lattes, a younger colleague of ours who had participated actively in the discovery of π mesons in the $\pi \rightarrow \mu + \mu_0$ decay, in Bristol. ... We were excited since the puzzle resulting from the Conversi-Pancini-Piccioni experiment[13] would be solved if the π was the meson involved in nuclear forces, say, the Yukawa meson. But why did the μ meson have a weaker interaction with nucleons? I proposed then that μ and μ_0 were not Yukawa mesons, but spin-½ particles forming an iso [*spin*] doublet. The μ capture, $\mu^- + p^+ \rightarrow n + \mu_0$, then did not have to go fast. It was objected that the cosmic-ray mesons decayed as $\mu \rightarrow e + \nu$ and not as $\mu \rightarrow \mu_0 + e + \nu$, as I was suggesting. ... I must mention that I was not aware, at the time, of the Bethe-Marshak paper ... on the two-meson hypothesis which preceded [*the*] π discovery.[14] I had already been accepted as a graduate student at Princeton to work with John A. Wheeler starting in January 1948. So I stopped thinking about this problem and, at Princeton, I became involved with graduate courses and in General Relativity research with Wheeler.[10]

Tiomno however had not abandoned his idea of a Fermi type of interaction. After hearing Wheeler's talk, he quickly estimated the coupling constant for muon capture by nuclei based on that assumption, and again obtained a value close to that from β decay. "In the midst [*of the work on General Relativity*], I was thinking about problems concerning mesons and had an idea – did some calculations on an interesting aspect and showed the results to Wheeler, and he was interested—so it seems that from now on, I will be working on mesons, let's see…", he wrote to Elisa[15] (Fig. 6.2).

In fact, this initiated a lively collaboration between Wheeler and Tiomno, which resulted in five publications in major journals within the year, and set the tone for Tiomno's Masters research. Tiomno and Wheeler felt themselves

Fig. 6.2 Wheeler and Tiomno in Princeton, probably around the time of the AAAS meeting in September, 1948. *Source* Private, archive [JT]

to be on the trail of a *universal* Fermi interaction, and set out to make a clear-cut case. Wheeler reported on their first results at the Pasadena meeting, while Tiomno started working with great intensity. "I am in a phase which is not too good for studying, since I am having a great number of ideas (mostly stupid ones), and so I have a tendency to be easily distracted", Tiomno wrote to Elisa.[16] In fact, his ideas were not stupid, and it was no doubt this period that prompted Wheeler to say later that Jayme Tiomno was one of the three students who worked hardest and with whom he could work best.[17] (The other two were Richard Feynman and Robert Marshak, both of whom made major contributions to weak-interactions theory).[18]

Tiomno wrote to his family in Brazil,[19]

I have been working hard in the past weeks. I am writing two articles right now. One of them is on the meson decay—it was already nearly finished some 20 days ago, but then I discovered that a few simplifying assumptions used by Wheeler were not valid, so I had to redo the whole calculation using more precise assumptions (having to go against relativistic effects). The theoretical calculations are now finished, but the numerical computations are immense, since I am considering all the possible cases. I hope to finish by next week (this work is a collaboration with Wheeler).

The other article is on the capture of mesons by atomic nuclei—my idea, but under Wheeler's guidance. The preliminary results that I sent to Leite [*Lopes*] refer to this project, and he used them in the work that he presented to the *Academia*. However, there is still a bit (a lot) to do before it will be finished.[19]

The two articles that Tiomno mentions were both published jointly with Wheeler in 1949, in *Reviews of Modern Physics*. The first of them ([JT5], '*Energy spectrum from mu-meson decay*') contained Wheeler's calculation of the electron spectrum from muon decay, supplemented by Tiomno's work; and the second ([JT6], '*Charge exchange reaction of the mu meson with the nucleus*') presented their calculation of the rate of muon capture by (nuclear) protons, the 'charge exchange reaction'. Two more works were presented as papers at the New York meeting of the APS in January, 1949, on the coupling and on the particle spins in the $\pi - \mu$ decay chain; the abstracts of those papers were published in the Proceedings of the meeting in the *Physical Review*[20]; and finally, they published a review article on particle physics, which appeared in the *American Scientist*.[21]

Aside from the report by Wheeler at the Pasadena meeting, he and Tiomno made the first public announcement of their ideas on weak interactions at the

September 1948 *Centennial Meeting of the AAAS* (American Association for the Advancement of Science) in Washington/DC (Fig. 6.3).

Their work was also noted in the September 1948 issue of *Science News Letter*, with a photo of Tiomno and Lattes (who had come from California for the meeting), together with John Wheeler, Isidor Rabi and Willis Lamb:

Tiomno also published two papers in 1949 as sole author, one on his work on proton-proton scattering at USP,[22] and another, in the *Physical Review*, on the spin of the muon (his Masters research work).[23] As he wrote ironically, "I am even convinced that the three most important abilities needed by a fellowship holder here are (1st), initiative; (2nd), initiative; and (3rd), taking a maximum of initiative"; and he continued, "Naturally, it is also important to know German".[24]

Fig. 6.3 A contemporary magazine report of the AAAS meeting in September, 1948. *Seated*: Lattes, Wheeler; *standing behind*: Tiomno, Lamb, Rabi. *Source* Science News Letter, Sept. 25th, 1948, p. 199. Reproduced with permission from the publisher. https://www.sciencenews.org/archive/trumans-aaas-address

The Tiomno Triangle

In the second Tiomno-Wheeler article in the *Reviews of Modern Physics* ([JT6], early 1949), Tiomno included a diagram that expressed the idea of the *universal Fermi interaction* (UFI) in a very graphic way. At that time, the only processes known which might be described by the Fermi (Weak) interaction were nuclear beta decays (including negative-electron decays, positron decays, electron capture and the free neutron decay), and the pion-muon decay chain and reactions (pions decaying to muons or electrons, muons decaying to electrons, muon capture by nuclei). It had been Tiomno's idea while still in Brazil, after hearing César Lattes' seminar on the pion decay, that all of these processes might be mediated by the same type of force: Fermi's 'four particle, two-current' interaction, first suggested in 1933/34—thus, a *universal* Fermi interaction.

Fermi's original paper on that theory, which he submitted to the prestigious British journal *Nature*, was rejected on the grounds that it was "too speculative". This was a classic error on the side of scientific conservatism, as later admitted by the *Nature* editorial staff. He published it instead in 1933 in Italian in the obscure journal *La Ricerca Scientifica*, and in 1934 in *Il Nuovo Cimento*, as well as in German in the *Zeitschrift für Physik*. Its refusal by *Nature* prompted him to switch to experimental physics, where he had great success with neutron capture experiments, discovering many radioactive isotopes (and stimulating Otto Hahn and Lise Meitner to begin the research program that eventually led to the discovery of nuclear fission). Fermi was awarded the Nobel prize for physics in 1938 for that work. He later returned to theory.

For the 1949 *Reviews of Modern Physics* paper on muon capture, Tiomno provided a diagram that he had drawn to illustrate graphically the principle of the UFI. The first person to see it was Elisa Frota-Pessôa, in a letter on October 18th, 1948, as reproduced in Fig. 6.4.

Below is the printed version as it appeared in the original 1949 article[25] (Fig. 6.5).

Many years later, John Wheeler made a simple hand-drawn version, similar to that in Tiomno's letter to Elisa (above), for his autobiography[26]; it was important to him to emphasize its origin.

The three vertices of the triangle represent the three groups of particles whose currents interact via the Fermi-type interaction: N, P are the *nucleons* (neutron, proton) of nuclear beta decays and the neutron decay, while μ, μ_0 are the muon and its neutrino (now termed ν_μ), and e, ν are the electron and *its* neutrino (now termed ν_e). The lower two vertices contain what are now

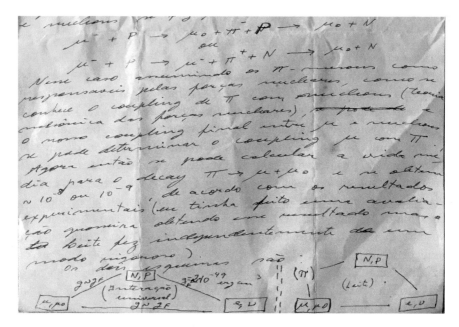

Fig. 6.4 Part of Jayme Tiomno's letter to Elisa on October 18th, 1948, showing his early versions of the 'Tiomno triangle' (at the bottom of the page). *Source* Letter from JT to EFP, Oct. 18th, 1948. Private archive, [EFP]

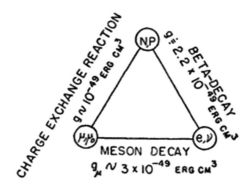

Fig. 6.5 The 'Tiomno (-Wheeler-Puppi) triangle': here the printed triangle as it appeared in the Tiomno-Wheeler article in *Rev. Mod. Phys.* **21**, 144 (1949).[25] Reproduced with permission.

called *leptons* (light particles), and these were the only leptons known in the 1940s (there are now known to be three 'generations' of leptons, the heaviest being the tau particle or *tauon*, τ, and its neutrino, ν_τ).

The upper vertex in fact represents *all* of the strongly-interacting particles, now termed *hadrons*. The existence of the *Strong interaction*, presumed to be responsible for the nuclear binding force, had been supposed even before the

Weak interaction was suggested. The nucleons would clearly be subject to this interaction, in addition to the Weak and Electromagnetic interactions; but then so would the recently-discovered *pion*.

The division of the hadrons into two subclasses, according to their masses, was already apparent in the late 1940s: the *baryons*, or heavy particles (in 1948, only the *nucleons* n and p); and the *mesons*, or particles of intermediate mass (in 1948, only the *pion*—K mesons ('V particles') had been seen but not yet identified).

The mass of the electron, given as an equivalent energy according to Einstein's famous equation $E = mc^2$, and expressed in *electron volts*, eV (the energy that an electron gains when it passes through a potential difference of 1 V), is 511,000 eV (or 511 keV). The nucleons are almost 2,000 times more massive: The proton's mass is 938.3 MeV (million electron volts), and the neutron's is 939.6 MeV. The π meson has an intermediate mass, equivalent to 139.6 MeV, about 275 times the electron mass.

The classification as 'baryons' or 'mesons' is still used, but it now refers not to the masses, but instead to the elementary-particle constituents within the hadrons: baryons contain *three* (valence) *quarks*, and mesons contain *two* (as quark-antiquark pairs). Thus, in its more general form, the triangle's upper vertex should represent all of the *hadronic* currents. (It has been mentioned, for example by J.J. Sakurai in his book '*Invariance principles and Elementary Particles*',[27] that the triangle could be extended by a fourth vertex to include the *hyperons*, another subclass of hadrons, namely those baryons that contain 'strange' quarks, i.e. quarks of the second generation. It would then become a tetrahedron).[27] Another suggestion for modifying the triangle was made by Tiomno's fellow student and later colleague of many years, José Leite Lopes, who proposed in a review article to add the pion in the center of the triangle, with additional diagonal lines indicating the nature of the couplings.[28] The paper containing the original Tiomno triangle became a classic, and was later republished in a series of selected papers by the Physical Society of Japan, in a volume on the '*Weak interactions in particle physics*',[29] which included the most important works on that topic.

The story of this triangle and its reception in the physics literature is a somewhat unfortunate chapter, an example of the inaccurate and confused attributions which seem to have occurred often in the history of the Weak interactions (and of course in many other fields as well). One reason for this is the well-known phenomenon that ideas whose 'time is ripe' often occur about the same time to different scientists, who report them in different ways. Later, it is often difficult to sort out just who had 'priority', and to give credit where credit is due. The history of the UFI, including the lack of recognition for Tiomno's early contributions, was summarized later in a review by Leite Lopes,[28] and also in Tiomno's paper from the 1984 Racine Conference.[10],[19]

Wheeler himself wrote in his autobiography[26]:

> [*on the Tiomno triangle*] ... We found it to be a very beautiful representation of a pattern in nature. It doesn't answer the question, 'Why the muon?', but it compactly summarizes a lot of particle physics. I always thought that this triangle should be called the 'Tiomno triangle'. He got there first. But a few months after our paper appeared, Giampietro Puppi published similar ideas in an Italian journal. He, too, saw the great simplicity of a common interaction among nucleons, electrons, muons, and neutrinos. As luck would have it, the Tiomno triangle is now known to everyone as the 'Puppi triangle', even though Puppi did not include a diagram in his paper.

We will return to this question later on (cf. Chap. 15).

During his graduate-student years at Princeton, Tiomno had the opportunity to meet and collaborate with a number of well-known physicists. Aside from his first mentors, John Wheeler and later Eugene Wigner, another of those was *Wen Yu Chang*, who was studying meson decays using a cloud chamber. Wheeler asked Tiomno to aid in analyzing the results—"I am Chang's technical advisor!", he wrote facetiously to Leite Lopes. He analyzed a case in which a meson emits an electron of high energy, finding that it could have an energy of up to 25 MeV, but with a very low branching ratio of only 0.8%. But this remained a unique observation and could not be verified. In any case, it was a good way to get to know Chang.[31]

Another collaborator of Tiomno's in Princeton was **Chen Ning Yang** (born 1922). Yang, who was in Princeton (at the IAS) during Tiomno's graduate-student years, was born in Hefei, in the *Anhui* Province in southeastern China. His youth and later school years were complicated by the Japanese invasion of China in 1937, and his family left Beijing, where he had been attending school, and returned to Hefei, moving in 1938 to Kunming in the *Yunnan* Province in the southwest of China. There, he was able to attend the National Southwestern Associated University, receiving his Bachelor's degree in physics with a thesis on the interpretation of molecular spectra using group-theoretical methods. He continued graduate work at the Tsinghua University, which had moved from Beijing to Kunming during the war, receiving his MS there in 1944 with a thesis on statistical mechanics. He obtained a fellowship for graduate study in the USA, which he was able to begin only in 1945, at the University of Chicago. He originally intended to work with Enrico Fermi, but because the latter was involved in military work at the nearby Argonne National Laboratory, Yang, as a non-US national, remained in Chicago, working with Edward Teller, one of the famous Hungarians who had emigrated to the USA in the 1930s.[32]

Yang, who used the first name 'Franklin' as his Western appellation, received his Ph.D. at Chicago in 1948 and then spent a postdoctoral year as assistant to Fermi, before accepting an invitation to the Institute for Advanced Studies (IAS) at Princeton, then directed by J. Robert Oppenheimer. He remained at the IAS for 16 years, moving to the State University of New York (SUNY) at Stony Brook in 1965. At Princeton, he began his collaboration with Tsung-Dao Lee, with whom he proposed parity violation by the Weak interactions in 1956, leading to their joint Nobel prize the following year. Yang became famous for his introduction, with Robert Mills, of the class of theories now known as Yang-Mills theories (1955), and for his suggestion of parity violation with Lee (1956); Tiomno played a certain role in the latter event.

In the summer of 1949, after John Wheeler, his previous thesis advisor, had departed on sabbatical for France, Jayme Tiomno began collaborating with Yang. Their work together led to a joint paper in the *Physical Review*[33] in 1950, entitled '*Reflection Properties of Spin-½ Fields and a Universal Fermi-Type Interaction*'. In it, they consider the description of 'parity-even' and 'parity-odd' quantum objects, and they introduce the term 'Universal Fermi Interaction'. Although the actual conclusions of the paper were not very influential, it is important for historical and precedence reasons (cf. Tiomno (2005) and Yang (2005)).

Yang himself said of this article (in his commented publications, from 2005[34]) that it had been inspired by discussions with A. Wightman, another young theorist who was at Princeton at the time, leading him and Tiomno to raise the question of possible phase factors in the parity operator for a Dirac particle. He notes that the scheme proposed in the paper was later found to be irrelevant for describing experimental findings; however, the experience of working on the paper proved useful to him later, in 1956, when he proposed parity violation in the weak interactions together with T. D. Lee. He also mentions that Fermi was very much interested in this paper. At a conference in Chicago in September, 1951, he arranged a session to discuss it. The proceedings of that conference show that Fermi asked a (for him) typical question: he inquired as to how the different classes of particles might be experimentally distinguished.[35] Fermi's emphasis on the experimental meaning of theoretical constructs was an important characteristic of his style, according to Yang.

We can presume with some certainty that the experience of working on this paper with Yang was useful for Jayme Tiomno, as well. Yang was two years his junior, but his academic career had not been so seriously interrupted by the 2nd World War as had Tiomno's, and he had already published a number of articles by the time he went to Princeton. We note that at the time when Tiomno collaborated with Yang, the latter had already published, together with T. D. Lee and M. Rosenbluth, an article on '*Interactions of Mesons with Nucleons and Light Particles*'.[36]

After more than 30 years, when the early history of the UFI was under discussion, Yang wrote to Tiomno,[37] affirming that their joint work in 1950[33] was important for his later parity work with Lee, and it was cited in their 1956 parity paper Lee & Yang (1956).

The paths of Tiomno and Yang crossed again several times, and they later collaborated on topics in particle physics, for example briefly in Rio de Janeiro in 1960,[38] but did not publish any further joint articles. Yang had a distinguished career at Stony Brook, where he was director of the Institute for Theoretical Physics which now bears his name. He retired there in 1999 and returned to China, where he is a Distinguished Professor at the Tsinghua University in Beijing and at the Chinese University of Hong Kong.

The Michigan Summer School

As is also usual in many American universities, Tiomno was first required to obtain a Masters degree before entering the Ph.D. program; this meant passing a series of examinations and carrying out research (even though his previous education and scientific experience were already more than equivalent to the MS). He performed the required research work for his Master's under Wheeler's supervision (Fig. 6.6).

Tiomno's original fellowship to Princeton, from the U.S. Department of State, ended in January 1949 and could not be renewed (since he had missed the deadline for a renewal application, in February 1948, while he was traveling to Princeton). He was, however, able to obtain another fellowship, from the Rockefeller Foundation. His application was supported by the Physics Department at USP, where he still had an assistantship (from which he was on leave). In June of 1949, Tiomno finished his MS degree, and around the same time, John Wheeler departed with his family for France, where he was to spend the academic year 1949/50 in Paris on a Guggenheim Foundation fellowship. Wheeler's sabbatical year was originally planned for 1950/51, but was moved forward when he was granted leave earlier by the University.[3]

Fig. 6.6. Jayme Tiomno in his cap and gown, after the graduation ceremony for his Masters degree, June 1949. *Source* Private archive, [JT]

This coincidental circumstance led to Eugene Wigner's becoming Tiomno's 'doctor father', rather than John Wheeler.

Tiomno's new fellowship permitted him to attend the course on Field Theory at the Michigan Summer School (which by then had established a long tradition). The Summer 1949 physics course was given by Richard P. Feynman, a former Wheeler doctoral student, and Tiomno found it very useful[39]:

> ...in the summer of 1949, still with the sponsorship of the Rockefeller Foundation, I participated in the post-doctoral course on field theory at the University of Michigan, under the orientation of R.P. Feynman ... , which had great benefits for my scientific education.

This was probably his first serious interaction with Feynman,[40] although they had met earlier in Princeton and in New York. This relationship would continue for many years, in Brazil as well as in the USA. The 1948 and 1949 summer school courses were given by two of the founders of modern quantum electrodynamics (QED), Julian S. Schwinger (1948) and Richard P. Feynman (1949). Schwinger and Feynman later shared the 1965 Nobel prize in physics with Shin'ichiro Tomonaga of Japan for their 'fundamental work on quantum electrodynamics'.

Their courses are recalled by the elementary-particle theorist Robert Finkelstein[41] in his memoir:

> During this period I attended two sessions of the Michigan summer school at which Schwinger (1948) and Feynman (1949) described their respective re-formulations of QED. Schwinger's was deeper and more complete, while Feynman's was easier to use but at that time incomplete. ... Bethe at that time described this period as the most exciting in physics since the great days of 1925–1930, when quantum mechanics was being discovered. ... By this time the 'Universal Fermi Interaction' (UFI) had been proposed by Wheeler and Tiomno. Adopting the UFI, Mal Ruderman and I calculated (1949) $\pi e/\pi \mu$ in the same way, i.e. pre-Schwinger, that I had previously calculated the $\pi \gamma$ rates. ...[42]

We can see that the introduction of the UFI, which Tiomno had considered on his own while still in Brazil in 1947, had not gone unnoticed by other physicists. Finkelstein attributes the UFI to 'Wheeler and Tiomno'; in fact, the phrase 'universal Fermi interaction' was first used in the paper by Tiomno and Yang,[33] but the concept had already been introduced in a striking form in the second of Tiomno and Wheeler's papers,[25] and had been suggested by some other authors[43] in different forms about the same time. Finkelstein's young colleague Mal Ruderman also attended the summer school, where he and Jayme Tiomno were the only participants who had not yet completed their doctorates.[44]

Tiomno's Ph.D. Thesis

After the summer break in 1949, Jayme Tiomno turned all of his attention to finishing his doctoral thesis, on the general topic of '*The theory of the Neutrino and double Beta Decay*'. Wigner, who would be his thesis advisor, was by now a famous theoretician. While Tiomno and Wigner had no joint publications,

Wigner certainly had an influence on the younger physicist at an important juncture of his career, just when he was writing his doctoral thesis.

Eugene Paul Wigner (1902–1995) was born in Budapest, Hungary.[45] He studied chemical engineering at the Technical College (now *Technische Universität*) in Berlin, and completed his doctorate there in 1925 under the supervision of Michael Polanyí, another famous Hungarian, who was a physical chemist. Wigner met Albert Einstein and Leó Szilárd while in Berlin; Einstein impressed him and confirmed his intention of beginning a career in physics, rather than the more prosaic chemical engineering preferred by his family. Szilárd was also Hungarian, three years older than Wigner, and he likewise later emigrated to the USA, after an interlude in England. Szilárd collaborated with Einstein on practical problems while in Berlin, and was later instrumental in calculating the details of nuclear chain reactions, together with Wigner.

Wigner attended the colloquia of the *Deutsche Physikalische Gesellschaft*, organized by Max von Laue in Berlin, and this heightened his knowledge of and his interest in physics. During his third year in Berlin, Wigner worked as an assistant at the *Kaiser-Wilhelm-Institut* (KWI) in Berlin-Dahlem. The KW Physics Institute was headed by Albert Einstein, and Max von Laue was its vice-director, but it still had no building of its own.[46] He instead worked in the facilities of the KWI for Physical and Electrochemistry, headed by Fritz Haber (now called the *Fritz-Haber-Institut*), where his mentor Michael Polanyí had a research group, and Einstein also had an office during his early years in Berlin.

After finishing his doctorate in 1925, Wigner remained for a year in Berlin as assistant to the theoretician Richard Becker at the *Technische Hochschule*, and then went to Göttingen as assistant to the mathematician David Hilbert, who by that time was however already rather old and no longer scientifically productive. Wigner worked on his own and developed applications of group theory to quantum mechanics, which were reflected in his publications '*Gruppentheorie und Quantenmechanik*' in 1929, and '*Group Theory and its Application to the Quantum Mechanics of Atomic Spectra*' in 1931. They established his reputation as an innovative theorist. He completed his *Habilitation* (academic teaching qualification) in Berlin in 1928 and became Adjunct Professor for Theoretical Physics at the *Technische Hochschule* in 1930.

The following year, he moved to the USA, where he worked first as a Fellow at Princeton, then in 1936/37 as Professor at the University of Wisconsin, finally returning as Professor of Mathematics to Princeton University, where he spent the rest of his career, retiring in 1971. Together with Leó Szilárd, Wigner developed the theory of nuclear chain reactions, and, along with Einstein, the three of them played a role in initiating the US nuclear weapons program in the early 1940s. He received the Nobel prize in physics in 1963 for his contributions to nuclear physics and his applications of symmetry principles to physics, sharing the prize with Maria Goeppert-Meyer and Hans Jensen, who had independently developed the Nuclear Shell Model to explain the energy-level structure of many atomic nuclei. In his later years, Wigner worked on fundamental physics and interpretations of quantum mechanics (e.g. the '*Wigner's Friend*' thought experiment (1956), a complement to Schrödinger's Cat; and his essay on '*The unreasonable effectiveness of mathematics in the natural sciences*' (1960)).

Leó Szilárd, Eugene Wigner, John von Neumann and Edward Teller, all Hungarian immigrants who worked in modern physics and mathematical physics, were sometimes called 'the Martians' by their American colleagues, due to their apparently other-worldly intellectual abilities.

Tiomno described an early encounter with Wigner as his 'doctor father' in a much later conversation (in the 1970s), as recounted by A.L.L. Videira[47]:

> ... having asked [*Tiomno*] about the topic of his thesis, Wigner posed some questions: 'Have you seen this article? That other one? Do you know the work of Majorana? ' ...
>
> [*and Tiomno answered*], with the security that comes of strength, or rather, with the intrepidness of those who have yet to discover Wigner's way of being, '—I have already read all of Majorana's published work on neutrinos.' '—Ah! Very good!', replied Wigner '—Then you can explain to me certain aspects which I was never able to understand. Given that in Majorana's theory, there are only two neutrino states, and that in Dirac's theory, there are four, then the specific heat of the vacuum according to Dirac must be greater than the specific heat of Majorana's vacuum?' This was followed by still another question that the young student was likewise unable to answer. 'Ah, good! I see that you *also* haven't understood Majorana's theory, and that studying it could be a good way to begin your thesis.'

Ettore Majorana was an Italian theoretician who developed a fermion theory analogous to the Dirac theory, but whose solutions are two-component spinors rather than four-component spinors. They thus describe particles (evidently without electric charge !) which are their *own antiparticles*. There has been much speculation as to whether neutrinos are Dirac particles or Majorana particles.

Tiomno's thesis topic may have been a kind of last resort; after it was clear that Wheeler would be leaving in June, 1949, he no doubt wanted to finish his thesis as quickly as possible, and chose topics which would be of current interest but could be treated fairly rapidly. On the other hand, it was somewhat daring of him, as a young, beginning theoretician, to propose to hatch not one, but *several* new theories.

Neutrino physics has been a subject of continual interest over the past 85 years, ever since Pauli's suggestion of the neutrino's existence, and is showing no signs of flagging. Since the 1960s, it has become an important experimental topic as well as one for theorists, and now motivates the construction of enormous and expensive detectors. Double beta decay was first suggested in 1935 by Maria Goeppert-Mayer, who reportedly consulted with Wigner before publishing her hypothesis.

In its simplest form, double beta decay is simply the simultaneous decay of two neutrons or two protons within an unstable nucleus, accompanied by the emission of two electrons or two positrons and, correspondingly, two neutrinos. This can be the preferred decay mode if the adjacent nucleus (at $Z \pm 1$) is *less* strongly bound, so that no energy gain would occur through single beta decay, but a neighboring nucleus *twice removed* (at $Z \pm 2$) is *more* strongly bound, favoring a double decay. Double electron capture can also occur.

The more exotic form, up to now not observed experimentally, involves the emission of two electrons or positrons, or double electron capture, but *no* emitted neutrinos—this would be possible if neutrinos are Majorana particles. This version is called 'neutrinoless double beta decay'. The characteristic decay times for double beta decay are very long, making it difficult to observe in practice. As of the present date, 13 examples of 'ordinary' double beta decays have been detected, with half-lives of the order of 10^{18} years. A lower limit for the half-life of neutrinoless double beta decay has also been established: it is 10^{25} years!

Tiomno's young colleague at Princeton, Edward L. Fireman (1922–1990), had carried out the first experimental study of double beta decay, using a Geiger counter, for his own Ph.D. thesis, completed in 1948 with John Wheeler as thesis supervisor. Tiomno used Fireman's data in his thesis, and he may have been inspired to write about double beta decay by the latter's work. He no doubt met Fireman, still in Wheeler's group, soon after arriving at Princeton.

Jayme Tiomno's Ph.D. thesis,[48] dated September 1950, is unusually clear and well-structured. It reflects his lifelong penchant for didactically transparent writing. He refers to it several times as 'this paper', as though he were writing a scientific article. However, he did not find time to publish it immediately—several articles were published some years later, containing some of the new ideas developed during the writing of his thesis.[49]

Tiomno begins with a brief Introduction, in which he sets out his goals and briefly describes the history of neutrino theories and of double beta decay up to that date. This is followed in Part I by a general summary of quantum field theories, still in a state of flux at the time, which distinguishes in particular *local field theories* and *non-local Schrödinger projection theories*. In Part II, he discusses Lorentz invariance (i.e. invariance under Lorentz transformations, which signals conformity with Special Relativity, a requirement for any modern theory in which relativistic velocities or energies occur). He also discusses the time-reversal invariance of quantum electrodynamics (QED), and restrictions on the possible forms of Hamiltonians (quantum-mechanical operators for the total energy of a system) that are imposed by time-reversal invariance. Finally, in this part, he deals with the Lorentz transformations of spinor fields, which he will use to describe the leptons.

In Part III, Tiomno continues with the theoretical underpinnings of his proposed models, by discussing the Hilbert space underlying the spinor field operators. Hilbert spaces are a mathematical construct, named for the German mathematician David Hilbert, which generalizes three-dimensional Cartesian space to arbitrary dimensions. The operators of quantum mechanics, which represent observable physical quantities, operate on a Hilbert space, and the 'state vectors', defining the state of a quantum-mechanical system, are vectors in the Hilbert space.

Subsequently, in Part IV, he describes and classifies theories of neutral spin-½ particles (e.g. neutrinos), and discusses neutral particles that are their own antiparticles (Majorana particles, in a general treatment). The second section of Part IV then introduces the formulation of possible neutrino theories and their application to a Fermi-type interaction. Tiomno restricts himself to local field theories (based on the results of Part I), and proposes to describe all fermions (except the neutrinos) as Dirac particles. He then distinguishes three types of theories, depending on the description of the neutrinos as Dirac particles; or as Dirac particles but with a projection operator in the interaction; or alternatively as Majorana particles. The interactions are written using the five ranks of current operators (S, V, T, A, P),[50] expressing them in terms of the Dirac matrices γ_i.

In this section, Tiomno also develops a simple form of the 'lepton numbers' and 'baryon numbers', originally suggested by Fermi and later used to classify particle reactions in terms of 'conservation of lepton number', etc. Tiomno refers here to 'conservation of particles'. He then discusses the classification of the various theories in terms of Lorentz invariance and of the structure of the underlying Hilbert space. He further considers theories without 'conservation of particles' (this applies to neutrinos as Majorana particles, since then neutrinos and antineutrinos are identical and the 'lepton number' concept fails); this class could describe neutrinoless double beta decay. He also discusses 'projection theories', in which a part of the Dirac neutrino wavefunction is projected out in the interaction, so that only one kind of neutrino (or antineutrino) actually participates in the interaction. Finally, he considers explicitly Majorana-type theories, where only one kind of neutrino exists a priori; they are the most likely candidates for neutrinoless double beta decay.

In the last section of the thesis, Part V, he discusses the consequences of the various theories for experimental results from double beta decay, computing coupling constants and spectrum shapes in a long and mathematically challenging procedure, comparing them for the different current-operator ranks, and deriving predicted decay lifetimes. In the third section of Part V, he compares these results with available experimental data (still rather limited at that time). He concludes, in the final section, that Majorana-type theories give results which are not in disagreement with the data, but this does not justify the conclusion that the neutrino is a Majorana particle. He underscores the need for more data, in particular spectrum shapes and angular correlation data (very hard to obtain, given the exceedingly long decay lifetimes of double beta decays).

Tiomno's fellowship was scheduled to end in February, 1950, and he thus would not have had time to finish writing his thesis before leaving Princeton. Therefore, his plan was to finish the thesis back in Brazil, as his colleague Walter Schützer had also planned to do. Wigner was however angered by Tiomno's intention of an early departure, and insisted that he ask the Rockefeller Foundation for an extension of his fellowship. Wigner had been dissatisfied at Schützer's premature departure, and was not keen on having Tiomno repeat that, wanting him to stay until his doctoral thesis was completed. Contrary to Tiomno's expectations, the Rockefeller Foundation granted the extension without problems; but his leave from the Faculty in São Paulo presented an obstacle. Normally, the University (USP) would grant a leave for at most two years; furthermore, the Department was seriously lacking in teaching capacity. After considerable negotiation and a letter from Wigner to the Chairman of the Department at USP, he was allowed to extend his leave until September.

Tiomno had the support of two young colleagues in Princeton. The first of them was *David Bohm* (1917–1992), then still an assistant professor at Princeton, who in fact served as reader for Tiomno's thesis, since Wigner was traveling at the crucial time. The second was *Arthur Wightman* (1922–2013), another Wheeler student, who completed his doctorate in 1949, and was interested in pion physics (he was later involved in the formulation of axiomatic quantum field theory).

In July, 1950, when Tiomno was concluding his Ph.D. thesis, Wigner went to the University of Wisconsin to give a summer school course on group theory, and suggested that Tiomno attend, as well. He could participate in the course and also work together with Wigner on finishing his thesis. Both of them however had too little time to finish it: Wigner was scheduled to travel to Europe in September, and Tiomno intended to return to Brazil.

Finally, in September 1950, Tiomno defended his thesis. Wigner considered it to be generally in order, except for some of the English formulations. The only point that he found lacking was a complete discussion of double beta decay in what he called mixed theories.[51] In his Acknowledgments, Tiomno of course thanks Eugene Wigner for his supervision of the thesis, and also Arthur Wightman and David Bohm for 'helpful discussions'.

Tiomno was only the third Brazilian physicist to formally earn a doctorate.[52] The first was José Leite Lopes, who received his Ph.D. in Princeton/USA in 1946, and the second was a woman, Sonja Ashauer, who obtained her DPhil at Cambridge/UK in 1948.[53]

Returning Home

Tiomno's final months at Princeton in 1950 were very busy. He was working quite intensively, first on formulating, and then on actually writing and editing his thesis. His letters to Elisa became less frequent. He also began typing them, apparently for practice typing which would increase his speed for writing up the final version of the thesis. Having finished his Ph.D. at Princeton in late September, he prepared his return to Brazil.

On October 7th, 1950, Tiomno left the USA, bound for Brazil, to resume his post at USP. This time, he did not experience an adventurous flight; the Rockefeller fellowship paid for a passage by ship, and he took a Uruguayan steamer which stopped at *Santos*, the port city of São Paulo. His plan was to spend 2 or 3 years there, and then join the ambitious project that was underway in Rio de Janeiro: establishing the Brazilian Center for Physical Research (CBPF). There had been some confusion in Brazil earlier in the year because of a rumor that Tiomno would return directly to Rio rather than to São Paulo, where he still had an assistantship. The offer from USP was indeed attractive; his colleagues were eager to keep him there: He would be first assistant and would receive a supplement as a full-time employee. Furthermore, he had obtained the extension of his leave abroad at the price of promising to return to São Paulo, in spite of the attraction of the newly-founded CBPF in Rio. For Tiomno, his promise to the University and his loyalty to the Department in São Paulo were more important than the favorable offer.

Back in Brazil, even in São Paulo, he would also be able to see Elisa again. Since the Brazilian summer of 1947, when they began their intimate relationship, they had had few opportunities to be together, first because of Tiomno's stay in São Paulo, and then his nearly three years at Princeton. Now, the time had come to meet with her personally and to decide whether their relationship had a future.

In his luggage, Jayme Tiomno carried seven papers published in prestigious journals, his Ph.D. thesis, contacts and collaborations with several well-known physicists,[54] and an electric refrigerator—a gift to his mother, to replace the wooden icebox that the family had used for years.

Notes

[1] Jayme Tiomno to Elisa Frota-Pessôa on Feb. 10th, 1948; [EFP].

[2] Fornazier & Videira (2018).

[3] See Wheeler & Ford (1998), Wheeler's autobiography.

[4] Jayme Tiomno to José Leite Lopes, Feb. 18th, 1948; [JLL].

[5] Jayme Tiomno to José Leite Lopes, April 14th, 1948; [JLL].

[6] Infeld & Schild (1949).

[7] Jayme Tiomno to Elisa Frota-Pessôa, April 11th, 1948; [EPP]; cf. also Tiomno (1966a).

[8] See the website https://todayinsci.com/W/Wheeler_John/WheelerJohn-Quotations.htm.

[9] The definitive scientific biography of Wheeler has yet to be written (his autobiography is available, cf. [3]); but there are various scientific-biographical works describing different aspects of his life and science. See for example Klauder (1974); Christensen (2009); Ciufolini (1995); Misner et al., (2009). The dissertation by Terry M. Christensen, 'John Archibald Wheeler: A Study of Mentoring in Modern Physics' Christensen (2009) gives a detailed discussion of Wheeler's guidance of his many students; it is available as a pdf at. https://ir.library.oregonstate.edu/concern/graduate_thesis_or_dissertations/3r074x13x.

[10] Quoted from 'The early period of the universal Fermi application', originally contributed in 1984 to the Proceedings of the Racine Conference on Fifty Years of Weak Interactions; cf. Tiomno (1984/94). See also Chapter 15, section on 'The Weak-interactions interlude'.

[11] Jayme Tiomno to José Leite Lopes, Nov. 27th, 1947; [JLL].

[12] Bruno Touschek, *'Excitation of Nuclei by Electrons'*; cf. Touschek (1947).

[13] Conversi, Pancini & Piccioni (1947)—the 'CPP experiment'. The authors compared the rates of decay and capture of negative muons in carbon and iron plates. It could be inferred from their data that the muons were produced by the fast decay of another particle (the *pion*). See also Hans Bethe and Robert Marshak—the 'Bethe-Marshak paper' Bethe & Marshak (1947). They postulated two mesons: One, strongly interacting (now known to be the pion), decaying to the second, weakly interacting (the muon). Tiomno knew of the CPP result but not the two-meson postulate in 1947.

[14] Bethe & Marshak (1947).

[15] Jayme Tiomno to Elisa Frota-Pessôa on June 9th, 1948; [EFP].

[16] Jayme Tiomno to Elisa Frota-Pessôa on June 15th, 1948; [EFP].

[17] Bassalo & Freire (2003). Their article is entitled *'Wheeler, Tiomno, and Brazilian Physics'*.

[18] This quote may be inaccurate, since Marshak was not a student of Wheeler's, nor did they work together, although they were certainly well acquainted. It is quoted in Bassalo & Freire (2003) as a statement made, presumably by Elisa, in an interview with Jayme Tiomno and Elisa Frota-Pessôa in August, 2003, and recorded by the authors and by Sérgio Joffily (their Ref. [8]).

[19] Jayme Tiomno to his parents and siblings, August 20th, 1948; [STT].

[20] See [JT8] and [JT9].

[21] The journal of *Sigma Xi*; cf. [JT11].

[22] See Chapter 5, and [JT7].

[23] [JT10].

[24] Jayme Tiomno to Elisa Frota-Pessôa, Sept. 24th, 1948; [EFP].

[25] See [JT6]; this is the article containing the 'Tiomno triangle', based on the talk at the Pasadena meeting in June, 1948.

[26] Wheeler & Ford (1998); the triangle drawing is on p. 175, his description on p. 176.

[27] Sakurai (2015, pp. 167–168). The original suggestion of a 'Tiomno tetrahedron' was apparently made at the 1956 Rochester Conference on High-Energy Physics by Murray Gell-Mann, who introduced the concept of 'strangeness'; cf. Leite Lopes in MacDowell et al., (1991).

[28] Leite Lopes (1996); cf. p. 14, Fig. 4. This article gives a good, compact summary of the development of weak-interactions theory. See also the essays in MacDowell et al., 1991. Note that this article is a revised version of a shorter essay, entitled *'The Principle of the Universal Fermi Interaction'* and published in the *Festschrift* for Jayme Tiomno's 70th birthday MacDowell et al., (1991). There, a number of alternative forms of the 'Tiomno triangle' are displayed, and the problem of the 'mutating name' of the original triangle is mentioned: "*This is only an example of the well-known tendency in the community of physicists of industrial countries for the rarity of quotation of articles the authors of which are out of one of the groups of these countries—thus the Tiomno-Wheeler triangle became known as the Puppi triangle (and in the*

process Wheeler was also pushed away)—although, of course, I can only be happy with this homage to Giampetro Puppi whom I know and admire." This somewhat ambivalent remark may have been poorly received by Jayme Tiomno, and perhaps led to the revised version of 1996.

[29] The Physical Society of Japan (org): *Series of Selected Papers in Physics,* volume on '*Weak Interactions in Particle Physics*'. Physical Society of Japan (1972).

[30] Tiomno (1984/94); See also Marshak, (1997) and Bassalo & Freire (2003), as well as Chapter 15, section on 'The Weak-interactions interlude'.

[31] Jayme Tiomno to José Leite Lopes, March 23rd, 1948; [JLL].

[32] Argonne National Laboratory was founded in 1946 in Lemont, Illinois, near Chicago, under the auspices of the University of Chicago, as the successor to the 'Metallurgical Laboratory', where the first experimental work for the Manhattan Project was begun and the first working nuclear reactor was constructed. It is operated today by the US Department of Energy and is one of the major national laboratories in the USA.

[33] C. N. Yang and J. Tiomno, in: the *Physical Review,* Vol. 79, p. 495, 1950; see [JT12].

[34] C. N. Yang's commented publications, collected in book form, Yang (2005).

[35] Quoted in the *Proceedings of the International Conference on Nuclear Physics and the Physics of Elementary Particles,* eds. J. Orear, A. H. Rosenfeld, and R. A. Schlüter (Mimeographed notes, University of Chicago, 1951), p. 109.

[36] Lee, Rosenbluth & Yang (1949).

[37] Chen Ning Yang to Jayme Tiomno, May 10th, 1985; [JT].

[38] cf. Chapters 9 and 10.

[39] Quoted from Tiomno (1966a).

[40] cf. Tiomno (1977), and Chapter 9.

[41] See Finkelstein (2016). Finkelstein worked for many years at UCLA, but he had previously been at Berkeley and at Princeton with Oppenheimer. He also worked on weak interactions at the same time as Tiomno.

[42] Finkelstein (2016). The calculation that he mentions in the quoted passage was later relevant to Tiomno's continuing work on Weak interactions (cf. Chapter 9). See also Rudermann & Finkelstein (1949).

[43] Puppi (1948); Pontecorvo (1947); Klein (1948).

[44] Videira (1980).

[45] See his autobiography, Wigner & Szanton (1992).

[46] It had been founded in 1917, during the 1st World War, when there were no funds to construct a building, and the situation was not much better during the Weimar Republic, while Wigner was in Berlin. The building was finally constructed in 1936 on the Berlin-Dahlem KWI campus with funds from the Rockefeller Foundation.

[47] Videira (1980) mentions Jayme Tiomno's work with David Bohm, and he quotes the anecdote with Wigner from a conversation with Tiomno; it was requoted by Fornazier & Videira (2018).

[48] See [JT13]. The thesis is available as a pdf from the CBPF [document CBPF-DH-001/18], and it has been discussed in detail by Fornazier & Videira (2018). Here, we give only a brief summary. Interested readers should consult the original (in English).

[49] See Chapter 9.

[50] See Chapter 4.

[51] Eugene Wigner to Jayme Tiomno, Sept. 14, (1950); [JT].

[52] See Preface, Backnote [1].

[53] César Lattes did, however, receive an *honorary* doctorate from USP in 1948.

[54] Besides the physicists already named, we should also mention *Abraham Pais* and *J. Robert Oppenheimer.*

7

Elisa

Early Times

Elisa Frota-Pessôa was a woman who was ahead of her time. She studied physics and worked in scientific research, when both physics and research were still in their infancy in Brazil. Both were uncommon activities even for men at the time, and all the more so for the few women who had acquired college degrees. For those few who *had* obtained a university degree, almost the only usual career was in teaching. Over and over, Elisa had to prove that, although a woman, she was competent and could do physics, as if the one were incompatible with the other. She also refused to surrender to conservative and traditional customs and rules that regarded women who had separated from their husbands, especially if they decided to remarry, with suspicion and disdain. In short, she was a remarkable woman.

She was born in Rio de Janeiro on January 17th, 1921, less than a year after Jayme Tiomno. Her parents were *Juvenal Moreira Maia* and *Elisa Habbema de Maia*, and she was given the name *Elisa Esther Habbema de Maia* (Fig. 7.1).

Elisa's interest in science and specifically in physics was awakened early, when she was in her second year at the high school *Escola Paulo de Frontin* in 1935/36, by her science teacher, *Plinio Süssekind Rocha*. Her father, who was a lawyer and of conservative inclinations, would have preferred to send her to the state Normal School in preparation for working as an elementary teacher, but she wanted to attend *Paulo de Frontin*. She was thus registered at both schools, causing some difficulties because of a new law prohibiting candidacy at two state schools at the same time. After an interview with the principal

© Springer Nature Switzerland AG 2020
W. D. Brewer and A. T. Tolmasquim, *Jayme Tiomno*, Springer Biographies,
https://doi.org/10.1007/978-3-030-41011-7_7

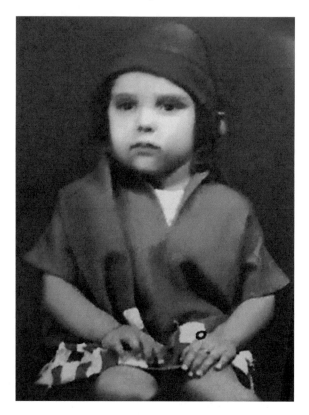

Fig. 7.1 Elisa at three years of age. *Source* Private, archive [EFP]

of *Paulo de Frontin*, Andrea Borges, who was favorably impressed by Elisa's entrance test results, her father used his legal skills to have her admitted there.

While her case was being decided, she worked at home, solving exercises provided by Süssekind, among others. She entered late in her high-school class, somewhat behind due to her delay in admission, and was immediately fascinated by Süssekind's physics course. Süssekind himself was at first incredulous that she—a girl—had been able to solve the exercises correctly by herself, and presumed that she had been helped by an older brother or some other male relative. She replied that she had solved them herself, and that it would be simple to verify that by asking her to come to the blackboard and answer any questions he wanted to pose. Shortly thereafter, she became an informal teacher for those in her class who were having trouble keeping up, with her own classroom and schedule.[1] She found this to be the best way to learn—by teaching.

Plinio Süssekind Rocha, in addition to teaching at the high school, was assistant professor at the *Universidade do Distrito Federal*—UDF. When Elisa was ready to begin university studies (Fig. 7.2), he suggested to her that she should not study engineering, as she initially intended, but instead should major in physics. She had thought that physics courses were offered only as supporting material for 'practical' subjects like engineering…

Many of her teachers had been educated at the UDF, including her future husband, *Oswaldo Frota-Pessôa*. Of Oswaldo, whom she encountered as her biology teacher and greatly admired, she says, "[*he*] was the best teacher that I ever met. He was always thinking of inspiring his students to also become good teachers themselves". Oswaldo came from an intellectual family that had arrived in Rio from the state of *Ceará*, in the northeast of Brazil. His father, José Getúlio da Frota-Pessôa, was an educator and administrator who was an associate of Anisio Teixeira, the founder of the UDF, and became Secretary of Education in Rio. He was referred to familiarly as 'Doutor Frota'. He believed strongly in the value of education and passed that inclination on to Oswaldo.

Elisa married very early, at 17. She had a 'boyfriend' while still in school, and there was even talk of marriage. But it soon became clear that he was very traditional, and would not consider allowing his wife to work independently. So she broke off the relationship.[2] No doubt she was in love with Oswaldo,

Fig. 7.2 Elisa's graduating class at her highschool. She is at the lower left in the picture. *Source* Private, archive [EFP]

her biology teacher, whose family lived in the same neighborhood and were friendly with hers. She had been able to convince her father to let her attend the school that she wished; but he was adamant that she should not study at a university, believing that "a woman's place is in the home". This conservative position of her father was in contrast to examples of independent women with their own professions from her mother's side: Elisa's French grandmother was a midwife, who had her own livelihood and lived independently. And her Aunt Esther, from whom Elisa received her middle name, had been a dentist with a flourishing practice, especially favored by women patients.[2] They may well have served as role models for Elisa, but there is no doubt that she herself had a great deal of perseverance, self-confidence and independence, without which she could never have overcome the obstacles put in her way.

Elisa was still a minor and had to be granted permission by her family to marry, which was given. Their wedding was held in a church, although neither of them was religious; however, her family were adherents of the—at that time still ubiquitous—Catholicism in Brazil.

Her husband Oswaldo (Fig. 7.3), although only four years older, was to some extent her guide for her own education, and he was supportive of her wish to study physics and enter research, in contrast to her own family. Elisa and Oswaldo had two children: Sonia, born in August 1942, and Roberto, born in April, 1944 (she completed her Bachelor's in 1942, just four months after the birth of Sonia, and was carrying her second child when she obtained her *licenciatura*—her teaching certificate); so she was an early example of a 'student mother'. Oswaldo's family went out of their way to help in every possible manner. His parents, Zezé and Dr. Frota, and his sister Regina lived in an adjacent house and took care of Sonia and Roberto, so that Elisa could study, work and even travel, when necessary. Oswaldo had majored in Natural History at the UDF and went on to study medicine at the *Faculdade de Medicina* of the UB, as a contemporary of Jayme Tiomno there. He completed the course in medicine in 1941, and later, in 1953, successfully defended his doctoral thesis in biology at the FNFi, while working as assistant to the Chair of Biology.

Elisa regarded the UDF very highly, having heard much about it from her teachers; but, like Jayme Tiomno, she could not herself study there, since it had been closed by the government, giving rise to the *Faculdade Nacional de Filosofia* (FNFi), which was founded as a faculty of the UB, losing many of its most talented professors in the process. In 1940, she took the entrance examinations to the FNFi, in spite of warnings that "physics is nothing for a woman" and predictions that she would fail the examinations; in fact she succeeded and entered in 1940, along with one other (male) candidate (they

Fig. 7.3 Oswaldo Frota-Pessôa, taken when he was about 75. From: *American Journal of Medical Genetics*, June 28th, 1996, Vol.63 (4), pp.585–602. Online at: https://doi.org/10.1002/(SICI)1096-8628(19960628)63:4<585::AID-AJMG13>3.0.CO;2-E. Reproduced with permission (photo cropped)

were put in a class with the mathematics students, who followed the same curriculum for the first two years).

University Years in Rio de Janeiro

As a student of physics at the FNFi, Elisa met *José Leite Lopes*, *Jayme Tiomno*, *Mauricio Matos Peixoto* and *Leopoldo Nachbin*, as we have seen. Three of them, Elisa, Leite Lopes, and Tiomno, became very close friends, and one of their common features was their passion for physics. As she said many years later, "[*It*] was a group that really wanted, above all else, *to do physics*" [1] (our emphasis).

One of her mentors and role models during this period, in addition to Plinio Süssekind Rocha and her husband Oswaldo, was Bernhard Gross, whom she calls the "Father of Physics in Rio de Janeiro", since he had been publishing research results and encouraging students to do research since 1935. But she also believed [1] that the overall "Father of modern Physics" in Brazil was Gleb Wataghin, who had founded the Department of Physics at USP in 1934 'by pulling on his bootstraps', and who educated a whole generation of research physicists and teachers.

Elisa (Fig. 7.4), like Jayme Tiomno, was critical of the courses offered at the FNFi, in particular those of the Italian professor of experimental physics, *Dalberto Faggioani*, who gave introductory courses during her first year as a student. He belonged to that group of academics who were sent from Mussolini's Italy to promote fascism; and he apparently was more interested in political agitation than in teaching physics.[1], [3] Unlike Tiomno, she did not fall under the spell of *Luigi Sobrero*, perhaps because she was inclined towards experimental physics rather than theory.

In her second year at the Faculty, her supervisor was *Joaquim da Costa Ribeiro*, an experimentalist. By then, the Italian professors were leaving, due to Brazil's involvement in the war. Even before graduating, she began research in the group of Costa Ribeiro, although in an interview she notes, similarly to Tiomno, that her mentor did not give her much guidance; he himself had been mainly self-taught, and evidently held that to be the best model for his

Fig. 7.4 Elisa, around the time that she finished her studies at the FNFi. *Source* Private, archive [EFP]

students. Elisa studied on her own, using books suggested by Oswaldo and by her fellow students.

She began research in 1942 with measurements of the radioactivity of minerals, a topic of some strategic importance at the time, and of interest to Costa Ribeiro because he had discovered the thermodielectric effect in samples with radioactive impurities. She sometimes had to work in a spare room in the *Faculdade de Medicina* at *Praia Vermelha*, and she remembered leaving in the late evenings, walking by candlelight through the dissection rooms. She took her children as infants to the laboratory while she was nursing them, keeping them in a baby buggy. She worked as a student assistant to Costa Ribeiro in 1942/43, and then as an official, paid assistant from 1944 on. Elisa was the second woman to obtain a physics degree in Brazil (together with Sonja Ashauer, who graduated from USP in the same year, 1942). Sonja Ashauer, who went on to complete her doctorate under P.A.M. Dirac in Cambridge, unfortunately passed away at a very young age, in 1948, soon after returning to Brazil from England. The *very* first woman physics graduate in Brazil was Yolande Monteux, who completed her degree at USP in 1938. She worked for around twenty years at the *Instituto de Pesquisas Tecnológicas* in São Paulo, until in 1960 she moved to France, where she continued working at the *Bureau International de Poids et Mesures*. Yolande Monteux died in 1998 in France. Thus, Elisa was effectively the first *practicing* woman physics researcher in Brazil. However, like many Brazilian physicists of her generation, she never submitted a doctoral thesis; in her own words, her goals were [1]:

...To establish a [*research*] school and provide orientation to various students for their research, to publish the results... with them; and later, those students would guide other students – after a certain time – and continuing onwards... At the FNFi, there was a 'doctoral course' which consisted of writing a thesis and then defending it; however there were no real [*graduate*] courses as there are today. To obtain the doctorate, you chose a topic and wrote your thesis. You kept a notebook, where you recorded all of your activities. I thought of writing my thesis on radioactivity, which was the field that I was working in. In the end, I published various works on radioactivity at the CBPF, and never wrote the thesis. I found it much better to simply continue publishing my work, since at that time, most people didn't write a thesis, they just published their results. The first doctorate in physics granted in Rio was conferred by the CBPF in the 1960's... I believe that the first real course for doctoral students, the first course for post-graduates, was set up in Rio at the CBPF, set up by Jayme when there was already a lot of research going on at the *Centro*.

During her time as assistant in Rio, Elisa was in close professional contact with Jayme Tiomno, who was also contracted as an assistant to Costa Ribeiro. When Jayme Tiomno went to São Paulo in 1946, they began a correspondence,[4] at first as colleagues.

The Relationship with Jayme Tiomno

During nearly seven years (1940–46), Elisa and Jayme were in almost daily contact, first as fellow students and later as assistants to Costa Ribeiro in Rio de Janeiro, but it was in the first half of 1947, when she had separated from Oswaldo and Jayme returned from USP after his fellowship there ended, that they began an intimate relationship.

Oswaldo and Elisa separated as husband and wife, although they continued living in the same house and remained outwardly a couple. Their relations with each other and with Jayme Tiomno were still cordial and friendly. For both Tiomno and Elisa, their 'first love' was perhaps *physics*—but they could share that with each other, along with all the other aspects of their lives. Both were worried about the reactions of their families, so they were not given to speaking about their emotions and inner lives, even with their own families, and only some very close friends were aware of their relationship.

When Jayme Tiomno moved back to São Paulo in mid-1947, now as assistant to Mario Schenberg, while she stayed in Rio, still working as assistant to Costa Ribeiro, they continued their correspondence,[4] now much more intense, which they kept up while he was in Princeton in 1948–1950. However, their letters were still more like communications between colleagues than love letters.[5]

Their plan was to get a fellowship for Elisa at USP for the following year, as Tiomno had; that way, they could be together in São Paulo. Tiomno began to talk with his colleagues at USP about arranging a fellowship for Elisa, and was quite surprised and outraged when Wataghin made rather disparaging comments about the presence of women in physics.[6] But he went ahead, obtaining the fellowship for Elisa. Costa Ribeiro, for whom she was working at the FNFi in Rio, agreed to grant her leave for a period of one year, beginning in March 1948.

However, as we have seen, Tiomno traveled to Princeton at the beginning of February of that year, in a hurry due to the deadline for presenting himself to the University there. In fact, in 1948, Tiomno and Elisa 'exchanged places'; she went (in early March) to USP on a fellowship, working there with

Marcelo Damy de Souza Santos (an experimental physicist) in the department founded by Gleb Wataghin, while Jayme left São Paulo and began his graduate work at Princeton. While she was in São Paulo, Elisa's children, Sonia and Roberto, then 6 and 4, stayed with their father Oswaldo and his extended family. Oswaldo, as well as his parents Zezé and Dr. Frota, and his older sister Regina, volunteered to help her. They lived in the *Santa Tereza* district in Rio, overlooking the famous imperial church, *Igreja de Nossa Senhora da Glória*, which was illuminated on the feast day of its patron, the Virgin Mary—on Assumption Day, August 15th—which happens to be Sonia's birthday. The children were well cared for there, and Elisa could visit frequently and could return to Rio quickly if needed. (The 400 km from São Paulo to Rio de Janeiro took around 8 h by bus in those days, but travel by air was possible in an emergency.)

Princeton was another matter, far away and rather out of touch from Brazil, as long-distance telephone calls were quite expensive and the mail was slow. Thus, the nearly three years that Jayme spent there were a real separation for him and Elisa. Tiomno's family still did not know of his relationship with Elisa, but he certainly knew that they would not approve of it. For Tiomno, and probably for Elisa as well, *physics* had the highest priority; she expected that their relationship would continue when he returned to Brazil, but they had not made any promises to each other. It is indeed possible that they both saw this separation as a 'trial period' to test their feelings and maintain a certain distance. If that is so, then their loyalty to each other passed the test with flying colors.

Elisa's arrival in São Paulo was not easy for her. The University (USP) at that time still did not have its own campus, and was without living quarters or accomodation for visiting researchers. She found a small house in the *rua Dr. Melo Alves*, not far from the faculty (FFCL). Apart from that, her fellowship, provided by the University, took several months to begin being disbursed on a regular basis, causing various difficulties for her in the new environment. The Department of Physics there was undergoing a stressful period, with numerous internal quarrels and a lack of funds, in particular for purchasing material for the laboratories. Even so, she managed to develop a research program. She worked with the experimentalist *Oscar Sala*, who at that time was busy constructing an accelerator based on a Van de Graaff generator, and was planning to do research on a series of aluminum isotopes.

Tiomno was very enthusiastic about the Princeton experience, and he was eager for Elisa to have a similar experience abroad. But there was a problem: She could not go to Princeton, because the University still did not accept women students at that time. Perhaps she could go to Columbia University, or to a university in Europe. Tiomno and Leite Lopes also tried, through their contacts, to obtain a fellowship for Elisa to go abroad. But in contrast to Tiomno's fellowship, in the case of Elisa they did not have the support of the Rockefeller Foundation, since it did not usually give scholarships to women. In a conversation with Harry M. Miller, representative of the Rockefeller Foundation for Brazil, Leite Lopes insisted that "she was a good element, was doing well and so on…, and that she was already married". According to him, apparently one of the reasons for not giving scholarships to women was that they might marry before finishing their studies, and then drop out of school.[7]

Their correspondence [4], [5] provides a good source of information on his scientific development and achievements while Tiomno was in Princeton (cf. Chap. 6); but in addition, it contains details of his 'extra-curricular' activities and his daily life there, as well as of Elisa's to some extent, and we mention some highlights in the following: In early March of 1948, Tiomno again went to New York City, visiting the cyclotron at Columbia University and the laboratories (among others, Willis Lamb's labs, where he had measured the Lamb shift a few years earlier, and also the labs of I.I. Rabi; both of them were Nobel prize winners, Lamb somewhat later). He sent a clipping from the *New York Times* on Lattes' work, to be shown to Wataghin and others in São Paulo. He also wrote about the crystals used to detect gamma rays at Princeton (Robert Hofstadter was developing gamma-ray detectors there at the time, and introduced the Na(Tl)I scintillation crystals that became standard as gamma-ray detectors for many years).

By April 5th, Elisa had moved to São Paulo, and Tiomno warns her not to let herself become too involved in teaching duties by her host, Marcelo Damy (she was there on a research fellowship). Later that month, he refers to the 'Lattes wave', the publicity surrounding the π-meson detections, and expresses his opinion that it will be good for Brazil and for physics there.

In May and June, he writes about Mario Schenberg's arrest (Schenberg was politically active and a Communist), and reports that Wigner, Wheeler, Dirac and Oppenheimer had sent letters to the Brazilian ambassador to protest the arrest. In a letter in mid-September, Jayme Tiomno describes the AAAS meeting in Washington, where Wheeler made the first real public announcement of the 'Universal Fermi Interaction', attributing the calculations to Tiomno. Lattes also attended, and since he and Jayme shared a room, they had ample time to talk. At that time, they were close friends as well as colleagues. Tiomno's letter of September 24th closes with more personal remarks: "Remembrances to everyone there, and promise that we will have the opportunity to hug each other a lot in the near future, here … P.S.—I heard the sorry news of the death of Sonja Ashauer from Walter [*Schützer*]; it is a lamentable loss. In any case, I am still waiting for your answer to a question that I asked several times of you and others there over the past 6 months, and to which I have still had no replies: What was the situation of Sonja in the Department since she returned from England? Normal, or was there ill will towards her?". In his final letter to Elisa for the year 1948, Tiomno reports that he has received the Rockefeller fellowship, and that two places are still available, for which Elisa (or someone else in their group of colleagues in Brazil) could apply.[4]

First Research at the CBPF

When Elisa's fellowship at USP ended, Wataghin invited her to stay on there. But Costa Ribeiro wouldn't agree to give her another year's leave to São Paulo, and so she had to return to Rio. At this time, a very important project for Brazilian physics was underway in Rio: the founding of the *Centro Brasileiro de Pesquisas Físicas* (CBPF)—an innovative project, establishing an institute completely dedicated to physics research and teaching. And she was of course also happy to be there for that important event.

Upon returning to Rio in early 1949, Elisa began setting up the Department of Nuclear Emulsions at the newly-founded CBPF, and also resumed her work at the FNFi. Her scientific supervisor at the FNFi, Joaquim da Costa Ribeiro, was also a founding member of the Center, as was Guido Beck, who returned from Argentina somewhat later to help establish the Department of Theoretical Physics there. Elisa began studying some emulsion samples that had been exposed to pions in Berkeley, brought back by César Lattes. They contained the tracks of pion decays, and she was attempting to identify the 'branching ratios' of different decay modes. This was important for developing the theory of the Weak interactions. She wrote to Costa Ribeiro, who was traveling in Europe:

> I initially planned to work on the determination of the maximum energy of the electrons from some decays. In August, I had the opportunity to start working on photographic plates, with the long-term goal of finding the percentages of positive mesons which decay via muons and directly to electrons. That determination, along with some other data which are already known, will give us knowledge of the mean lifetimes for each of the processes. This research seems to me to be much more interesting and necessary at the present time.[8]

Lattes at first discouraged her work, saying that she would not succeed where the Americans had failed, and prompting her to engage instead in discussions of the political situation of the Center. But Elisa continued studying the samples, together with her student, Neusa Margem, and was able to obtain sufficient statistics to show that the fraction of pion decays leading to electrons and electron neutrinos must be less than 0.5% of the main decay mode (to muons and muon neutrinos), if it occurred at all (cf. Chap. 4). This was an important result, which we will take up again in connection with Jayme Tiomno's work on the Weak interactions. She submitted the resulting article for publication to an American journal, and it was accepted, but the journal wanted a change in the text. Lattes, by now enthusiastic about the results, encouraged her to publish quickly, since this would be the first scientific result from the Center, and so it was published in Portuguese in the *Annals of the Brazilian Academy of Sciences*.[9]

This was indeed the first publication from the Center, but unfortunately it was little noticed by the international scientific community, and was later forgotten even by Jayme Tiomno, who at the time of its publication was still in Princeton—a serious oversight, as we shall see later. The result was later confirmed at CERN, and disclosed in a famous paper published in 1958,[10] which found the fraction of decays leading to electrons to be $(1.22 \pm 0.30) \times 10^{-4}$, by observing those decays directly.

Neusa Margem (known later in her career as *Neusa Amato*, her married name) was Elisa's first graduate student. She was educated at the *Universidade do Brasil* (now UFRJ), and was the fourth woman to obtain her degree in physics in Brazil.[11]

In spite of her successes at the CBPF, Tiomno continued to feel that Elisa should go abroad—"it is very important for your education and to open up new horizons".[12] But César Lattes held a different opinion, preferring that she stay at the *Centro*, contributing to the research work of the new institute. In the end, Elisa, Leite Lopes and Lattes arrived at the conclusion that at that time, the *Centro* was still in a phase of initial organization and could not develop its research program on a grand scale—in truth, the only research result at that point was Elisa's. Thus it would be more productive for her to make use of the time and to qualify herself abroad so that she could be of more help to the *Centro* when she returned. She consulted with Schenberg, who was in Belgium, asking about the possibility of joining his group there.[13] He however replied that it would be better to go to England and to work with Blackett in Manchester, where they were using Wilson cloud chambers and photographic-emulsion plates. She could try to obtain a fellowship from the British Council.[14]

Elisa felt that it would not be easy to get a fellowship for the UK, and she preferred to make contact with Leprince-Ringuet in France, who responded very favorably to her inquiry about spending a year in his group at the *École Polytechnique* in Paris.[15] His suggestion was that she should work with a cloud chamber, studying cosmic rays. The CBPF itself gave her a fellowship and she requested a year's leave from the *Faculdade Nacional de Filosofia* for her sojourn at the *École Polytechnique*.[16] Once again, Elisa and Tiomno were about to exchange places: Jayme would be returning to Brazil just at the time when she was to depart for Europe. Tiomno wrote in one of his letters from that period, "I hope that we manage to meet in September before your departure".[17] But in the end, although everything had been arranged satisfactorily, Elisa gave up the fellowship and decided to stay in Brazil. We cannot know if this was because of Tiomno, because of her two small children, from whom she would be separated for a long time, or for some other reason altogether. Quite probably a combination of factors played a role in her decision.

Life with Tiomno

Jayme Tiomno arrived back in Brazil in October, 1950, but in spite of his desire to participate in the project of establishing the CBPF and to be near Elisa, both in Rio, he had committed himself with his colleagues at USP to go instead to São Paulo on his return from Princeton. In fact, at the time of his return, it was not at all clear for Tiomno and Elisa whether they would—or should—continue their relationship and become a couple openly, with a long-term committment.

Tiomno traveled periodically from São Paulo to Rio. Elisa liked to tell an anecdote from that time: One day, not long after he returned to Brazil, he was visiting in Rio. He came to her and said, "You know, I am planning to get married…" She was rather shocked and taken aback, but to hide that, she replied, "How nice for you! Congratulations!"; and then he completed his unfinished sentence, "… to you!". During the next Brazilian summer vacation, on January 12, 1951, just five days before her 30th birthday, he and Elisa held an informal ceremony and thereafter considered themselves married (Fig. 7.5). Their decision about their mutual future was taken and sealed! Their marriage was unofficial, since Elisa could not marry again according to Brazilian law at that time. (Only after the law was changed to permit divorce could they formalize their marriage, which they did on September

27th, 1977. After more than 26 years of their being to all intents and purposes a married couple, that event was anti-climactic and little noticed by the rest of the family).

Both families were against their union: Elisa's family were opposed to her living with a man without being formally married to him, and Tiomno's family did not want him to live with a woman who was separated from her husband and had two children (Elisa later maintained that this did not play a serious role in their objections, however [2]). They would also have preferred that Jayme marry someone from the Jewish community, as had all of his siblings. He indeed had several girlfriends from that community in his younger days, but none had led to a more serious relationship. His family's objections created some initial strain in their relations to Elisa (and hers to them), although they later accepted her fully.

There was still another hindrance: Jayme Tiomno's father, Mauricio, was very ill and the family were afraid to break the news to him. Even Joaquim da Costa Ribeiro, who was the mentor of Elisa and Tiomno at the *Faculdade Nacional de Filosofia*, became involved in the discussion. He was a religious Catholic and opposed the relationship of the two.

Fig. 7.5 Jayme Tiomno and Elisa Frota-Pessôa, at the time of their informal marriage in 1951. *Source* Private, archive [EFP]

But in the face of all these difficulties, they continued to pursue their project of living together. They moved to the district of *Santa Tereza*, in the *rua Santa Alexandrina*, not far from the house where Elisa's children were living with Oswaldo and his parents. Although Tiomno spent most of his time in São Paulo, where he was working at USP, he considered that address in Rio to be his principal residence. Over time, he assumed some responsibility for raising Sonia and Roberto, while always maintaining close contacts to Oswaldo and his family. The children remained mainly in the care of Oswaldo's family until the mid-1950's.

It is notable that Elisa Frota-Pessôa and Jayme Tiomno had no children of their own, although they were both relatively young (she was 29, he 30) when they began their conjugal life. She was small and delicate, and doctors feared that a further pregnancy could be dangerous to her health. Nevertheless, she became pregnant while they were living in the *rua Santa Alexandrina*, but lost the baby, an event of great sadness to both of them. Although they would have liked to have children, it was not to be. In the course of time, Jayme Tiomno came to regard her children with Oswaldo, Sonia and Roberto, as his own family.

From the date of their informal marriage onward, when they committed themselves to a lifelong partnership, their lives and careers were intertwined, and that continued for the rest of his life, another 60 years. Jayme Tiomno's story cannot be told without including Elisa's story as well.[18], [19]

During the writing of this chapter, we learned of Elisa's death from pneumonia, in late December 2018, just a few weeks before her 98th birthday. This marks the end of an era —with her, the last of the immediate founders of the CBPF, and one of the last of the 'founding physicists' of her generation in Brazil—and a pioneering Woman in Science—has passed away.

Notes

[1] Elisa Frota-Pessôa (1990): Interview with Ana Elisa Gerbasi da Silva and Lizete Castro Pereira Nunes, held in Rio de Janeiro for the *Projeto de Estudos de Educação e Sociedade* at UFRJ, on March 29th, 1990.

[2] Personal communications from Sonia Frota-Pessôa, 2018/19.

[3] See Tiomno (1977).

[4] Jayme Tiomno and Elisa Frota-Pessôa donated their papers and correspondence to be kept in archives at the Museum for Astronomy and Related Sciences (MAST) in Rio de Janeiro (archives [JT] and [EFP]). Following Elisa's death in 2018, their correspondence from the years 1946–1950 was

found (unfortunately only his letters to Elisa, and not hers to Tiomno), and is now also in the archive [EFP].

[5] For more details, readers should consult the originals in the archive [EFP].

[6] Jayme Tiomno to Elisa Frota Pessôa, December 4th, 1947; [EFP].

[7] José Leite Lopes to Jayme Tiomno, November 1st, 1948; [JT].

[8] Elisa Frota-Pessôa to Joaquim da Costa Ribeiro, October 27th, 1949; [EFP].

[9] Frota-Pessôa and Margem (1950).

[10] Fazzini et al. (1958).

[11] Neusa Margem had been preceded as a 'Woman in Science' in Brazil by Yolanda Monteux and Sonja Ashauer at USP and by Elisa Frota-Pessôa at the UB. She was initiated into science in a similar manner as Elisa: She obtained her secondary education at the Colégio Rivadavia Correa, where she was, like Elisa, also a pupil of the physics teacher Plínio Süssekind Rocha. She was the daughter of Lebanese immigrants, and, again like Elisa, she did not have the support of her family for embarking upon a scientific career. According to Ligia M.C.S. Rodrigues, in a text available on the CNPq Portal, "When she was nearing the end of her secondary education, she prepared to look for employment and to begin working, in view of the fact that the financial situation of her family was not satisfactory. Plinio, who had taken note of the fact that his brilliant pupil Neusa was not planning to continue her studies because she needed to work to help her family, gave her incentives and convinced her to take the entrance exams for the physics course at the Faculdade Nacional de Filosofia (FNFi) of the Universidade do Brasil. In order to aid her and other interested pupils, he gave a series of free classes during the vacation period". See https://memoria.cnpq.br/web/guest/pioneiras-view/-/journal_content/56_INSTANCE_a6MO/10157/902836. Neusa was five years younger than Elisa and obtained her Bachelor's in physics in 1945, and her *licenciatura* (teaching certificate) the following year. During her scientific career, Neusa published 116 articles. For more details on the professional life of Neusa (Margem) Amato, see '*Neusa Amato dos Vinte aos Oitenta*' ('Neusa Amato from the Twenties to the Eighties'), by Alfredo Marques Marques (2007).

[12] Jayme Tiomno to Elisa Frota-Pessôa, Dec. 18th, 1949; [EFP].

[13] Elisa Frota-Pessôa to Mario Schenberg, January 19th, 1950; [EFP].

[14] Mario Schenberg to Elisa Frota-Pessôa, February 07th, 1950; [EFP].

[15] Elisa Frota-Pessôa to Joaquim da Costa Ribeiro, March 27th, 1950; [EFP].

[16] Louis Leprince-Ringuet to Elisa Frota Pessoa, March 18th, 1950, and Elisa Frota-Pessôa to Mario Schenberg, April 19th, 1950; [EFP].

[17] Jayme Tiomno to Elisa Frota-Pessôa, June 21st, 1950; [EFP].

[18] Additional information on Elisa's life and scientific career can be found in the archives at the Museu de Astronomia e Ciências Afins (MAST) in Rio

([EFP]), and on the website of the Brazilian National Science Research foundation CNPq, as well as in the article written in honor of her 80th birthday (the 'homage'): Silva Lima et al. (2004).

[19] An interesting, sometimes controversial comparison of her 'public image' as presented by Silva Lima et al. (2004) and her own 'self-image' (or personal narrative) as found in her interview Frota-Pessôa (1990) is given in a recent article Linhares and Silva (2017) from the proceedings of a conference held in Florianopolis, Brazil in 2017. Compare: Silva Lima et al. (2004) and Linhares and Silva (2017).

8

The CBPF: In Pursuit of a Dream

Return to São Paulo

On returning to São Paulo in late 1950, Tiomno encountered a Physics Department which was quite different from the department that he had experienced prior to his 2–½ years at Princeton. When he left USP, in February, 1948, it was considered to be the principal physics center in Latin America, and an important nexus for attracting talent to the field; but by the time of his return, it was without leadership, and full of intrigues and internal conflicts. This was mainly due to the loss of its earlier academic and intellectual leaders.

The first of them to be lost was Mario Schenberg. At the end of the 2nd World War, Getulio Vargas was forced to abdicate his office as President of the Republic, and the Brazilian Communist Party had been rehabilitated. Mario Schenberg joined forces with the Party and ran for election to the Legislative Assembly in São Paulo (the equivalent of the state legislature or parliament) in 1946, becoming a substitute member. At the end of 1947, when a previously-elected member retired, he advanced to full membership in the parliament of São Paulo, becoming a very active member. Shortly thereafter, in January, 1948, as the Cold War began to heat up, the Party was again declared illegal—Schenberg lost his mandate and was arrested.[1] His parliamentary mandate had lasted all of 48 days! Various prominent persons in Brazil and abroad interceded with the authorities to have him released, sending letters for example to the Brazilian ambassador in the USA, asking him to intervene in favor of Schenberg.[2] In April, 1949, he was allowed to leave the country to attend the International Congress of Intellectuals for Peace, in Wrocław, Poland. He

© Springer Nature Switzerland AG 2020
W. D. Brewer and A. T. Tolmasquim, *Jayme Tiomno*, Springer Biographies,
https://doi.org/10.1007/978-3-030-41011-7_8

traveled with his wife Julieta and their young daughter Ana Clara, and after the Congress they took refuge in Paris. A few months later, he was hired by the Free University of Brussels, where Giuseppe Occhialini was also working; Schenberg remained there for three years.[3]

Approximately a year after Schenberg's arrest, in May of 1949, César Lattes, who had again taken up his assistantship at USP on returning from the USA, asked to be released from his position there, and transferred to Rio de Janeiro. He wished to dedicate his efforts exclusively to the CBPF project in Rio and, beyond that, Leite Lopes and Costa Ribeiro had managed to establish a chair in Nuclear Physics expressly for him at the *Universidade do Brasil*. His decision to leave USP was also due in part to a serious dispute with another professor of experimental physics at the Department there, *Marcelo Damy de Souza Santos*. Lattes' request for release from his contract at USP was accompanied by a letter to the Rector, denouncing Damy. The latter, for his part, went to the press and publicized Lattes' departure, detailing the internal conflicts within the Department.

Finally, in September of the same year, Gleb Wataghin, the founder of the Department at USP, returned to Italy, where he accepted a professorship at his *alma mater*, the University of Turin. The departure of Wataghin left a gaping vacancy in physics at USP.

Even in early 1948, shortly after Tiomno had left for Princeton, the internal quarrels in the Department at USP had intensified, culminating in an accusation of misappropriation of funds against its Director, Marcelo Damy, and the Vice-Director, Hans Stammreich. The latter was a physical chemist, educated in Berlin, who had arrived in Brazil at the beginning of the war as a refugee, and had set up a laboratory for Raman spectroscopy at USP. Because of the conflicts in the Physics Department, he moved to Chemistry. Meanwhile, the difficulties in Physics continued to escalate over time. Walter Schützer, who arrived back from Princeton at the end of 1949, described the terrible situation to Tiomno, who was still in the USA. Schützer also associated the internal problems of the Department with the departure of Wataghin.[4]

One by one, other students and assistants were migrating to the new *Centro* which had been founded in Rio, causing a drain on the Physics Department at USP. This also contributed to creating a climate of competition between Rio de Janeiro and São Paulo. By the end of 1949, Marcelo Damy had quit as director of the Department at USP, and that position was taken on by *Abrahão de Moraes*, a professor at the Polytechnic School of USP, who tried to resolve the crisis and to re-establish the Department. One of his first objectives was to search for a renowned physicist to replace Gleb Wataghin in the

Chair of Theoretical Physics. He invited Guido Beck, but Beck had already agreed to go to the CBPF on a long-term basis. He also considered Giancarlo Wick and Bruno Ferretti, and even invited Giampietro Puppi and Felix Villars. But without success; none of them was available to go to Brazil at that time.

Tiomno, still in Princeton, felt an obligation to aid in re-establishing the Physics Department in São Paulo. For one thing, the Department had maintained his position (paying a salary while he was on leave) throughout the whole period while he was in the USA, and he felt indebted for that. And in addition, he was convinced that Brazil should have at least two strong centers of physics research and teaching. The country couldn't allow itself the folly of letting such an important location for physics research slide into irrelevance. His intention to return to USP was balanced by his desire to participate in the project of establishing the *Centro Brasileiro de Pesquisas Físicas*, and by his relationship with Elisa, who had already returned to Rio and was involved in setting up a laboratory and working at the *Centro*. From Princeton, at the end of 1949, Jayme wrote to Elisa: "I accepted [*the invitation to return to USP*], I think that this will be a step towards restoring some equilibrium there".[5] His intention was to remain for two or three years and then to go on to Rio de Janeiro.

Tiomno returned to the Department of Physics at USP in October of 1950. He arrived there full of enthusiasm and ideas. One of his first activities was to write an article with Schützer on the S-matrix. While still in Princeton, shortly before his departure for Brazil, he had examined the MS thesis that Walter Schützer had written under Wigner's guidance. Tiomno found some inconsistencies, and Wigner agreed with his observations, suggesting that he write a paper together with Schützer, containing the correct formulations. This gave rise to their article '*On the Connection of the Scattering and Derivative Matrices with Causality*', which appeared in Volume 83 of the *Physical Review* in 1951.[6]

This paper harks back to a suggestion of Wigner's that a 'causality condition' must determine the analytic properties of the S matrix. *S*-matrix theory was originally introduced by Wheeler in 1937—the name stands for 'scattering matrix'—and it was proposed as a simplified way of describing scattering events between two particles. Werner Heisenberg generalized it in 1943 to a basic principle of particle interactions, and it was regarded for a time as a substitute for quantum field theories in particle physics. It was the subject of considerable theoretical effort in the 1950s and 60s, but was supplanted by a field-theoretical description of the Strong interaction (*quantum chromodynamics*, a cornerstone of the Standard Model) in the 1970s. However, it is still used in theories of quantum gravity.

The article by Tiomno and Schützer treats the case of scattering of waves from an atomic nucleus. The causality condition consists in maintaining that the amplitude of the scattered wave depends only on the amplitudes of the incident wave at earlier times and at points on the nuclear surface. Strict causality implies that there is no scattered wave before the incident wave impinges on the nucleus. The considerations of Tiomno and Schützer were originally non-relativistic, but have been generalized to the relativistic case (necessarily, since the *S* matrix is presumed to be relativistically correct, as well as analytic and unitary). Their articles attracted considerable attention, as S-matrix theory was beginning to be important in the early 1950s.

Some years later, in 1957, Wheeler remarked that this was the first paper to derive a connection between the properties of the *R* matrix and the *S* matrix. He also explained that the difficulty they pointed out in the paper, "… there is a connection between causality and the properties of the *R* matrix; however we are not able to derive all the properties of the *R* matrix from causality alone…", defined a problem which challenged researchers for some years. Only in 1957, six years later, was it solved by Nicola N. Khury,[7] who cited Schützer and Tiomno's paper as a "foundation stone for this more recent progress".[8a] Later, Marvin Goldberger, during the celebration of '15 Years of Dispersion Theory', mentioned the article as one of the precursors of that theory.[8b]

In this connection, the historians of science José Maria Bassalo and Olival Freire Jr.[9] affirm that "This article for the first time resolved the question of how to introduce a condition of causality into the mathematical formalism and to verify its implications for the *S* matrix", and that it has been referred to as a 'cornerstone of elementary-particle physics'. It is called the 'Schützer-Tiomno causality condition'.

In a letter to Wataghin, Abrahão de Moraes reported the good news: "Tiomno is very enthusiastic, he is giving seminars and working on topics associated with the *S* matrix".[10] For his part, Moraes made an effort to create the best possible conditions for Tiomno's work. He obtained a full-time position for Tiomno from the Faculty, increasing his salary by 50%, and attempted to increase it still more by 70%.[11] During a reorganization of the Department, Tiomno remained in the area of Rational and Analytic Mechanics (Classical Mechanics), set up by Abrahão de Moraes himself, along with Mario Schenberg (whose return was still anticipated) and Léo B. Vieira (the latter as assistant).

To fill the position previously held by Wataghin, Tiomno suggested the theoretical physicist *David Bohm*, who had been an assistant professor at Princeton when Tiomno arrived there. They had worked briefly on a formulation of the Dirac theory which would be conformally invariant[12]; that however had not been completed at the time. Tiomno would return to this project only a dozen years later (cf. [JT36]). Bohm was also reader for Tiomno's doctoral thesis, since both Wigner and Wheeler were away from Princeton at the critical time.

David Joseph Bohm (1917–1992),[13] who was born in the USA, in Wilkes-Barre, Pennsylvania, studied physics at Pennsylvania State College (now the Pennsylvania State University), graduating in 1939 (Fig. 8.1). He began graduate work that year at Caltech, in Pasadena/CA, where he met J. Robert Oppenheimer, who was dividing his time as professor during the late 1930s between Caltech and UC Berkeley. After a year in Pasadena, Bohm moved to Berkeley and continued his graduate work in Oppenheimer's group there. Some of its younger members were active in the American Communist Party, as were Oppenheimer's wife Kitty and his brother Frank, and Bohm also joined the party. In late 1942, Oppenheimer became scientific director of the Manhattan Project, and he went to Los Alamos in the spring of 1943 to begin the active development of a working nuclear weapon. He had invited Bohm to work there also, but Bohm was refused a security clearance because of his political affiliation, and even his thesis work—on proton-deuteron scattering—was deemed significant to the war effort and classified as 'secret', so that Bohm could not even access his own work to write his Ph.D. thesis. Through Oppenheimer's intervention, he was nevertheless able to complete his degree in 1943, remaining in Berkeley and working on plasma physics and the principles of synchrotron operation. He also carried out calculations for the electromagnetic separators ('calutrons') used to enrich uranium for making nuclear weapons during the Manhattan Project.

Bohm went to Princeton University in 1947 as assistant professor, at the invitation of John Wheeler and on the recommendation of Oppenheimer, who by 1947 had become Director of the Institute for Advanced Study there.

David Bohm became acquainted with Jayme Tiomno when the latter arrived in Princeton as a graduate student in early 1948. A year later, Bohm was subpoenaed by the 'House Un-American Activities Committee' (HUAC; its chairman at the time was Joseph McCarthy—'the McCarthy Era'), owing to Bohm's communist background. He refused to testify (i.e. to denounce his former colleagues), and he was therefore charged with 'contempt of Congress' in 1950, and arrested. The University suspended him from his faculty position, and even though he was later acquitted of the charges (in early 1951),

Fig. 8.1 David Bohm, probably around the time of his graduation from Penn State in 1939. *Source* David Bohm Society (DBS); see https://www.dbohm.com/david-bohm-pictures.html. Photographer unknown, public domain. Digitized version used with permission of the DBS

Princeton's president refused to reinstate him as a faculty member. Surprisingly, both Wheeler and Oppenheimer also failed to support Bohm's efforts to be reinstated. Oppenheimer recommended to him that he emigrate instead, and aided his efforts to move to the University of Manchester, in England. This was however unsuccessful.

Tiomno was aware of Bohm's need to leave the USA, and found that his accepting the open position in São Paulo would be a great help in reconstructing the Physics Department there. His suggestion was endorsed by Allen Shenstone and by Einstein himself, and they sent letters of recommendation for Bohm to Abrahão de Moraes.[14a] The invitation from USP was for Bohm to remain for (at least) three years as tenured Professor of Physics and Mathematics, receiving an annual salary of approximately 8500 dollars. In the invitation, it was emphasized that his name had been suggested by Tiomno, and that Tiomno and Walter Schützer would be very interested in collaborating with him on theoretical physics.[14b]

Bohm accepted the invitation. He was of the opinion that there were many interesting problems that they might work on together.[15] Furthermore, he could go to Europe in case he proved unable to adapt to Brazil. He arrived in São Paulo in early October, 1951, but was unhappy there from the start. Soon after his arrival, the American Consulate in São Paulo confiscated his passport, making it difficult for him to travel. In the end, Bohm remained at USP for four years before he moved on to Israel in 1955.

Bohm was one of the more controversial figures in physics in the second half of the twentieth century, both because of his political history and his resulting 'wanderings', and also because of his work in fundamentals of quantum mechanics, which was at odds with the conventional picture (the 'Copenhagen Interpretation'). He was undoubtedly also one of the most brilliant physicists of his time. During his tenure at Princeton, Bohm had often had discussions with Einstein, and that no doubt had stimulated his interest in fundamental quantum mechanics, which became his main research topic in his later career. His contributions to a deterministic quantum theory were regarded with skepticism in the early period of his career, but later generated three schools of fundamental physics (in Paris, London and Munich), and are now recognized as important impulses to the field of the interpretation of quantum physics, which has become increasingly active in the past 40 years. His theory is now generally referred to as the *de Broglie-Bohm Theory*, since its basic idea, of a 'guide field' (originally called a 'pilot wave') that gives rise to a deterministic trajectory for quantum–mechanical particles (e.g. electrons), was originally suggested by the French theoretician *Louis-Victor Pierre Raymond de Broglie* (Nobel prize winner in 1929 for his work on the wave-particle duality in quantum mechanics).

After Bohm's first publications on the subject, in the early 1950s, de Broglie resumed his interest in the deterministic theory, which he had abandoned following the Solvay Conference of 1927 in favor of the majority view, the 'Copenhagen Interpretation'. In the early 1950s, de Broglie's assistant *Jean-Pierre Vigier* visited Bohm in São Paulo, and they collaborated on the deterministic quantum theory and wrote several joint papers, culminating in Vigier's *Thèse d'Etat* in 1954 and their article in the *Physical Review* the same year, based on the hydrodynamic model for a deterministic quantum theory. Interestingly, Vigier's Marxist leanings may have contributed to his enthusiasm for a deterministic version of quantum mechanics, and no doubt also played a role in the mutual sympathy between him and David Bohm. Vigier developed his own version of the theory, and continued publishing on the topic, initially together with Louis de Broglie, through the early 1990s. This represents the 'French school' of the de Broglie-Bohm theory.

During Bohm's tenure in Brazil, in 1953, his doctoral student from Princeton, David Pines, also visited him in São Paulo, and they developed the *random phase approximation*, important in condensed-matter theory, and the concept of *plasmons*, quantized excitations in a plasma (an electrically-charged gas phase; Bohm had already worked on plasma physics while still at Berkeley). Bohm himself became disillusioned with the scientific possibilities in Brazil, and, although he had taken out Brazilian citizenship after his U.S. passport was confiscated,[16] he moved to Israel in 1955, working at the *Technion* in Haifa.[17]

Tiomno's single joint paper with Bohm and R. Schiller, '*A causal Interpretation of the Pauli equation*', appeared in *Il Nuovo Cimento* the same year that Bohm left Brazil, and made use of the hydrodynamic model originally suggested by Erwin Modelung.[18] Tiomno described the results of that paper,

> ...the extension of the causal Schrödinger equation developed by Bohm to the Pauli equation for particles with spin was carried out. We obtained classical equations of motion for a fluid of particles with spin and turbulence, with Hamilton equations in complete agreement with those introduced by M. Schenberg in his thesis for the Chair at the FFCL (USP), except for the 'quantum potentials and tensions'.[19]

At the Physics Department of USP, Tiomno also gave courses, organized seminars, and took on a first graduate student, *Paulo Saraiva de Toledo*, to work on field theory. He in addition helped to organize the '*Symposium on New Techniques for Physics Research in Latin America*' (held in Rio in 1952), and in gaining the support of UNESCO to boost research in physics in Brazil. At USP, besides the group in theoretical physics to which Tiomno belonged, there were several other groups working in experimental physics. One of them, led by Oscar Sala, was completing the construction of a Van de Graaff accelerator with an energy of 3–4 MeV; another, that of Marcelo Damy, was working with a betatron producing 28 MeV electrons; and a third was continuing the work on cosmic rays initiated by Wataghin, in particular on penetrating showers.

In spite of the fact that his work in São Paulo was going well, Tiomno's heart wished to return to Rio permanently; there he could participate in the CBPF project, and furthermore, the separation from Elisa created a stress factor which was increasingly disturbing to the couple.[20] Thus, in March of 1952, Tiomno asked to be dismissed as First Assistant to the Chair of Rational and Celestial Mechanics at USP, turning over the leadership of his research group to David Bohm, and transferred to Rio. Quite possibly, Bohm was one of the people who suffered most from Tiomno's move from São Paulo to Rio, as one can see from his comments in an interview more than 30 years later:

> Not long after that [*when Bohm's passport was confiscated by the US Consulate in Brazil*], this fellow Tiomno came to me and said he's moving on to Rio because he likes it better, his wife is there and so on. He was one of the main people I would be able to work with. There was another fellow, Walter Schützer. I was able to work with him later, but to a much smaller extent. This was a big blow in a way that I was sort of left alone.[21]

Early Plans and Dreams

Finally, Tiomno was able to join the Center that he had helped to establish. The idea of founding the *Centro Brasileiro de Pesquisas Físicas* (CBPF), the Brazilian Center for Physics Research in Rio de Janeiro, dated from the early 1940s. It was an old dream, most likely born out of discussions among the 'Three Musketeers': Jayme Tiomno, Elisa Frota-Pessôa and José Leite Lopes, while they were studying at the FNFi of the *Universidade do Brasil*. Each of them was full of enthusiasm for pursuing physics research and teaching, and they all had been disappointed by the forced closing of the *Universidade do Distrito Federal* (UDF) in Rio in 1939, preventing them from studying there. The UDF had been an attempt to establish a research university with high standards of excellence, but it had fallen victim to the fears of the Vargas government. The *Universidade do Brasil* (UB), which took over the faculty from the UDF, was neither equipped nor inclined to encourage research— and thus also not the teaching of graduate students in the physical sciences, for which research facilities are indispensable. This became even clearer to the three when they finished their undergraduate work and began research at the UB. Another serious problem was the resistance of the University to providing full-time positions for its professors. The great majority of them had other professional activities and went to the University only to give their courses. In contrast, the CBPF founders wanted their center to be a space dedicated to science, where the professors would apply all of their time to both teaching and research at the same institution.

Jayme Tiomno explained concisely the necessity of creating a new institution:

> The *Centro* was founded because the universities wouldn't permit research in their own institutions; or rather, they created all [*possible*] obstacles to research.

This was general in Brazil, with the exception of São Paulo. All the Brazilian universities created the maximum of difficulties for research … In all the Brazilian universities, when individuals went abroad and became researchers, and returned to try to carry out research at a university, they suffered negative reactions from the universities [*which were*] directed by bachelors. This was an obstacle for all of our students who went back to their [*home*] states and had to deal with reactions that were really violent, including campaigns of demoralization ['*mobbing*']. And the fundamental reason is this: They wanted to carry on research, and the bosses of the universities realized that, as soon as they would manage to establish research at a university, they would begin to have power within that university, and would thus take power away from the bosses.[22]

At first, they expected that such a space dedicated to research could be housed at the *Fundação Getulio Vargas* (FGV). This foundation had been set up by President Getulio Vargas himself in 1944, to be a center for the education of public servants and for applied research in the interest of the government, such as the establishment and monitoring of economic indicators, with its scope oriented towards economics and administration. In October of 1945, after the end of the war, Vargas was deposed and the direction of the organization changed; it was now led by Paulo Assis Ribeiro. He reoriented it, opening the scope of the FGV to scientific areas with emphasis on stimulating research. Departments for physics, mathematics, geology and biology were set up, directed respectively by José Leite Lopes, Lélio Gama, Othon Leonards and Antonio da Silva Mello, all of whom had a strong interest in research. Each of these departments had its own scientific journal: *Summa Brasiliensis Physicae, Summa Brasiliensis Mathematicae, Summa Brasiliensis Geologiae* and *Summa Brasiliensis Biologiae*. Among them, the first to establish its own structure was mathematics. It was this Foundation which provided the fellowship for César Lattes to go to Bristol, where he participated in the pion discovery.

However, a short time later, a group associated with Getulio Vargas and his original concept for the Foundation, led by Luiz Simões Lopes, put an end to the scientific departments and dismissed their directors, with the intention of restoring its original structure.[23] This was a severe blow to the group that had dreamed of a space devoted exclusively to physics research. "But we must not lose hope", Tiomno wrote to encourage Leite Lopes. And they indeed weren't about to give up easily. They considered making the Physics Department at the UB into a semi-autonomous institute, separate from the FNFi. To that end, they needed to attract more people to the department in Rio. Stimulated by the positive results from the presence of Wataghin in São Paulo, Tiomno also found the idea of trying to attract a well-known physicist to Rio to teach Advanced Physics there appealing; that physicist should stay in Rio for at least two years.[24]

Their new institute would be created within the university itself, but completely dedicated to physics research and (graduate) teaching. In 1947/48, Tiomno and Leite Lopes shared this idea with the young César Lattes, who at that point was on his way to Bristol to participate in pion research there. He was also quite enthusiastic about the proposal. They initially tried to obtain support from the Rockefeller Foundation, which had a representative in Brazil and was interested in fostering the development of physics in that country. They soon found, however, that some major event would be necessary to stimulate sufficient support, perhaps a notable discovery by a member of their group, which would give prestige and visibility to their proposal.[25]

In early 1948, having participated in the discovery of pions the previous year, Lattes went on to Berkeley (following his marriage in Brazil), under a fellowship from the Rockefeller Foundation. There, the 184″ cyclotron was the only machine available worldwide at the time that could provide a sufficiently energetic particle beam to produce pions artificially. It had been financed originally by the Rockefeller Foundation, specifically for meson production, but had been used during the 2nd World War as a prototype for the magnetic separation devices ('calutrons') constructed to obtain enriched ^{235}U for the weapons program, in Oak Ridge, Tennessee.

The projectile energy for particle production has to be greater than the equivalent mass of the particles to be created, about 140 MeV for the pion. Taking the relativistic mass increase of the projectiles, as well as momentum conservation with the stationary target into account, the required kinetic energy of the alpha particles in the laboratory frame is around 290 MeV. This was near the maximum energy of the 184″ cyclotron at the time.

Lattes collaborated with Eugene Gardner at the UC Radiation Laboratory (now Lawrence Berkeley National Laboratory), using high-energy alpha particles striking a carbon target to produce pions (the elementary process is a proton-proton collision). Pions had indeed been produced by the cyclotron since at least late 1946, after it was again put into operation at the end of the war; but no suitable detectors were available to identify them until Lattes arrived with his knowledge of how to observe their tracks in the nuclear emulsions. Their successful detection (on February 21, 1948) was quickly reported in *Science*.[26]

Lattes, at 24, became famous overnight within the physics community, and among the public at large as well, thanks to press coverage of the discoveries. He was invited to address conferences at many North American institutions: the Universities of Rochester, New Mexico, Harvard and Princeton, and Brookhaven National Laboratory. He gave a presentation at the Annual Meeting of the American Physical Society. He was also written up in the *New York Times*.

"That boy is a miracle!", exclaimed Leite Lopes in a letter to Tiomno. His achievements were also given ample coverage in the Brazilian press, where he was hailed as a young genius, a prime example of a successful Brazilian, and a national hero. He looked (and acted) the part. This was precisely the event that they needed to make possible the establishment of a Physics Center in Rio de Janeiro. At a time when both Lattes and Tiomno were in the USA, Leite Lopes gave interviews to the Brazilian press, explaining the significance of the pion experiments and the importance of producing artificial pions in the laboratory. The Brazilian military were also interested in the new experiments, since they were in the area of nuclear physics, a field which they held to be of strategic importance. Leite Lopes was invited to give lectures at the Army Technical Academy and even to the General Staff of the Army. He wrote enthusiastically to Tiomno:

> The news hit here like a bomb. All the newspapers carried articles about Lattes, declaring his results to be a great step forward for science, with telegrams from the USA, etc. The impression that I have is that now is the moment to apply pressure and get our Institute of Physics in Rio with all the trimmings.[25]

At the same time, Tiomno and Lattes, both in the USA, made use of their time to discuss the project for the new Center in Rio de Janeiro. Lattes and Tiomno first conferred in May 1948, when Lattes went to Princeton to give a talk, and then later in September, during the AAAS Washington meeting.

Lattes returned to Brazil for a brief stay in December of 1948. His arrival there was a great success. He had become a public figure, and was received by President Eurico Gaspar Dutra, by several Ministers and by various other authorities. While at Berkeley, Lattes met another young Brazilian, *Nelson Lins de Barros*, who was studying there and was working as secretary at the Brazilian consulate (in San Francisco). Lins de Barros was a member of a large and influential family (he had 21 siblings). A much older brother, João Alberto Lins de Barros, had been a Minister and a close confidant of Getulio Vargas. Although by this time he was only a city councilor in the Capital, he had maintained many political and economic connections in Brazil.

César Lattes and José Leite Lopes were invited to spend some days at João Alberto's country estate in order to talk about the planned new institute. During the conversations, João Alberto changed their initial idea; instead of creating an Institute of Physics within the University of Brazil, they would establish a new foundation dedicated just to physics, autonomous, with private and public funding, outside Rio and São Paulo, but collaborating with both. The foundation would have a research institute, and Lattes himself would be its scientific director, carrying out research, educating new researchers and teachers for the universities, and supporting them through fellowships. From the beginning, the new foundation would cooperate with the universities in Rio and São Paulo, stimulating the existing centers. It seemed to them that it was not reasonable for each location to have its own betatron or accelerator for the restricted use of the local researchers. Rather, such instruments should be shared within the whole emerging physics community. The new institute would not be understood as opposing or competing with the one or the other group, but as supporting physics as a whole.[27]

Some years later, Lattes commented on this sojourn at João Alberto's estate and his decision to abandon other promising opportunities and join the adventure of setting up the CBPF:

> I returned and committed a crazy act, as only young people will do … I came to Rio with a contract to give two seminars per week at the FNFi [*Universidade do Brasil* (*UB*)], and to be the Scientific Director of a Center which was only a foundation filed in the public registry … When I arrived in Brazil, I met Nelson and British [*Henry British Lins de Barros*], another brother who was an officer in the Navy. To everyone's surprise, when we got to João Alberto's house, we learned that his wife was a half-sister of Leite Lopes. They hadn't seen each other for 20 years. This was how the CBPF was born. Everything was in the family.[28a]

However, the plan was not based solely upon family relationships. Nuclear physics was becoming a great opportunity in itself. The founders of the *Centro* obtained the support of entrepreneurs and industrialists worried about a future shortage of electrical energy, which might threaten their projects for industrial development, but could be alleviated by turning to nuclear power. At the same time, the military believed a high level of competence in nuclear physics to be of great strategic importance, as had been so well demonstrated at the end of the 2nd World War. In this respect, the influence of Rear Admiral *Álvaro Alberto da Motta e Silva*, a chemist and the ex-president of the *Academia Brasileira de Ciências*, was a great help. Finally, there was a growing scientific community with an interest in research, which would support the project of establishing an institute dedicated to science; it included Costa Ribeiro, occupant of the Chair of Experimental Physics at the FNFi, and Carlos Chagas Filho, professor and director of the *Instituto de Biologia*, as well as the former participants in the project of the *Fundação Getulio Vargas*, such as Paulo Assis Ribeiro, Lélio Gama and Othon Leonards, among others. As for the small group of founding scientists, composed of César Lattes, Jayme Tiomno, José Leite Lopes, and Elisa Frota-Pessôa—what they really wanted was simply to pursue physics. And even more: they wanted to establish physics research in Brazil.

César Lattes had already been invited to work at Harvard University, and the other members of the group would sooner or later have been able to obtain good positions at research institutions abroad. But, as Lattes himself later explained, "At that time, no-one wanted to go there [*i.e. to go abroad*] with the idea of starting a career outside Brazil. No-one wanted to stay abroad. We all thought, let us say in somewhat patriotic language, of improving things in Brazil".[28b]

The official registration of the foundation took place on January 15th, 1949, making João Alberto Lins de Barros its President, Álvaro Alberto da Motta e Silva its Vice President, and César Lattes its Scientific Director. One hundred and sixteen persons signed the registration form as founding partners, among them scientists, members of the military and industrialists. In fact, Lattes did not participate in that initial assembly; he had to return to the USA in January, 1949 to finish up his work at the Radiation Laboratory in Berkeley, and had left a power of attorney with Leite Lopes. Back in Berkeley, he received notice by telephone that the Center had been registered with the authorities as a foundation, and that he was its Scientific Director. He was 24 at the time, and had a certain amount of experience in scientific research, but none as an administrator or research coordinator.

Lattes was planning to return to Brazil in March to participate in the project together with Elisa and Leite Lopes, but nothing was going to be easy for the group of founders. Without expecting such a rapid development in the process of setting up a new Center outside the university, Leite Lopes had accepted a fellowship from the Guggenheim Foundation and was programmed to go to Princeton for a year beginning in February, to work with Oppenheimer. He traveled to the USA in early 1949, shortly after the formal creation of the CBPF as a foundation. In March, just before Lattes returned to Brazil, he was in Princeton to participate in a seminar on meson physics, together with the Japanese theoretician Hideki Yukawa, who was a guest professor there at the time. Tiomno, Leite Lopes and Walter Schützer were also there, and Hervásio de Carvalho, who was studying in Washington/DC, went to meet his colleagues. There, at Princeton University, the final strategy for establishing the *Centro* was hatched out (Fig. 8.2).

Many scientists wonder how to explain the rapid progress of physics in Brazil, a country with practically no scientific tradition. For Tiomno, it was

Fig. 8.2 A group photo taken in Princeton. In the front row are Hervasio de Carvalho, José Leite Lopes and Jayme Tiomno; standing behind are César Lattes, Hideki Yukawa, and Walter Schützer. This photo was taken in March, 1949, during Yukawa's tenure as visiting professor at Princeton in 1948/49. *Source* Private archive of Jayme Tiomno [JT] and Leite Lopes archive [JLL] (public domain)

a manifestation of collective enthusiasm, mainly by a group of idealists, who had Gleb Wataghin as their tutor.[29]

When his fellowship ended in March, Lattes returned to Brazil and went to USP, but as we have seen, after two months he asked to be dismissed from his position there, and transferred to Rio de Janeiro in May, 1949. While Tiomno and Leite Lopes were still in Princeton, Lattes joined forces with Elisa, who had also returned from São Paulo, and with Ugo Camerini, who had been a student at USP, along with several assistants, to take the first steps to begin research at the CBPF.

José Leite Lopes, who by now had a secure professorship at the FNFi, had a more relaxed attitude towards the *Centro*. In an interview, he says the following about the early period at the CBPF:

What did we have, then, at that time, at the CBPF? Lattes, Camerini, young people such as Elisa Frota-Pessôa and other students, who were instructed in turn by Lattes and Camerini, who was an experimentalist. And Lattes began to apply the technique of nuclear emulsions, special photographic plates that were exposed to cosmic radiation. He sent them to accelerators, they were developed in the lab and examined with a microscope, and that permitted research [*on those particles that left tracks*]. That group was set up rapidly in Brazil, with Lattes, Camerini and Occhialini, who came to help out Lattes and some younger people. In theoretical physics, there was myself; I called on Tiomno and Guido Beck, and they stayed there [*at the Centro*].[30]

In fact, the actual establishment of the Center was carried out by the experimentalists: Lattes, Elisa, Camerini, and Occhialini, since Tiomno and Leite Lopes were in the USA and Beck had not yet joined the project. However, as Lattes was mainly engaged in raising funds and organizing the administration, it fell to Occhialini, Camerini, and Elisa to establish the first scientific activities at the *Centro*, which led her to publish the first paper there. Although Leite Lopes was 3 years older than Elisa and already had a professorship at the FNFi, they were part of the same group of students, and in 1950, she was not a young student: She already had a position at the University as assistant to Costa Ribeiro, and was a senior researcher at the *Centro*.

The original idea of the group of founders, that the *Centro* would be established neither in Rio nor in São Paulo, proved not to be feasible. They realized that they didn't have the resources to give full-time salaries to the professors, nor even to guarantee them long-term contracts. They would have to continue working at the universities. Apart from that, the proximity to students was of fundamental importance. It was only at the universities that they could encounter talented young people and recruit them to work at the *Centro*.

The relation between the *Centro* and the University, and their physical proximity, found a consensus among all the founders. Leite Lopes made the following remarks about it:

My preoccupation was to make a connection between the *Centro* and the *Faculdade*, since, in my view, an isolated *Centro* would not be viable. The Faculty had an element, its young students, who could be recruited.[30]

Tiomno had a similar view:

We had to remain within the *Faculdade de Filosofia* in order to find the best students and take them to the *Centro*, where we could train them as physicists... otherwise, we would have remained isolated from the *Faculdade*. Thus, we accepted secondary positions there, since that was the only way to acquire students.[22]

Elisa was the most active in bringing physics students to the *Centro*. She set up a system to find housing for the students, tried to obtain scholarships for them, and helped them to adjust to life in Rio. Together with Tiomno, she was pleased to welcome them into her home. Elisa also shared similar opinions to those of Tiomno and Leite Lopes on the role of the CBPF and its connection to the Faculty:

> The courses at the *Centro* were finally recognized by the *Faculdade*, but we never managed to merge the two. What I did manage to do, and it was one of the things that I most enjoyed doing in my life, was to take my students from there [*the FNFi*] to the *Centro*.[31]

In fact, the ideal of the unity of teaching and research that Jayme Tiomno upheld with such energy was not new; it was the philosophy of *Wilhelm von Humboldt*, the *Humboldt'sches Ideal*. It had been practiced in Europe for generations, at institutions such as the *Institut Pasteur*, the *Institut Pierre et Marie Curie*, the *Robert-Koch-Institut* and the institutes of the *Kaiser-Wilhelm-Gesellschaft*, which were all associated with universities and combined research with graduate teaching. The national laboratories in the USA, established after the 2nd World War, practiced a similar combination of research and graduate teaching, in general linked with a nearby university. But the concept was new to Brazil, and the challenge for the *Centro* was to make this model work in the face of resistance from the universities, in particular the FNFi at the *Universidade do Brasil*, and to find stable financing to make the *Centro* viable over a longer term.

Lattes initially consulted with USP about the possibility of establishing the CBPF there, but after hearing the opinion of the Department of Physics, the Rector Lineu Prestes replied that they had no interest in such a project, since they already had a department that was dedicated to physics and supplied what was needed from the University. The Rector of the UB, Leitão da Cunha, in contrast greeted the idea of establishing the *Centro* and its proximity to the University. The UB provided some land on its campus to construct the first building of the *Centro*, and a donation from the banker Mario d'Almeida enabled the construction of that building, a pavillion on the old UB campus at *Praia Vermelha*, near the Urca district.[32] (Its modern building now stands at the back of the Urca campus of the University). The UB also later gave a university mandate to the CBPF, allowing it to offer instruction in certain disciplines which then would be recognized by the University.

According to this model, students would take the first two years of courses at the University and would then attend lecture courses both at the University and at the CBPF. In return, the professors from the University could make use of the laboratories at the *Centro*. Leite Lopes returned from the USA in January of 1950 and was impressed by what had been accomplished during that first year:

> The *Centro* is going well. Lattes' prestige is enormous and opens doors everywhere, from the press to the Senate. He is really working selflessly to stabilize the *Centro*, instead of resting on his laurels. As a result, I think that we will grow properly. He has energy, administrative vision, and makes decisions as he should. The building for the *Centro* [*the pavillon*] should be finished by April.[33]

Indeed, it wasn't just Lattes' prestige; the time was simply ripe, although not for any of the reasons that motivated the little group of founders. At the very beginning, a significant amount of financial support was made available from a secret fund managed by Evaldo Lodi, President of the Industrial Federation, intended to combat communism. Lattes was initially not aware of the origin of the money; however, he never had to sign receipts or to submit reports for the use of those funds.[34]

Also at the beginning of 1949, shortly before the founding of the CBPF, the *Conselho Nacional de Pesquisa* (CNPq, National Research Council) was established, fulfilling an old aspiration of the scientific community in Brazil to have a central agency that would finance scientific research. This initiative, however, could be put into effect only with support from the central government, in the interest of directing science and especially nuclear technology. An indication of this was the signal presence of military officers in the CNPq: Its President would be Rear-Admiral Álvaro Alberto and its Vice President was General Armando Dubois, satisfying the demands of the Army and the Navy. We should recall that Álvaro Alberto was also Vice-President of the CBPF. During the year 1949, the Soviet Union successfully detonated its first nuclear device at its testing grounds in *Semipalatinsk*, Kazakhstan. This definitively demonstrated to the Brazilian government the importance of investing in nuclear physics. For the military officers at the CNPq, the CBPF would be the ideal place to develop the required knowledge. Therefore, in the first years of the decade of the 1950s, a major portion of the resources of the CNPq were invested in the CBPF. In the period 1951–1954, biology was the area that benefited most from CNPq funding, followed by physics; and within the area of physics, the CBPF received 75% of the financial support. In addition, it received a large amount of funds for special projects.[35]

The establishment of the *Centro* was however not unanimously supported in Brazilian academic circles. *Plinio Süssekind Rocha*, professor of Rational and Celestial Mechanics at the FNFi, was skeptical with regard to the research center. He felt that it was wrong to expend so much effort outside the University. The thinking of the experimentalist Armando Dias Tavares concurred with this; in spite of having participated in the initial seminars, he preferred to remain only at the University and not to join the *Centro*. Apart from them, physicists in São Paulo eyed the new project with distrust, both because of the loss of important researchers from their own institution and because of the competition for financial support.[36]

Adding to that, there was skepticism about the sustainability and continuity of the project. During a trip back to Brazil, while he was staying in Rio de Janeiro for a few days, Mario Schenberg was impressed by the enthusiasm of the group. Nevertheless, he found them too idealistic; and furthermore,

> ...they completely underestimated the difficulties of setting up a physics institute. It appeared to them to be only a question of money. Naturally, it is necessary to plunge into the task with enthusiasm, but I fear that the *Carioca* temperament [*natives of Rio*] brings with it a tendency to become quickly discouraged when the inevitable difficulties arise. It appears to me that in less than three or four years, it will not be possible to set up a good laboratory in Rio, even if all goes well.[37]

In fact, their enthusiasm was contagious. Facing up to difficulties was (and perhaps still is) what motivates physicists most strongly. In 1950, the experimentalist *Roberto A. Salmeron* moved from USP to the CBPF; the following year, Guido Beck left Argentina to go to Rio, strengthening the team of theoreticians there. Richard Feynman, who had already given lectures at the CBPF during his first visit in 1949, shortly after the founding of the *Centro*, returned in August, 1951, this time to stay for nearly a year, during his sabbatical; and finally in March, 1952, Tiomno, who had been visiting Rio frequently during 1951, officially left the FFCL at USP, and transferred to the CBPF and the FNFi/UB, to a less-important position there than he had in São Paulo.

The Golden Age of the *Centro*

The *Centro* initially opened in three rented rooms in the *Cinelândia* area, in a business district, surrounded by banks and office buildings, while waiting for its own building to be completed. With donated books for the library, and borrowed apparatus, the first small group of physicists began working there. The infrastructure was rather precarious; the work of Elisa Frota-Pessôa and Neusa Margem with the emulsion plates irradiated by Lattes in Berkeley, which led to the publication of the first scientific article from the CBPF, was carried out using microscopes borrowed from the Institute of Agricultural Chemistry and the Technical Division of the Police. But although the technical resources were limited, the desire to work was strong. It was the fulfilment of the dream of doing physics in Brazil and of building a Center which would aid in educating new generations of researchers and sowing the seeds for physics departments at the new universities that were beginning to spring up throughout the country.

The CBPF bought a Cockroft-Walton particle accelerator, manufactured in Holland, and E.O. Lawrence, in Berkeley, offered Lattes a 21″ cyclotron with an energy of 4 MeV at a low price. It would be built at the University of Chicago and installed in the city of *Niterói*, just across the bay from Rio de Janeiro. It would serve to train personnel and to produce radioisotopes. In addition, they started to negotiate with the Dutch Philips Corporation for the

Fig. 8.3 A festive lunch organized by the CBPF in Rio during Richard Feynman's second visit to Brazil, in 1952. Feynman is sitting on the right, fifth from the front. Just behind him is Elisa. Jayme Tiomno is sitting three places further back, wearing a white suit. In front of Feynman is Neusa Margem, Elisa's student. *Source* CBPF, and *Acervo* Jayme Tiomno [JT]

purchase of a 72″ cyclotron. The negotations stalled, and the President of the CNPq suggested that they order a 170″ cyclotron with an energy of 20 MeV, which was to be built in Chicago and brought to Brazil, to the Navy Arsenal in Rio de Janeiro, with the support of scientists and technicians from the University of Chicago. It would permit research in high-energy physics.[38] At that time, accelerators with maximum energies in the range of several GeV were being constructed in the USA and the USSR.

Officially, the contract with Philips was never signed due to technical requirements that the Brazilian government could not accept. Later on, Alfredo Marques, at that time an assistant of Lattes', revealed that the president of the CNPq, Álvaro Alberto, had explained to Lattes that the agreement with the Dutch company was causing diplomatic problems in the relationship with the United States. The purchase of the cyclotron from the University of Chicago, which would allow a degree of control by the American government, would demonstrate that Brazil's intentions were peaceful and did not aim at the development of nuclear armaments.[39]

Lattes also set up a research laboratory on Mount *Chacaltaya*, in Bolivia, where he had originally exposed the plates that led to the initial pion discovery. He wanted to use a Wilson cloud chamber, which was to be donated by Marcel Schein of the University of Chicago, for studying the 'V particles' that had been discovered by Rochester and Butler using a similar chamber at sea level.[40] The *Centro* entered into a formal collaboration with the University of Chicago to support this project.

Complementing experimental physics, the *Centro* provided adequate support for research in mathematics and theoretical physics. A Department of Mathematics was established and it took over the publication of the journal *Summa Brasiliensis Mathematicae*, which had been initiated by the *Fundação Getulio Vargas*. Starting from that department, *Leopoldo Nachbin* (the *d'Artagnan* of the 'Three Musketeers'), together with Lélio I. Gama and Mauricio Matos Peixoto, established the *Instituto de Matemática Pura e Aplicada* (considered today to be among the 10 best mathematics institutes in the world).

Tiomno set up the teaching laboratories at the CBPF and founded its collection of preprints, '*Notas de Física*', still in active use today. He later also organized the Graduate School of the CBPF, establishing a new standard for post-graduate education in Brazil. Until then, the doctorate was obtained, if at all, by simply writing a thesis and defending it, without any particular guidance. Tiomno created a graduate curriculum, including the study of recent scientific developments, preceding and accompanying work on the doctoral project. With the support of Feynman (Fig. 8.3), Tiomno and Leite

Lopes tried to modernize physics teaching in Brazil, substituting the long and detailed textbooks in use previously by more modern and specialized texts, in an attempt to replace the stultified, over-erudite instruction common at the time by a curriculum which would take the most recent developments into account. At the same time, Tiomno and Leite Lopes attempted to acquire teaching materials in Portuguese which would make physics accessible to a greater number of students.

The University was a *vivarium* where the most talented students in physics could be identified and encouraged to continue their studies at the *Centro*. In general, they were supported by being able to do research there and by receiving scholarships from international foundations, from the CNPq or from the CBPF itself. Elisa and Jayme Tiomno would invite their students into their own home to give them a feeling for that world of research. Tiomno also made good use of his scientific contacts abroad to send his best students to excellent foreign universities. The CBPF in addition invited students from various regions within Brazil and from other countries, mainly in Latin America. During the first 10 years of the Centro, more than 100 scholarships were granted to Brazilian students and others from Argentina, Bolivia, Peru and even from the USA.

The *Centro* developed research projects mainly in elementary-particle physics, cosmic rays (using nuclear emulsions), and theoretical nuclear physics; but there was also some research on photo-induced fission, atmospheric radioactive contamination, and radiochemistry. Researchers from the Center published more than 150 papers in recognized scientific journals during its first decade. The *Centro* also gave support to many universities that were being established at that time in Brazil, to help build up their physics departments; these included Recife (State of *Pernambuco*), *Minas Gerais*, and *Rio Grande do Sul*. From the beginning, it followed a policy of bringing prominent physicists from abroad to give courses and lectures, participate in seminars, and take part in specific research work. As examples, we could mention J. Robert Oppenheimer, Isidor I. Rabi, Leon Rosenfeld, Emilio Segré, Chen Ning Yang, Richard P. Feynman, Peter Bergmann, Giampietro Puppi, H. L. Anderson, S. de Benedetti, and Eugene P. Wigner, in addition to many others.

Meanwhile, the Department of Physics at USP gradually recovered from its crisis. Schenberg returned from Belgium, and became the head of the Department; Bohm was coordinating a research group; and both the betatron and the Van de Graaff generator had been completed and were operating. Furthermore, relations between the two centers, in Rio and São Paulo, had improved. Leite Lopes spent some time at USP and promoted several collaborations between the two centers.

In contrast to the professorial system at the University (UB), in which the holder of a Chair dominated the corresponding field, the CBPF had a very democratic structure: It had a President, an Executive Director, a Scientific Director, and a Secretary; but major decisions were taken by, or approved by, the Scientific-Technical Council, which represented all of its senior researchers, faculty members and technical staff. And there was also the Assembly, where only the founders had a vote. This structure was intended to make it resilient, even in times of political and economic difficulties. It did, however, suffer a considerable reverse in the mid-1950s:

At that time, the Executive Director of the CBPF, a certain *Álvaro Difini*, who was also the treasurer of the CNPq, diverted funds from the CBPF (originally intended for the construction of the 170″ cyclotron there)—rumor has it that he lost the money gambling. His family quickly raised the sum to replace the lost money, and the matter might have been settled quietly, but César Lattes, fearing that the affair would arouse curiosity among the press and that he would end up being held responsible, made the story public himself, causing an open scandal. Lattes made many accusations against the CNPq, which led to a reduction and finally the suspension of further financial support for the cyclotron.

In the end, support from the CNPq had to be completely withdrawn, and furthermore, the amount that had been diverted, some six million *old cruzeiros*, was not repaid until a dozen years later, when it was practically worthless due to inflation.

The financial scandal was amplified by another political disturbance which occurred in connection with radioactive minerals, in particular thorium-containing ores that were plentiful in Brazil. Álvaro Alberto supported the practice of using these rich ores as an exchange medium for financing Brazil's technological development, while many political figures believed that this practice was undermining Brazil's relations with the USA. Álvaro Alberto's political enemies used the scandal at the CBPF to have him removed from his office at the CNPq. The CBPF was caught in a crossfire: the right wing considered that it had acted against the national interest by supporting Álvaro Alberto's political positions. On the other side, the left wing considered the CNPq to be an institution aligned with American McCarthyism, and believed that it gave support only to people and institutions considered trustworthy by the government, such as the CBPF.

Furthermore, the *Centro* itself was divided. Professors Oliveira Castro, Ugo Camerini and Hervásio de Carvalho supported Lattes' actions, while Jayme

Tiomno, José Leite Lopes and Leopoldo Nachbin held a critical view of his public denouncement. As Tiomno later said, Lattes was morally justified in exposing the scandal, but the practical consequences of his actions for the Center were devastating, as were also the effects on its internal atmosphere. He explained that,

> The *Centro* was divided 50–50 during this period. There was one group who were in agreement, the other was not, in resolving the problem in this manner, even though, morally, we had nothing against the fact of [*publicly*] denouncing an irregularity.[22]

As a result of this crisis, the construction of the 21″ cyclotron, which was half finished, was stopped for a time and restarted only much later, although technicians from the University of Chicago were already in Rio and working on it. The construction of the planned 170″ cyclotron was never begun; and the strong financial support from the CNPq was lost. In addition, many people were greatly distressed by the situation: Guido Beck left the CBPF, and Tiomno experienced a period of increasing depression. Lattes suffered a breakdown. This was probably the first sign of his psychiatric disorder, which sometimes caused swings from a state of complete apathy to one of manic, frenetic activity, which, at extreme moments, went so far as to cause him to lose all contact with his feelings and rational thoughts.[39] He himself would later explain to some of his colleagues that the psychiatrists had diagnosed him as schizophrenic, and he was taking psychopharmica for that.[41] Although everyone involved was aware of the situation, it has seldom been a subject for public consideration, even today.[42]

Lattes went to the USA in 1956, ostensibly to work at the University of Chicago, where Enrico Fermi had recently died, leaving his group orphaned. He used the invitation to allow the situation at the CBPF to cool down in his absence. He was also undergoing psychiatric treatment during his stay in the USA. After his return in 1957, he continued working at the CBPF for a time, but finally left Rio for good in 1959, returning to USP in São Paulo, where he stayed for 8 years before moving on to the newly-founded university in Campinas ('*Unicamp*'). There he remained for the rest of his career.

With the passage of time, the crisis was overcome. Guido Beck returned to the *Centro*, and it found other sources of financial support to replace those lost from the CNPq. A member of the lower house of the Brazilian Congress (*Câmara Federal*), *Juscelino Kubitschek de Oliveira*, who later became President of the Republic, supported a project to provide a secure source of funding for the *Centro* from the Brazilian Federal government. Initially (1955), it amounted to 10 million old *cruzeiros* annually, which permitted the survival of the Center.

Tiomno[22] says of the subsequent period,

> ... the *Centro* developed rapidly right from its beginnings. Even during the period of this crisis, in 1954, it already had some international standing. With the new phase of development, when funds voted directly by the *Câmara Federal* became available, the *Centro* really achieved great international prestige and came to be considered as one of the major centers of physics research in the entire world.

However, one of the most dramatic consequences of this crisis was the alienation between Tiomno and Lattes which it provoked; never again were they able to restore their earlier close friendship.

Still another tragedy was a fire which occurred at the CBPF in May of 1959; it destroyed a major part of the library and the nuclear emulsion laboratory, which had been set up by Elisa for her work. In spite of the misfortune caused by the fire, it had a positive aspect: It mobilized physicists and institutions in various countries to help to restore the assets, especially books and journals, which had been lost by the CBPF. The mishap also provided a motivation to construct a new building, larger and with a better infrastructure for the laboratories, which was finally completed only in the 1970s.

In the mid-1950s, a number of physicists in Rio and São Paulo, many of them theoreticians, including Mário Schenberg, Jayme Tiomno, and José Leite Lopes, as well as the experimentalist César Lattes, began to feel the importance of developing other areas of physics research in Brazil besides the nuclear and particle physics which dominated their work at that time. In particular, solid-state physics, which was evolving rapidly in North America and Europe and promised important technical applications, needed a 'kick-start' to become established in Brazil. The CBPF offered an ideal platform for that effort, and many of the newer researchers there were later among the founders of the departments dealing with condensed-matter physics. They included *Jacques Danon* (1924–1989) and *Luis Marquez*, who subsequently made use of the newly-discovered *Mössbauer Effect*, as well as *Micheline Claire Levy Nussenzveig*, and later *George Bemski*, who came to Rio from São José dos Campos in the early 1960s. Tiomno was interested in and supported the establishment of groups working in condensed-matter physics in Rio and São Paulo, although this engendered some friction with the other experimental groups, who feared an unfair competition for the limited research funds.

By the end of the 1950s, the CBPF was fully established. The Department of Experimental Physics had a number of laboratories: Nuclear Emulsions, the Cockroft-Walton accelerator, Solid State Physics, and Cosmic Rays. There was also a Department of Physical Chemistry and a Department of Nuclear Chemistry, as well as the Department of Theoretical Physics and the Department of Mathematics. The Teaching Department founded by Tiomno, in charge of organizing courses and teaching general and experimental physics, played an important role in attracting students to the *Centro*, and its Graduate School provided a basis for their further education.

The Department of Theoretical Physics was led by Jayme Tiomno, José Leite Lopes and Guido Beck. According to a report written in 1960 by a mission sent by the International Atomic Energy Agency, in both the theoretical and experimental physics sections of the Center, "one finds outstanding scientists who enjoy not only national but international prestige and, of course, in their special subjects, mainly in the theoretical field, they are possibly the most eminent scientists in Latin America".[43]

The decade of the 1950s ended with the CBPF consolidated as an institute of physics; it was internationally recognized and active in helping to found and implement physics departments and institutes in various Brazilian universities. Those young idealists, its Founders, had managed to make their dream come true. But they had as yet no idea of the difficulties and problems which would face them in the coming years.

Notes

[1] See Chap. 7, section on 'The Relationship with Jayme Tiomno'.
[2] Jayme Tiomno to Elisa Frota-Pessôa, June 9th, 1948; [EFP].
[3] On the relationship of Mario Schenberg to the Communist Party, see Kinoshita (2014).
[4] Walter Schützer to Jayme Tiomno, December, 1949; [JT].
[5] Jayme Tiomno to Elisa Frota-Pessôa, November 7th, 1949; [EFP].
[6] cf. [JT14].
[7] Khuri (1957).
[8a] John Wheeler to Guido Beck, May 24th, 1957; [JWA].
[8b] Marvin Goldberger, 1969, at the commemoration of '15 years of Dispersion Theory', quoted by Bassalo and Freire (2003); see also Zichichi (1970).
[9] Bassalo and Freire (2003).
[10] Abrahão de Moraes to Gleb Wataghin, March 8th, 1951; [IF/USP].
[11] Abrahão de Moraes to Mario Schenberg, December 12th, 1950; [IF/USP].
[12] cf. [JT36]; compare also Videira (1980).

[13] See Peat (1997) for a biography of Bohm.

[14a] Abrahão de Moraes to David Bohm, August 20th, 1951; [IF/USP].

[14b] Abrahão de Moraes to David Bohm, April 27th, 1951; [IF/USP].

[15] David Bohm's reply to Abrahão de Moraes, May 1st, 1951; [IF/USP].

[16] Bohm's passport was confiscated, without legal justification, by the US Consul in São Paulo, making it impossible for him to travel outside Brazil (as intended by the US authorities). César Lattes intervened with Admiral *Álvaro Alberto*, who had been an ambassador and had connections with the Diplomatic Corps, to obtain a Brazilian passport for Bohm. See Leite Lopes (1988). He used it to travel to Israel and later to England. His US passport was restored only much later, in the 1980s, after he instituted legal action. Bohm never returned to the USA to live, but did visit there in his later years.

[17] There, he met his future wife, Sarah Woolfson, and they married in 1956. With one of his students from Haifa, *Yakir Aharonov*, he re-examined the famous 'EPR article' of Einstein, Podolsky and Rosen (1935), and reformulated its thought experiment (the 'EPR paradox') in a simpler form, which became the paradigm for testing the completeness of quantum theory, and was used by John Stewart Bell in formulating his inequalities. It is now termed the 'EPR-B' (for 'Bohm') experiment; see e.g. Friebe et al. (2018).

Bohm moved on to England after two years in Israel, working first at the University of Bristol, and then, from 1961 on, as Professor of Theoretical Physics at Birkbeck College of the University of London. In Bristol, working with Aharonov, who had accompanied him to England, he discovered the *Aharonov-Bohm effect*, which demonstrates the physical reality of the electromagnetic vector potential, originally considered to be only a mathematical construct (and thus unobservable). In London, he collaborated for many years with his colleague *Basil Hiley*, with whom he developed the de Broglie-Bohm theory further, establishing the 'London school' of that theory.

The third school dedicated to the deterministic quantum theory was established in Munich, Germany, by *Detlef Dürr* of the *LMU München*, whose group studied 'Bohmian mechanics' for a number of years. Bohm was not enthusiastic about their different interpretation, nor about the name they gave it.

Bohm renounced his political commitment to Communism after the suppression of the Hungarian uprising in 1956. In London, he became interested in other fields besides fundamental quantum physics, and made contributions to the theory of consciousness and the structure of the brain, as well as to philosophy and sociology. After his retirement from the faculty of Birkbeck College in 1987, he remained active in all of those fields until his death in October, 1992. Hiley continued the work on the deterministic quantum theory, and it remains today an active field in physics and philosophy.

[18] [JT21]; see also Madelung (1927).

[19] Tiomno (1966a).

[20] Personal communications from Sonia Frota-Pessôa, 2018/19.

[21] Bohm (1986); see also Freire (2005).

[22] Tiomno (1977).

[23] cf. D'Araújo (1999).

[24] Jayme Tiomno to José Leite Lopes, September 6th, 1946; [JLL].

[25] José Leite Lopes to Jayme Tiomno, March 12th, 1948; [JT].

[26] Gardner and Lattes (1948). Unfortunately, Lattes' collaborator at UCRL, Eugene Gardner, was suffering from berylliosis due to his exposure to beryllium dust during his work for the Manhattan Project, and he died prematurely in 1950. His early death may have prejudiced the hope of a Nobel prize for the artificial pion detection; cf. Vieira and Videira (2014), and also https://www2.lbl.gov/Science-Articles/Research-Review/Magazine/1981/81fchp4.html#06.

[27] José Leite Lopes to Jayme Tiomno, January 12th, 1949; [JT].

[28a] Lattes (1995a, b).

[28b] Lattes (1995a, p.81).

[29] Tiomno (1957); cf. [JT75].

[30] Leite Lopes (1977).

[31] Frota-Pessôa (1990).

[32] Leite Lopes (2004).

[33] Jose Leite Lopes to Jayme Tiomno, February 1st, 1950; [JT].

[34] Lattes (1995b).

[35] Statement made by Joaquim da Costa Ribeiro, Scientific Director of the CNPq, in the Council Plenary Session on February 28th, 1955; [EFP].

[36] Lattes (1995a, p. 86).

[37] Mario Schenberg to Jayme Tiomno. October 14th, 1949; [JT].

[38] Statement made by Joaquim da Costa Ribeiro, Scientific Director of the CNPq, in the Council Plenary Session on February 28th, 1955; [EFP]. See also Silveira, Joel: *Física*, in *Diário de Notícias*, February 26th, 1955, p.2; [EFP].

[39] Marques (2013).

[40] Rochester and Butler (1947). These were K mesons, or Kaons, whose characteristic V-shaped tracks in the cloud-chamber photos gave them their original name.

[41] From interviews with Mario Novello carried out for this book by ATT, August/September 2019.

[42] Vieira and Videira (2014).

[43] Quoted from the project application submitted to the United Nations Special Fund for the Organization of a Graduate School in Physics by the CBPF, 1960; [JT].

9

Research at the CBPF. Weak Interactions and Particles

Back in Rio de Janeiro

When Jayme Tiomno arrived back in Brazil from Princeton in October 1950, he was already a mature scientist. At 30, he had worked in several fields of theoretical physics and was in a position to carry out independent research and guide graduate students. As we have seen, he had worked on gravitational theory in São Paulo[1] and in Princeton,[2] and started a collaboration there with David Bohm,[3] which he pursued after Bohm came to Brazil at Tiomno's suggestion.[4] Upon arriving back in Brazil, Tiomno continued his research in particle physics and the Weak interactions. In São Paulo, he set up a research group and collaborated with Walter Schützer, also recently returned from Princeton. After about 18 months, Tiomno left USP to go to Rio and join the newly-founded CBPF—he returned to Rio definitively in March, 1952, leaving the group that he had started in São Paulo in the hands of David Bohm.

Elisa, informally Tiomno's wife since January 1951, moved together with Jayme soon after his return to Rio, to an apartment in *Grajaú*, a quiet residential area at some distance from *Santa Tereza*, where they had shared an apartment since early 1951, and from the CBPF.[5] Her first husband Oswaldo Frota-Pessôa, the father of Elisa's children, and his own family lived in *Santa Tereza*, and the children had been staying with them. Oswaldo submitted his doctoral thesis to the FNFi in 1953, and after obtaining his PhD in biology, he went for a postdoctoral stay with a fellowship to Columbia University in New York City, where he remained for over a year. He took the children Sonia and Roberto along, as well as his mother Zezé, who could look after

© Springer Nature Switzerland AG 2020
W. D. Brewer and A. T. Tolmasquim, *Jayme Tiomno*, Springer Biographies,
https://doi.org/10.1007/978-3-030-41011-7_9

them while he was working. After a year, the children went back to Rio, while Frota-Pessôa remained in the USA briefly until going on to São Paulo to assume a post at USP. This must have been an exciting interlude for the children, although the adjustment to the different climate, culture and language was perhaps difficult at first. Back in Rio in 1954, Sonia and Roberto lived with their mother and Tiomno[5] One of Jayme Tiomno's hobbies was relaxing by scribbling equations in the margins of his newspaper, which the children called "fifi". The term persisted for years as a synonym for *relaxing*. It was so natural within the family that until very recently (when both Sonia and Roberto were already over 70), they had not realized that it was an abbreviation for *física* (*physics* in Portuguese) (Fig. 9.1).

In March of 1952, Tiomno was elected to Permanent Membership of the *Academía Brasileira de Ciências* (ABC), the Brazilian Academy of Sciences, in Rio de Janeiro.[6a] His qualifications justifying that election were listed by the Election Committee[6b]:

He has published works on the radiation field of the electron, on the theory of the S-matrix, on proton-proton collisions, the decay of the μ meson and its capture by atomic nuclei, the description of μ mesons in terms of the theory of pairs of spin-1 particles, and on the universal interaction among spin-½

Fig. 9.1 Roberto Frota-Pessôa, Jayme Tiomno, and Sonia Frota-Pessôa at their house in *Grajaú*, mid-1950s. *Source* Private archive, [EFP]

Fig. 9.2 Jayme Tiomno and Elisa Frota-Pessôa in the mid-1950s. *Source* Private archive, [EFP]

particles. His thesis on 'Theories of the neutrino and double beta decay' is one of the most important theoretical studies of these particles.

In December of that same year, Elisa (Fig. 9.2) was also elected to Associate Membership of the ABC.[7]

Tiomno began to receive recognition from outside the country, as well. In June of 1952, he was invited to spend the following year as a visiting professor at Vanderbilt University (USA). However, he could not accept the invitation, since he was just beginning his work at the *Universidade do Brasil* and at the CBPF, and wasn't in a position to be away for a longer period of time in the next few years. Aside from that, as he explained, "We are now working hard in order to prepare a number of students for post-graduate work. This is very important for the development of physics in this country".[8]

Fundamental Physics and Particles

During this period, Tiomno continued to pursue his research in fundamental physics and particle physics, including the Weak interactions (the Universal Fermi Interaction, UFI, was at the time their predominant, but incomplete model)—a topic of great interest to physicists in the mid-1950s. Between 1951 and 1955, Tiomno published 9 articles, some alone, some with his students, and two with colleagues.[9] Several of Tiomno's papers from the early 1950s elicited strong responses from the scientific community. Four of those publications were 'general theoretical works', dealing with fundamentals of quantum mechanics. As described in the previous chapter, two were on '*S-matrix theory and causality*'[10], co-authored with Walter Schützer, and one was on '*A Causal Interpretation of the Pauli equation*', with David Bohm and R. Schiller.[11]

The fourth of Tiomno's general papers, written after his return to Rio, was his article on the '*Invariance of field theory under time inversion*'[12]; it was taken from his doctoral thesis ([JT13]). Tiomno himself mentions[13] that it was a "reformulation of Part II of [*his*] doctoral thesis, with some additional details. In particular, attention is called to the γ_5 transformation and to the conclusion, obtained in the thesis, of the necessity of anti-commuting different spinor fields which appear linearly in the Hamiltonian". The introduction of the γ_5 operator was important for the development of the theory of Weak interactions, and it was later used in the formulations of the Universal V–A Interaction by Sudarshan and Marshak[14] by Feynman and Gell-Mann[15] and by Sakurai.[16] This was the predominant theory of the Weak interactions from 1958 until the electroweak unification in the early 1970s.[17] Tiomno's article was however not published, but instead was made available only as a preprint in the CBPF's *Notas de Física*. He sent it to several of his colleagues, including Arthur Wightman, who took note of it and made some comments.[18]

His other five publications in the first half of the 1950s were in the field of particle physics in the wider sense. The first of those, co-authored with *Gabriel E. de Almeida Fialho*, was entitled '*Gamma radiation emitted in the $\pi \rightarrow \mu$ decay*' ([JT15], 1952). This paper treats an early example of a *radiative decay*, in which a beta decay or other weak decay is accompanied by electromagnetic radiation ('*inner bremsstrahlung*', here called 'gamma radiation'). This is the radiation that accompanies the acceleration of a charged particle—an electron, muon, etc.—during the decay. The branching ratio for such radiative decays, as compared to the non-radiative 'normal' decay, is usually very small, as they are second-order processes in the Electromagnetic interaction. Tiomno and Fialho calculated the spectra of muons and photons (gamma rays) in such radiative pion decays, as well as their branching ratios. The original article, initially published in the *Annals of the Brazilian Academy of Sciences*, was republished in a collection of selected articles by the Japanese Physical Society.[13,19] The undetected presence of this radiative decay branch was able to explain an 'anomaly' that had been observed in the $\pi \to \mu$ decays.

Tiomno also published articles on a '*Non-relativistic equation of charged particles with spin 3/2*', with Adel da Silveira[20]; on a '*Non-relativistic equation for particles with spin 1*', with the Argentine physicist Juan José Giambiagi[21]; and as sole author on a '*Relativistic theory of spinning point particles*'.[22].

At the beginning of July, 1954, Tiomno and Leite Lopes departed for Europe, to participate in the *Glasgow Conference on Nuclear and Meson Physics*, organized by the International Union of Pure and Applied Physics. They took advantage of their trip to make contact with a number of groups working in theoretical physics and to visit various laboratories in order to learn about the current state of the art with regard to synchrocyclotrons, thinking of purchasing one for the CBPF. They traveled from Rio to Lisbon, continuing on to Paris and then to London, where they visited the Department of Physics at the Imperial College, headed at that time by Patrick Blackett. Its equipment made quite an impact on them: "We were greatly impressed, since we had already forgotten what a true center of physics research is like".[23]

In Great Britain, they encountered Gleb Wataghin, John Wheeler and Leon Rosenfeld, among others. After the conference, they continued on to Manchester, Liverpool and finally Birmingham, visiting there what was considered at the time to be the most important English group in theoretical physics, and viewing the new synchrotron, with a beam energy of 1 GeV. They returned via Paris, where they visited the physics groups at the *Sorbonne*, and also the group of Leprince-Ringuet at the *École Polytechnique*.

Following the Glasgow Conference and his discussions in England, Tiomno returned to his doctoral thesis and extracted from it an article on '*Mass Reversal and the Universal Interaction*', which he published as sole author in *Il Nuovo Cimento* in 1955 ([JT20]).

[The term 'mass reversal' or 'mass inversion' was initially used, since this operator changes the apparent sign of the mass of a Dirac particle. Its physical interpretation was found to be simple; see e.g. [24]. Subatomic particles act like miniature gyroscopes[25] and their quantized 'spin' can take on only certain values and directions relative to a chosen axis in space, the 'quantization axis'. The *helicity* operator chooses the direction of motion (i.e. the particle's *momentum*) to be the quantization axis, and projects out the direction of the spin relative to it. For spin-½ fermions such as electrons, neutrinos, protons and neutrons, the spin can take on only two possible orientations—parallel ('spin up') or antiparallel ('spin down') relative to the quantization axis. A 'spin up' orientation can be represented by the curved fingers of the right hand, showing the direction of the spinning, while the outstretched thumb (e.g. pointing *upwards*) represents the momentum direction. The left hand then corresponds to 'spin down'; if the fingers curve in the same sense, the thumb will now point *downwards*. This quantity—helicity—is an intrinsic property of massless particles, such as the photon.[25] It was generalized to 'handedness' or '*chirality*' for particles with a finite mass, as expressed in Tiomno's 'mass-reversal' operation; from the Greek word for 'hand' (as suggested by Abdus Salam). A space inversion will reverse the handedness of all the particles in the system, since the spin is an *axial* vector, unchanged by space inversion, while the momentum is a *polar* vector, and changes its sign on space inversion. Chiral invariance relates directly to invariance under space inversion, or *parity conservation*. The 'parity operation' P also effects a mirror-reflection of a wavefunction or physical system. An experiment that is sensitive to the effect of the chirality operation can thus detect parity (non-) conservation. (A compact introduction to the chirality operation and chiral invariance is given for example by Assamagan (1995)).

Tiomno's 1955 paper treats the important γ_5 operator which was introduced in his thesis, here extended to quantum electrodynamics and meson theory. In fact, this was his first real publication on γ_5—since his thesis was unpublished, and his paper '*Invariance of field theory under time reversal*' (1954) was issued only as a preprint in the *Notas de Física*.

Tiomno lists three consequences of this work[13]: (i) the electromagnetic interaction should not show anomalous magnetic moments; (ii) interactions of π mesons should be of purely pseudoscalar (P) or axial-vector (A) character; (iii) the Universal Fermi Interaction (UFI) should be described by the operator combination S + P − T, or else A–V, if it is invariant under permutations of the interacting fields and mass inversion. (This latter assumption became obsolete the following year, when T.D. Lee and C.N. Yang[26] published their conjecture that parity is *not* conserved in the weak interactions, a suggestion which initially met with great skepticism).

Around this same time, the 'scandal' at the CBPF and the resulting upheavals that took place in 1954/55, as well as the subsequently changed climate at the *Centro*, had left Jayme Tiomno somewhat depressed. In January 1956, during the summer vacation period in Brazil, he spent a longer time in *Lambari*, a spa in the south of Minas Gerais, trying to relax a bit and recover his equilibrium. Nevertheless, by September he was traveling again, this time to the USA, to attend the Seattle *Conference on Theoretical Physics*, in Washington State. He had gone first to Washington, DC, and then to the University of California in Berkeley. There, he shared a room with the

Pakistani physicist *Abdus Salam*; they had been introduced by Leite Lopes, who had already met Salam at a conference in Geneva the previous year. From then on, Tiomno and Salam maintained a collaboration in physics and a strong friendship. The visit to Berkeley also provided an opportunity to converse again with *Emilio Segré*, whom Tiomno had met earlier when Segré had visited the CBPF. The flight from San Francisco to Seattle was like a preview of the conference there, carrying numerous physicists who were going to attend the meeting. In a letter to Elisa, he commented that this conference was "much more useful than the one in Glasgow for localizing my own work and verifying it in comparison to what my former colleagues and contemporaries are doing, and seeing what is being discarded along the way..." He concluded with the remark, "Undoubtedly, the context exercises a tremendous influence on scientific work".[27]

The mutual interest of Abdus Salam and Jayme Tiomno in neutrino theory probably brought them together after they met on the way to the *Conference on Theoretical Physics* in Seattle. They wrote a joint paper dealing with the masses of the elementary particles which was published in *Nuclear Physics in 1958*.[28] About this paper, Tiomno writes[13]:

> [*We*] speculate on the mechanism of the origin of the masses of particles, with an eye to explaining the empirical ratio of the masses of the proton and the electron, which is the same as that of the squares of the coupling constants of pions and the electromagnetic field.

[Their speculations seem rather quaint in view of the modern theory of the Strong interaction, which binds together the quarks that make up the proton. The observed mass of the proton is not simply the sum of the masses of its constituent quarks, which are determined by the Higgs field; Yang et al. (2018).

Rather, it results from the binding energy required to localize (confine) them within the proton. This binding energy makes itself felt in the observable world as a major contribution to the rest mass of the proton, as a result of the Einsteinian energy-mass equivalence. Other contributions come from a similar term for the gluons, as well as quantum-mechanical effects].

After the conference in Seattle, Tiomno traveled on to Princeton. There, he met with Wigner, Wheeler and various other colleagues. Wigner expressed the opinion that Tiomno ought to publish those remaining parts of his thesis which were still original, while Wheeler suggested that they could work together on an article about the K mesons, and that they should also write another article on π mesons, mainly to include an *erratum* to their earlier work on the meson decays. But Tiomno stayed in Princeton for only two weeks; and furthermore, he wanted to use the time to attend some seminars and to visit the library to update his knowledge of recent publications. Apart from that, he had begun to redirect his interests. "I think that I have an idea for a new research topic, and I have already discussed it with Wigner, who found it interesting".[29] He was starting to consider a new way to formulate the Universal Fermi Interaction.

By the beginning of 1957, Tiomno's depression had returned, but in spite of that he again traveled to the USA, to New York City and Rochester/NY, where he attended the 1957 *Conference on High-Energy Physics*. He also noted that the Rochester Conference had considerably more participants than the Seattle Conference, with both theorists and experimentalists, and "…an enormous quantity of results".[30] This time, for a change, he was not traveling alone; he was accompanied by his mother. Annita had a tubercular abscess in her lungs, and they wanted to hear the opinion of a specialist abroad. She made use of her son's trip to New York to have medical examinations performed there, and to consult with some American physicians. The logistics were not simple: making appointments with the doctors, taking Annita to the consultations and examinations, traveling to the conference, then more doctors, etc. Finally, they recommended lung surgery, although it was not urgent. She decided to return to Brazil and consider the problem there, postponing the surgery; in the end, it was never performed, and she remained in good health.

Tiomno and Elisa still intended to spend a longer stay at a research institution in the 'outside world', and he wished to take the opportunity to look for a place where he could continue his own research and where Elisa would also have a chance to work in a different environment[13] (and perhaps obtain material for her PhD thesis). To that end, he visited the group working with nuclear emulsions at Columbia University, directed by Gerard G. Harris, who had been his colleague at Princeton.

From the UFI to the Universal V–A Interaction

Indeed, the period from 1956 to 1958 was exciting, and decisive, for Weak-interactions theory. The excitement had begun at the previous Rochester Conference in April, 1956 and continued at the Seattle *International Conference on Theoretical Physics* in September, which Tiomno attended. In Rochester, the experimentalist Martin Block, via Richard Feynman, asked the apparently naïve question as to how important parity conservation in the Weak interactions really was. And Lee and Yang argued that there was no solid experimental evidence for the conservation of parity in the Weak interactions.[31] In Seattle, they announced their forthcoming paper[26] on parity non-conservation. Tiomno also mentioned his $1 \pm \gamma_5$ projector[32] to Abdus Salam there, and Salam subsequently wrote a paper, 'On Fermi Interactions', which he circulated as a preprint but never published see Ali et al. (1994). It contained some of the ideas that soon led to the Universal V–A theory.[33] Lee and Yang's announcements rapidly motivated several specific experimental tests[34a, b], all published in 1957, and they verified that parity was indeed violated to a maximal degree in the Weak interactions. If parity is *conserved*, that is if physical laws are unchanged by a space inversion, then in a process like beta decay, the particles emitted must show equal amounts of 'left handed' and 'right handed' chirality. A space inversion would then just interchange these, and the result would leave the overall process unchanged. In contrast, if parity is *not* conserved, there would be different amounts of the two chiralities; in the case of *maximal* parity violation, only *one* chirality would be observed. The Wu experiment[34a] showed indirectly that in the beta (–) decay of ^{60}Co, only *right-handed antineutrinos* are emitted (by measuring the angular distribution of the emitted electrons with the spin direction of the parent nucleus fixed, and considering conservation of linear and angular momenta). A later, similar experiment[34a] on the beta (+) decay of ^{58}Co showed that the emitted *neutrinos* are only *left-handed.*

This led to a new two-component theory of the neutrino—not in the sense of the Majorana theory, in which neutrinos and antineutrinos are identical, but rather in the sense that each has only *one possible spin orientation* (helicity). This theory, in turn, is now obsolete, since it has been observed that the three 'generations' of neutrinos (electron neutrino, mu neutrino, tau neutrino) can interconvert; this implies that they have a finite mass (thus they are not massless, fully relativistic particles, traveling at the speed of light, as originally thought). They then do not have pure helicity states, as proposed by the two-component theory.[25]

Wolfgang Pauli, who had initially scoffed at the suggestion of parity violation, was motivated to send out a card in the form of a death notice after it had been verified by experiments involving nuclear beta decay[34a] and meson decays.[34b] (Fig. 9.3).

The confirmation of parity non-conservation provoked a complete reshuffling of the cards of the Weak interactions, and was soon followed by the 'ultimate' *Universal V–A Theory*, which Jayme Tiomno narrowly missed, as we shall see.

In the second half of the 1950s, Tiomno published three articles on the Universal Fermi Interaction and seven on particle physics and field theory.[35]

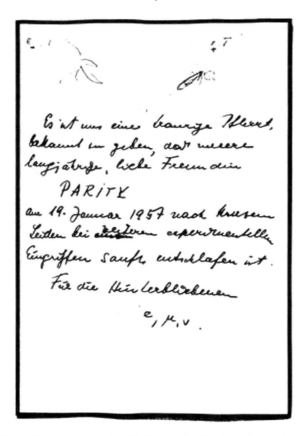

Fig. 9.3 Death notice card sent by Wolfgang Pauli after the verification of parity non-conservation in the weak interactions. It reads: "It is our sad duty to inform you that our beloved friend of many years, PARITY on the 19th of January 1957, following a brief affliction, after repeated experimental interventions, has passed away peacefully. For the bereaved, e, μ, ν". *Source* Pauli archives, CERN. Reproduced with permission

T. D. Lee and C. N. Yang received the physics Nobel prize in 1957 for their suggestion of parity non-conservation in the Weak interactions. The establishment of parity violation, however, had still not solved the problem of how to formulate the fundamental interaction in a way that would be consistent with all the experimental data. That task was complicated by the fact that some of the experiments were found to be incorrect a few years later. Tiomno proposed a solution in his paper '*Non-conservation of Parity and the Universal Fermi Interaction*', published in *Il Nuovo Cimento* in 1957 ([JT28]). In it, he used his $1 \pm \gamma_5$ projectors to show that the experimental data would be in agreement with either S + P − T or A–V as operator combinations, *with* violation of parity. He had arrived at similar conclusions using the γ_5

operator in his earlier work on '*Mass Reversal and the Universal Interaction*' ([JT20]), but there assuming parity *conservation*. Before the Rochester Conference in April, 1957, Tiomno paid a visit to Columbia University, where he met T. D. Lee. "Lee and Yang have already done practically everything that I had planned to do on the subject of the interactions of π mesons, including the calculations that Erasmo is already working on", he wrote to Elisa[36] [this refers to *Erasmo Madureira Ferreira*, Tiomno's colleague at the CBPF; cf. Chap. 14, 'the PUC quartet'].

Based on an earlier experimental result of Mahmoud and Konopinski[37], Tiomno chose the S + P − T combination, which turned out to be wrong. It was by 1957 generally known in North America and Western Europe that the logic leading to the S + P − T current-operator combination was faulty, but Tiomno, isolated in Brazil, was unaware of that. He also failed to remember the article of Ruderman and Finkelstein,[38] who calculated the (branching) ratio R of pion decays leading to electrons or to muons, and showed that R would be about 1 for a pseudoscalar (P) interaction, roughly 10^{-4} for A, and 0 for S, V, or T. While the exact measurement of R appeared only in 1958,[39] Elisa had already set an upper limit on R in her first publication from the CBPF in 1950, which Tiomno had discussed at length with her at that time in his correspondence from Princeton; it already ruled out a P-type interaction, leaving A–V as the correct alternative. He had indeed not considered her work when he wrote his article, which was submitted on July 2nd, 1957. (Sudarshan and Marshak published their results at the Padua-Venice Conference on '*Mesons and newly-discovered Particles*', September 22–28, 1957, after failing to announce it at the Rochester Conference in April; and the Feynman–Gell-Mann paper was received by the *Physical Review* on September 16th, 1957; it appeared in early 1958). Tiomno wrote to Elisa from Rochester,

> The day before yesterday, I gave a talk on my work, following Schwinger and Gell-Mann, who proposed a theory with a certain similarity to mine. In his case, however, an even weaker interaction with K mesons is assumed, and that appears to be in better agreement with the experimental data than my hypothesis of interactions of comparable strength to those of the π mesons. Still, the experiments might yet be interpreted to show a stronger interaction. I discussed this subject with various people, and I have the idea of making calculations of several processes which could permit a choice between the two schemes.[30]

Had Tiomno made the connection to Elisa's 1950 results when writing his 1957 article, he would have been the first to publish the universal V–A interaction. He later mentions[40] that he spoke to Richard Feynman about this paper, while Feynman was spending some time at the CBPF in the summer of 1957. Feynman replied that he himself was also working on the UFI, using $1 + \gamma_5$ projectors and a combination of the Dirac and Klein-Gordon equations, and that newer experimental results favored V–A (a remark which Tiomno found unjustified at the time). Feynman was keeping in touch with the experimentalists at Caltech (e.g. Felix Boehm) via short-wave radio. The result of Feynman's work was the famous Feynman and Gell-Mann paper[15], often taken to represent *the* discovery of the universal V–A interaction.[41] Feynman initially felt triumphantly that this was "the only law of nature that I could lay a claim to", but later admitted that he was not its only—or even its first—claimant.[42]

Elisa later commented on this story:

> In order to compare his theory with experimental results, he [*Tiomno*] arrived back from the USA and looked for data to be able to choose which theory [*i.e. which operator combination*] was correct; he used the results of some famous experimentalists at the time,[37] quite reasonably. In the end, my results indicated the opposite choice, and it was the correct one; but he had already published...[43]

The development of the V–A theory has an interesting history in itself, treated in many review articles and books[41]—and Jayme Tiomno made significant contributions to that story. However, as later emphasized by John Wheeler, his contributions were not always adequately recognized by the scientific community, like those of several other actors in this saga. That has been variously attributed to his being a physicist from a 'third-world country'.[44] While this may have played a certain role, one can see from the history of science that such a lack of recognition is by no means limited to scientists 'on the periphery' of the community.

Tiomno's other two published papers on the UFI in that period were 'π-*Electron Decay and the Universal Interaction*' (with Colbert Gonçalves de Oliveira, in the *Annals of the Brazilian Academy of Sciences*, 1958; [JT30]); and '*The Failure of the Space Reflection Principle*' (in the *Proceedings of the Mathematical Society of the University of Southampton*, 1959; [JT32])—both were conference reports, not much noticed; the second of them was an analysis of all the available experimental data on parity non-conservation.

The main portion of Tiomno's published work in the second half of the 1950s, aside from those three papers on the UFI, dealt with particle physics and included several articles on 'strange' particles, called *hyperons* ([JT23, 25, 27]).

[The term *hyperon* was introduced in the early 1950s, initially to denote particles that were more massive than individual nucleons but less massive than a heavy hydrogen nucleus (deuteron). In modern terminology, it refers to a baryon which contains one or more quarks with the 'flavor' *strange*. The K mesons also possess this property of *strangeness*].

About these articles, Tiomno remarks[13]:

These are contributions to the theory of strange particles which aroused the attention [*of the scientific community*], especially owing to the following points:

(1) They proposed a model of doublets, later taken up by [*Abraham*] Pais, which had repercussions for some time due to the interest in theories with global symmetry;

(2) They proposed a unification of the basic octet of baryons, now definitively incorporated as an octet of SU_3;

(3) They proposed a symmetry of rotations within an orthogonal, 7-dimensional space (O_7) of isotopic spin. This resulted automatically in a global symmetry G– and cosmic symmetry (Sakurai). Various works based on this model represented attempts to overcome difficulties in the theory of strong interactions. The fundamental work of Ne'eman started from my theory; it was published at the same time as that of Gell-Mann, and both finally introduced the SU_3 symmetry—today firmly established for strong interactions.

(4) They justified the fact that the mean lifetime of the π^0 meson was greater than theoretically predicted, by taking account of pairs of intermediate hyperons which appropriately reduce the amplitude of the decay.

Concerning the proposal of O_7 symmetry for the strong interaction, Videira[1] makes the following points:

At the Rochester Conference in 1957, Tiomno proposed a global supersymmetry O_7, and when Yang came on a visit to Rio in 1960, he suggested that they try to find a subgroup of O_7 which would be satisfactory, since that group indeed has too much symmetry, so to speak, giving rise to forbidden processes and conservation laws that are not observed. The two made the attempt during only a month, until Yang returned to the USA. [*Abdus*] Salam, who while visiting Argentina professed to be enchanted by O_7, gave the [*idea*] of that group to Ne'eman to study. He, in turn, went from O_7 to O_8—which has still more symmetry—but with an enviable advantage: it has SU_3 as a subgroup.

The *Moinho Santista* Prize

Tiomno did in fact receive significant recognition in 1957, in the form of the prize '*Premio Moinho Santista*' for the Exact Sciences—in Brazil, to be sure, and not from the international scientific community. The *Moinho Santista* Foundation was created in 1955, with the purpose of 'projecting great names in the sciences, in letters and the arts onto a national level' by means of an annual prize (accompanied at that time by the not-negligible sum of one million *cruzeiros*). Its first recipient, in 1956, was *Angelo Moreira da Costa Lima*, whose area of specialization was biology and physiology. In the following year, physics was chosen as the area to be awarded the prize, and it was bestowed on Jayme Tiomno (Figs. 9.4 and 9.5).

Guido Beck was scientific consultant to the selection committee, and he asked Wheeler's opinion on "Tiomno's position among other Brazilian theoretical physicists".[45] Wheeler could hardly have been more complimentary towards Tiomno. He wrote:

> I have the honor to report that Professor Tiomno's work cannot be surpassed for its simultaneous awareness, both of the experimental situation as regards weak particle interactions, and of the fundamental principles of symmetry, invariance and causality. His investigations have contributed in an important way to our present-day understanding of the light particle interactions and of dispersion theory. His work is well known all over the world. Moreover, he himself is widely respected for his energy, originality, ability and scientific integrity. It appears true, that in the whole hemisphere of the Earth South of the equator, Tiomno is the most distinguished scientist concerned with the theory of elementary particle transformations and causality. He is a physicist of whom Brazil, with its other distinguished physicists, can justly be proud".[46]

In taking their decision, the selection committee considered his activities and personality as basic characteristics of a good scientist, as follows:[1]

1. 'the ability to distinguish clearly what has been proven and what is a hypothesis;
2. the ability to analyze in detail the consequences of each hypothesis accepted,
3. always searching for experimental data to support the theories developed;
4. rationally [*critically*] analyzing the experimental results;
5. seeking opportunities to discuss his work with others, so that many of his projects are collaborative;
6. maintaining absolute scientific integrity; and
7. showing great energy and enthusiasm for scientific work'.

Fig. 9.4 Jayme Tiomno lecturing at the *Moinho Santista* prize ceremony in 1957. *Source* Private archive, [JT]

This last characteristic of great energy and enthusiasm for everything concerning his science constitutes, unequivocally, one of the most incisive in Tiomno's personality, and manifests itself in *all* the multiple directions and quadrants of his scientific activities".[1]

Fig. 9.5 Jayme Tiomno and his mother Annita at the *Moinho Santista* prize ceremony in 1957. *Source* Private archive, [JT]

With the prize money, Tiomno and Elisa were able to pay part of the cost of their first own apartment in the *Lagoa* region of Rio de Janeiro, not far from the Botanical Garden, where they moved (from *Grajaú*) in 1961.

Feynman's Visits to Brazil, 1949–1966

During the 1950s, there were frequent visits to Brazil and to the CBPF by the North American physicist *Richard P. Feynman* (an *iconic* physicist, in retrospect). They began in July of 1949, when Tiomno, still at Princeton, had arranged Richard Feynman's (and Cécile Morette's) visits to Rio. The relationship of Feynman to Tiomno and the CBPF was to continue for many years.

Richard Phillips Feynman (1918–1988) was one of the best-known and most successful of US physicists in the 20th century. He not only produced a large number of original and innovative works in theoretical physics, receiving the Nobel prize for physics in 1965 (together with J. Schwinger and S. Tomonaga); he was also one of the most famous 20th-century physicists among the general public, in the USA and in many other parts of the world. Several scholarly biographies have appeared since Feynman's death.[47,48] The most authoritative and scientifically complete among them is the book by Jagdish Mehra, '*The Beat of a Different Drum*'.[49] Feynman's visits to Brazil are also described in Leite Lopes' book on the history of Brazilian physics.[50]

Feynman's fame was the result of four factors: (i) his sparkling and ebullient personality, which made his lectures similar to the performance of a stand-up comic (except that he was usually speaking about physics, and not about everyday life); (ii) his introductory physics course given at Caltech in 1961–63, which was published as '*The Feynman Lectures on Physics*' with the help of his Caltech colleagues Robert B. Leighton and Matthew Sands[51] and quickly became a bestseller, a position which it maintains even today; (iii) his two semi-autobiographical books, '*Surely you're joking, Mr. Feynman*'[52a] and '*What do you care what other people think*'2[52b], written from informal interviews with his young friend Ralph Leighton (the son of physicist R.B. Leighton), which were also long-term bestsellers; and (iv) his successful participation in the Inquest Commission into the Challenger space shuttle disaster (1986), in particular a televised session in which he demonstrated the reduced resilience at low-temperatures of the essential rubber sealing rings, by using a glass of ice water.

Richard Feynman, following his undergraduate studies at MIT, carried out his graduate work at Princeton, obtaining his Ph.D. under the guidance of John Wheeler there in 1942. He spent the war years working in the Theory Group at Los Alamos, where the first nuclear weapons were designed; that group was led by Hans Bethe. After the war, he became a faculty member at Cornell University in Ithaca/NY, where Bethe also taught. He maintained contact to his earlier mentor Wheeler, and often visited Princeton during his tenure at Cornell. In early 1949, either after a seminar talk at Princeton[49], or at the New York APS meeting[52a], he met Jayme Tiomno and Walter Schützer, and Tiomno suggested to him—Feynman was already planning a visit to Latin America—that he should learn Portuguese and go to Brazil. Whether this occurred at Princeton or New York, or later at the Michigan Summer School at which Feynman lectured and Jayme Tiomno attended[50,53]—is not clear, but both Feynman[52a] and Tiomno[54] agree that it was Jayme Tiomno who first convinced him to go to Brazil. (Tiomno recalled much later that it was at the Summer School, but this is unlikely, due to the timing of Feynman's first Brazilian stay).

Feynman made his first visit to Brazil and the CBPF in July, 1949, soon after its founding, and remained there for 6 weeks. At that time, he was in the final phase of his 10-year effort to re-formulate quantum electrodynamics (QED), and if he worked scientifically while in Rio, it would have been on writing his final papers on QED, which appeared in 1950/51. It was for this work that he shared the Nobel prize in 1965.

Feynman's first Brazilian visit is described in '*Surely you're joking…*',[52a] and by Mehra[49] and Leite Lopes.[50] We note here that he made an intermediate stop in Recife on the way to Rio de Janeiro, where he was received by Lattes' parents-in-law[52a]; and on arrival in Rio, he was filmed being greeted by Lattes—two celebrities, surprisingly *not* politicians, soccer players, or movie stars, but *physicists*!

Jayme Tiomno was asked later in an interview[54] if the presence of Feynman had helped the CBPF, and he replied that the key to its early success had been its *structure*, rather than publicity or politics. Feynman gave a public lecture on QED at the Brazilian Academy of Sciences shortly before returning to the USA, which he delivered in Portuguese. He translated it in advance and had it corrected by some Brazilian students. He was surprised that the speakers who preceded him spoke in (broken) English, while he spoke in (broken) Portuguese—a case of politeness on both sides. He was later elected to external membership of the Academy.

Feynman was no doubt charmed and enchanted by Rio, still relatively unspoiled in those days, and on his return to the US, he mentioned to Tiomno that he would like to go back, this time for a longer stay. Tiomno replied enthusiastically to his suggestion, but he feared that the CBPF would not have sufficient resources to support a long visit by Feynman. When Feynman was granted a sabbatical year jointly by Cornell and Caltech, as a transition year between his former and his new university, Tiomno suggested that he could go to Rio again during that year, when he would be receiving a university salary from those institutions.[55]

Feynman had namely decided to move from Cornell to Caltech, in Pasadena/CA, as his colleague Robert Bacher had done shortly before him. He had a sabbatical year scheduled at Cornell, and Caltech agreed to let him take it in the academic year 1951/52, before beginning his duties there. However, in the event, Leite Lopes found a professorship at the FNFi which Feynman could occupy for a year as visitor. So in August of 1951, Feynman again went to Rio, this time for a longer stay at the FNFi, where he would have regular teaching duties, and at the CBPF (Fig. 9.6). This was an interlude for him, both professionally, since he was between universities, and scientifically, since he had finished his work on QED—his last two publications in that area appeared in 1950 and 1951—and he as yet had no new 'problem of interest'. He did, however, work on nuclear energy levels in light nuclei and on meson theory during his stay in Rio.

In 1951/52, during Feynman's second visit to the CBPF, Guido Beck and Jayme Tiomno arrived to complete the department of Theoretical Physics there. However, neither collaborated with Feynman, as far as is known. Leite Lopes did work with him, and they issued a joint article '*On the pseudoscalar meson theory of the deuteron*' as a preprint in the CBPF's *Notas de Física*. It was presented as a paper at the '*Symposium on New Techniques for Physics Research in Latin America*', held in Rio de Janeiro in July, 1952, of which Tiomno was one of the organizers. Feynman had already left Brazil by that time, so the paper was given by Leite Lopes.[56].

It is interesting to remark that Richard Feynman and David Bohm were both in Brazil at the same time, during nearly a year after Bohm's arrival in São Paulo in October (and Feynman's in Rio in August) of 1951, until Feynman left in mid-1952—and both were there thanks to Jayme Tiomno. They were nearly the same age, and had much in common—but they had very different personalities, and their reactions to Brazil were correspondingly diverse. Feynman embraced Brazilian spirit wholeheartedly, even playing with a Samba School in the Carnival of Rio in early 1952, while Bohm was disturbed and frightened by the social inequality, the noise, dirt and disorderliness of life there. A conversation between the two of them—real (but fictitious in the details of its dialogue), at a meeting of the Brazilian Society for Progress in Science (SBPC) in Belo Horizonte in November, 1951[50] is reproduced in the book '*The Age of Entanglement...*', by Louisa Gilder.[57] The details of timing are not quite perfect in that fascinating reconstruction, employing the 'docunovel' technique introduced by Gilder, since they mention Tiomno's departure for Rio, which in fact occurred only three months later; but Bohm's disappointment and feelings of abandonment due to that move come through clearly[58] (Tiomno may in fact have already informed

Fig. 9.6 Richard Feynman, during an early visit to the CBPF. *Source* Private archive, [JT]

Bohm of his planned move by that time). Leite Lopes[50] also reports that Bohm and Feynman on other occasions discussed Bohm's deterministic quantum theory, towards which Feynman (in contrast to most other physicists at the time) was very open-minded.

Shortly before leaving Brazil in 1952, Feynman gave a public lecture on the teaching of physics, in which he recounted his experiences with Brazilian students and decried the practice of rote learning. His later visits to the CBPF, in 1953, 1957, and 1963, are described in Leite Lopes' book[50], as are his efforts to help when the library of the *Centro* burned in 1959. In his lectures on 'physics didactics' in 1952 (and again in 1963), Feynman cited many problems related to science education in Brazil (such as poor salaries for secondary-school teachers, the predominance of 'mass education' in the public schools, and others at the university level). According to Ildeu de Castro Moreira[59], who has studied Feynman's lectures and their long-term results, many of them are still present. Feynman's 1952 talk had a powerful effect on the scientists who heard it, and it had repercussions over the whole scientific community in Brazil.

Oswaldo Frota-Pessôa wrote in his column '*Ciência em Marcha*' ('Science on the March') in the newspaper *Jornal do Brasil* from May 25th, 1952:

I just attended an astounding lecture. A physicist who is world-renowned in the field of quantum electrodynamics, Richard P. Feynman, from the California Institute of Technology (USA), explained to us, in clear, although rather truncated Portuguese (he has spent a year in Brazil), that, in truth, we are teaching science to our pupils and students, but they are not understanding it … And the worst part of it is that he is absolutely right. Feynman is a rare person: he says directly what he is thinking, and everything that he is thinking; and he says it with such enthusiasm and tenderness that it becomes at the same time cutting and enchanting. "Our children are valuable" and "We are teaching nothing" were the two unique ideas that Feynman claimed to develop, since the third—"how to teach successfully"—already exceeds his competence, he being a researcher and not a pedagogue.

But the truth is that he did say what is most important about how to teach. Science is the description of natural phenomena, he postulated, referring of course to the physical sciences (and not to mathematics). Teaching science is, therefore, putting the students into contact with the natural phenomena. The usual instruction, restricted to definitions and lifeless formulas, memorized—or even if understood—never manages to become genuine science teaching. Only when the students are carrying out their own research into real phenomena, which are effectively taking place right in front of them (and not virtually on the blackboard), only when they are investigating, sharpened by their curiosity and by their enchantment in the face of the mysterious, are they really learning science… The greatest value of science, said Feynman, is to feed our curiosity and to give us the incomparable pleasure of revealing the unknown. We should teach it in order to increase the circle of those who are able to participate in the delicious adventure of *knowing*.[60]

All together, Richard Feynman visited Brazil six times during the years 1949–1966. After his brief initial visit in 1949, and his longer stay in 1951/52, he returned in 1953 (for three months, working on his theory of superfluid helium); in 1957, when he discussed his theory of the Weak interactions with Jayme Tiomno, and prepared his part of the Feynman and Gell-Mann paper on the Universal V–A interaction[15]; and again in 1963, as lecturer at a summer school and to attend a conference on education, where he gave a second talk on physics teaching in Latin America. Finally, his 'carnival' visit, in early 1966, was however hardly related to physics. After that visit, he never returned again to Brazil or to the CBPF. Richard Feynman passed away on February 15th, 1988, three months before his 70th birthday.

The University of London

As to Tiomno's and Elisa's planned sabbatical, Abdus Salam suggested that rather than going to the USA, they could spend a year in England, where he had recently obtained a professorship at the Imperial College, London (in 1957). He was establishing a Department of Theoretical Physics there.

He also suggested that Elisa could work at University College London (UCL)—both divisions of the University of London, located on opposite sides of Hyde Park. At University College, there was a group working with nuclear emulsions, led by *Eric H. S. Burhop* (1911–1980). Elisa wrote him a letter inquiring about the possibility of going there to work, which Burhop welcomed.[61] She thus went to the UCL to work with nuclear-emulsion plates which had been exposed to high-energy particle beams.

Burhop was an Australian who had worked at the Cavendish Laboratory under Rutherford in the 1930s. He returned to Australia during the 2nd World War to contribute to the development of radar, and later joined the Manhattan Project in the USA during the final year of the war. Afterwards, he went back to England, where he was offered a position at the University College in London. There, he collaborated with the Universities of Edinburgh and Padua, setting up a high-altitude cloud chamber, and worked with Powell and Occhialini on detecting the K mesons. A few years after Elisa's stay in London, he spent a year at CERN, and later became leader of the bubble-chamber group at UCL. Somewhat disturbingly, Burhop was suspected of passing 'nuclear secrets' to communists in the USA and was investigated by the FBI there and by MI5 in the UK, although he was never officially questioned or charged; this became public only long after his death in 1980.

He received Elisa warmly in London and gave her some responsibility for teaching younger students, as well as a place to continue her research on cosmic rays and mesons. Her first report regarding her fellowship reads, "We are principally interested in the interactions of the K [*mesons*] with two nucleons, and in studying hyperons and hyperfragments. I have concentrated initially on making a first analysis of the 'stars' [*indicating nuclear disintegrations within the emulsions*], and on the calibration of the 'stack' [*of nuclear-emulsion and absorber plates*]. This work is a collaboration among the groups in London, Bristol, Padua, Milan and Brussels".[62] In a later interview, she says of her work at the UCL: "[*I*] worked at the University College in London, and enjoyed it immensely. They gave us every assistance. Once I had settled in there, the professor [*Eric Burhop*] divided up one of the classes and made me responsible for part of them—even though he hardly knew me...".[43]

In London, Tiomno collaborated with Abdus Salam, preparing an article with him on '*Isotopic Spin Relations in Hyperon Production*', which appeared in the journal *Il Nuovo Cimento* in 1961 ([JT34]). Tiomno[13] describes that work as follows:

> [*this is*] an analysis of the experimental results from reactions [*of the type*] π + p → K + Σ, verifying the properties of charge independence. [*We found*] not only good agreement with the isotopic spin theory, but also showed that peculiar circumstances occur at energies in the range of 1 GeV. Reactions which should satisfy a [*single*] relation between their amplitudes in fact satisfy two relations. The implications of this result are examined [*in this work*], although the problem is still open, apparently related to the 'false' asymmetry of the Σ's produced.[13]

This work also implied that neutral bosons (the K^0 mesons or neutral kaons) could have distinct antiparticles, a result which became important a few years later in the discussion of CP violation.

[This is true of *composite* bosons, like the mesons, since they contain quarks and antiquarks. It is *not* true of the *elementary* neutral bosons, such as the photon, the Z^0 particle (neutral intermediate vector boson, which mediates the neutral currents in the Weak interactions), or the Higgs boson].

The 'peculiar circumstances' referred to are an early example of a *resonance*, resulting from a short-lived reaction or particle.

Tiomno also collaborated there with *Susumu Kamefuchi*; their joint article[63] was published as a communication to the Brazilian Academy of Sciences in 1959. Kamefuchi, a Japanese theoretician eight years younger than Tiomno, was known for his work on field theory. He was visiting at the Imperial College as a Research Fellow at the same time that Tiomno was there, and he published a longer paper together with Abdus Salam in 1961.[64] The short paper with Tiomno, entitled '*Some Reflections in Quantum Field Theory*', deals with the definition of an operator for spatial reflection in the case that parity is only approximately conserved. It is a rather speculative treatment of a fundamental aspect of quantum mechanics. Tiomno's last paper on the Weak interactions, '*The failure of the space reflection principle*' ([JT32]), a review of the experimental evidence for parity non-conservation in the Weak interactions, was published with Tiomno as sole author during his London stay. It appeared in the *Proceedings of the Mathematical Society of the University of Southampton* in 1959; it was evidently submitted to a conference while

Tiomno was in England, and appeared in the conference reports. His year there was thus relatively productive, although it did not lead to any new long-term collaborations or dramatic changes of direction in his research. David Bohm was also in England by that time (initially in Bristol), but there is no evidence that he and Jayme Tiomno collaborated again.

During Elisa and Tomno's stay in London, the disastrous fire at the *Centro* in Rio de Janeiro broke out. It burned mainly the library, which was however located just one floor above Elisa's nuclear-emulsion laboratory. About this event, she made the following sad remarks[43]:

> … But I had an enormous misfortune.… I had set up a laboratory for nuclear emulsions at the *Centro*, completely equipped, and before leaving Brazil, I had requested a special microscope. I had been requesting it [*every year*] for six years from the *Conselho Nacional de Pesquisas* [*CNPq*], as it was very important for my work.
>
> While I was in England, they wrote to me, saying that the microscope had arrived. I was quite happy about that. Then, just three days later, I received another letter: 'The library has caught fire and burned!' The library was just above my laboratory, and all the microscopes were destroyed".[43]

While Elisa and Jayme were in London, Elisa's daughter Sonia, who was still in high school, lived with the family of one of her classmates, while her son Roberto, then 14, went to a boarding school, in *Nova Friburgo*, about 3 hours' drive from Rio. When Elisa and Tiomno returned from London, Sonia moved back to live with them, but Roberto preferred to remain at the boarding school, traveling sometimes to Rio.[5] His fellow pupil there, Sergio Joffily, later became a student of Elisa and Tiomno's. In his contribution to the homage to Elisa in 2004[65], Joffily recalled:

> At the *Colegio Nova Friburgo*, Center of Pedagogical Studies, we were both test subjects for experiments on teaching and learning. Frota, as he was called by his fellow students—he is now a renowned surgeon in Rio de Janeiro—enjoyed that boarding school so much that when his mother returned [*from England*], he stayed there until he had finished high school.

Tiomno and Elisa had both requested leave from the CBPF and the FNFi, initially for the academic year 1958/59, and had obtained a fellowship from the CNPq to support their stay in London. In May of 1959, Tiomno made an effort to extend the fellowship, with the intention of staying on in London for a second year, and applied to the Faculty in Rio for a corresponding extension of their leave of absence (for himself and Elisa). However, it proved impossible to extend the fellowship, so that after he had attended the Paris

Fig. 9.7 Jayme Tiomno and Elisa Frota-Pessôa in Paris, at a spot on the western end of the *Pont d'Iéna*, looking north along the Seine. This was probably at the time of the Paris *Conference on Relativity and Gravitation* in June, 1959, just at the end of their London sojourn. *Source* Private archive, [EFP]

Conference on Relativity and Gravitation in late June, 1959 (Fig. 9.7), he and Elisa returned to Brazil.[66] The fact that Tiomno wished to attend that conference shows that he had maintained his interest in field theory and gravitation, in which he had worked briefly over 10 years earlier, in São Paulo and Princeton. He had however published only a few scattered articles in that field in the intervening years; most of his work had been on particle physics. But his continuing interest in gravitation theory would be revived in the future.

An additional factor contributing to ending their sojourn in England must have been Elisa's health[5]—in the late 1950's, the dreaded London fogs were still a threat, although somewhat less so than earlier in the decade. Those fogs could descend upon the city for periods of a week or more during the winter months, trapping the fumes from thousands of coal heating fires and coal-fired power plants, and presenting a serious health hazard to London's inhabitants. It was not unusual for more than a thousand excess deaths to be registered within a 10-day period due to the fogs. The worst one had occurred in December, 1952, and is reported to have caused over 4000 excess deaths within 5 days. Elisa was accustomed to the mild tropical climate of Rio de Janeiro, and she apparently suffered from chronic bronchitis while in London, a condition which persisted even after warmer weather had arrived in the spring. This was certainly a factor which would have limited her ability to live there for a longer period.

Indeed, Elisa's planned Ph.D. thesis, which she intended to write after returning to Brazil, as well as her continued collaboration with the English group, both never came to pass. The reason was not only her efforts within a new project of César Lattes', as she suggests in a later interview, but also her health, which continued to be uncertain for several years after the London stay. This was due not just to the climate and the air pollution in London, which had affected her lungs, but in addition to a severe depression which was apparently induced by her relentless work during that year, compounded by the short, dark winter days in London and the general greyness of their life there, and by the psychological effects of the fire at the CBPF. Later, she was under treatment for several years and recovered completely only around 1963/64.[5]

Notes

[1] Compare Videira (1980). See also Leite Lopes (2004), and Chap. 5.

[2] His work there with Wheeler on the motion of infinitesimal masses in GR was superseded by Infeld and Schild (1949) before it could be published; cf. Chap. 6.

[3] See Peat (1997) for a biography of Bohm, and compare the section on Bohm in Chap. 8.

[4] A rather complete discussion of Bohm's stay in Brazil, as well as other aspects of his life and science, is given by Olival Freire Jr.; see Freire (2005).

[5] Personal communication from Sonia Frota-Pessôa, 2018/19; [EFP].

[6a] cf. http://www.abc.org.br/membro/jayme-tiomno/.

[6b] Records of the Membership Election Committee, ABC; in Archive [JT].

[7] cf. http://www.abc.org.br/membro/elisa-esther-maia-frota-pessoa/.

[8] Letter from Jayme Tiomno to David Hill at Vanderbilt University, June 25th, 1952; [JT].

[9] See [JT14]–[JT22].

[10] '*S-matrix theory and causality*', [JT14] and [JT17], with W. Schützer.

[11] [JT21].

[12] [JT19].

[13] See Tiomno (1966a).

[14] Sudarshan and Marshak (1958); see also R.E. Marshak, letter to Tiomno, Sept. 14, 1984; [JT].

[15] Feynman and Gell-Mann (1958).

[16] Sakurai (1958).

[17] For original articles on the electroweak unification, see e.g. Weinberg (1967), Salam and Ward (1959), Glashow (1959).

[18] This is the article [JT19] in Tiomno's publications list; cf. the letter from Arthur Wightman to Jayme Tiomno, Dec. 6th, 1955; [JT]. Wightman comments on the 'mass reversal operation' and expresses doubts as to its connection with time reversal invariance (the CPT theorem was not yet well established at that time).

[19] CBPF, *Notas de Física*, Vol. I, No. 1: online at http://cbpfindex.cbpf. br/publication_pdfs/NF00152.2011_11_28_11_55_33.pdf.

[20] [JT16].

[21] [JT18].

[22] [JT22].

[23] Jayme Tiomno to Elisa Frota-Pessôa, July 6th, 1954; [EFP].

[24] Ouchi et al. (1956), Pecina-Cruz (2005). See also Chap. 4.

[25] Chapter 4, Section on 'Beta decay', and Backnotes [6]–[8] there.

[26] Lee and Yang (1956).

[27] Jayme Tiomno to Elisa Frota-Pessôa, September 23rd, 1956; [EFP].

[28] [JT29].

[29] Jayme Tiomno to Elisa Frota-Pessôa, Sept. 28th, 1956; [EFP].

[30] Jayme Tiomno to Elisa Frota-Pessôa, April 21st, 1957; [EFP].

[31] cf. Mehra (1994), Chap. 21.

[32] In his talk for Jayme Tiomno's 60th birthday, Videira (1980) recounts that Tiomno, while writing his thesis, mentioned to C.N. Yang that the operator $1 \pm \gamma_5$ would violate parity conservation if used in the current-current interaction. Yang replied that he was happy to have studied under Fermi, who did not hold parity conservation to be a fundamental property of nature—a prophetic remark, pointing to Lee and Yang's later conjecture, and to their Nobel prize in 1957.

[33] See Chap. 10, section on Abdus Salam.

[34a] Wu et al. (1957). (^{60}Co); see also Postma et al. (1957) for ^{58}Co, and Hargittal (2012) for Wu's attribution.

[34b] Garwin et al. (1957), Friedman and Telegdi (1957).

[35] See [JT23]–[JT32].

[36] Jayme Tiomno to Elisa Frota-Pessôa on April 9th, 1957; [EFP].

[37] Mahmoud and Konopinski (1952). Tiomno himself, in a memoir written in 2005, later described how he was misled; see Tiomno (2005) and [JT107]. Cf. also the chapter by Allan Franklin, '*The Konopinski-Uhlenbeck Theory of ß Decay: Its Proposal and Refutation*', in the book *Wrong for the Right Reasons* (Buchwald and Franklin 2005), which describes another example of how a theory can go astray, also involving the Weak interactions and Konopinski.

[38] Ruderman and Finkelstein (1949); see also Finkelstein (2016).

[39] Fazzini et al. (1958).

[40] Tiomno (1984/94).

[41] For more details on the origins of the Universal V–A theory, see e.g. Wu (1964); the essays in MacDowell et al. (1991), particularly Leite Lopes' chapter; Leite Lopes (1996), Marshak (1997), and Tiomno (1984/94), as well as Mehra (1994) (who quotes extensively from Wu 1964). cf. also [JT68] and [JT102].

[42] See Mehra (1994), Chap. 21, p. 453. Feynman's elation at discovering a new law of nature is described there. R.E. Marshak quotes Feynman as later having realized that he was not the only physicist, nor even the first, to discover that law—see Chap. 15, 'The Weak-interactions interlude', and backnotes [19] and [21] there.

[43] Frota-Pessôa (1990).

[44] Compare the remark quoted in Chap. 6, backnote [28].

[45] Guido Beck to John Wheeler, April 9th, 1957; APhS Archives.

[46] John Wheeler's statement sent on May 24th, 1957 to the *Moinho Santista* selection committee; [GB].

[47] Gleick (1992).

[48] Gribbin and Gribbin (1997).

[49] Mehra (1994).

[50] Leite Lopes (2004): This is Leite Lopes' book on the history of physics in Brazil (in Portuguese). The parts about Feynman were recorded in a memoir in 1988 (in English), published in the CBPF *Ciência e Sociedade* series; see Leite Lopes (1988).

[51] Feynman et al. (1964–2006).

[52a] Leighton and Feynman (1985).

[52b] Leighton and Feynman (1988).

[53] cf. Chap. 6, section on the Michigan Summer School.

[54] Tiomno (1977).

[55] Jayme Tiomno to Richard Feynman, March 6th, 1950; [JT].

[56] Leite Lopes (1988); the paper was also published in the Proceedings of the Symposium (CBPF, June 16th–19th, 1952).

[57] Gilder (2008).

[58] cf. Chap. 8, section on David Bohm; cf. Chap. 8, p. 119.

[59] Castro Moreira (2018).

[60] Oswaldo Frota-Pessôa, in *Jornal do Brasil,* May 25th, 1952. His article is cited by Castro Moreira (2018), and it is reprinted and included within a note on the website of the Brazilian Physical Society, available as a pdf (in Portuguese) at http://www.sbfisica.org.br/fne/Vol14/Num1/fne-14-1-a14.pdf.

[61] Elisa Frota-Pessôa to Burhop, March 2nd, 1958 (this letter is erroneously dated as March 2nd, 1948); [EFP].

[62] Report by Elisa Frota-Pessôa on her first trimester of the fellowship, September 15th, 1958; [EFP].

[63] [JT31].

[64] Kamefuchi et al. (1961).

[65] Silva Lima et al. (2004).

[66] These details can be found in a report from Tiomno to the FNFi/UB from 1959, in which he explains his delayed return to Rio de Janeiro; [JT].

10

A New Decade: The Calm Before the Storm

Shortly after Jayme Tiomno and Elisa Frota-Pessôa arrived back in Brazil from their sojourn in London, a new decade began—the 1960s. That decade was a turbulent time in the lives and careers of Jayme and Elisa, reflecting wider events in Brazil and throughout the world in those years. As far as Tiomno's own activities are concerned, we can divide those ten years roughly into three periods. The first of those, described in this chapter, might be called the 'calm before the storm', while he was still working in Rio de Janeiro, but also spending some time abroad, as a visiting scientist in the USA and Italy, and before the turbulent times of the mid- and late 1960s surged in.

Academic Politics

On returning from London in July 1959, Tiomno and Elisa resumed their activities at the CBPF and the FNFi/UB in Rio. Elisa also began treatment for her health problems. A year later, on the 29th of July, 1960, Joaquim da Costa Ribeiro passed away. He was the occupant of the Chair of General and Experimental Physics of the FNFi faculty, *Universidade do Brasil*, where Tiomno and Elisa had begun their scientific careers. It was thus necessary to hold a new competition to fill the vacancy left by Costa Ribeiro's passing. One of Costa Ribeiro's assistants, the experimental physicist *Armando Dias Tavares*, considered himself to be a sort of natural heir to Costa Ribeiro, and thus the legitimate contender to occupy his now-vacant position. César Lattes supported Tavares' candidacy; however, by that time, Lattes had already transferred to USP. Tiomno, on the other hand, saw the opportunity to bring

© Springer Nature Switzerland AG 2020   **171**
W. D. Brewer and A. T. Tolmasquim, *Jayme Tiomno*, Springer Biographies,
https://doi.org/10.1007/978-3-030-41011-7_10

Roberto Salmeron back to Brazil from CERN/Geneva, where he had been working since obtaining his doctorate in England, having spent three years at the CBPF in the early 1950s. If Salmeron returned and occupied the vacant Chair at the University, he could also be contracted as professor by the CBPF.

The Department of Physics didn't want to nominate any of Costa Ribeiro's assistants as an interim replacement to his position, in order to avoid influencing the outcome of the competition. They thus decided to ask Tiomno to take over its functions temporarily, until the competition could be organized.[1] For Tiomno, taking over those functions was an additional burden, but he nevertheless agreed on a short-term basis, on the condition that the competition be carried out as quickly as possible and that all of those interested would be given an equal chance to participate in it. He also preferred that instead of a formal contract, he simply be designated temporarily as the person responsible for the Chair, in order to avoid the appearance that he was taking advantage of the situation. Aside from the fact that the new occupant of the Chair would have to be an experimentalist, he had no interest at all in becoming involved in administrative and university politics. He preferred to maintain the continuity of his research. Elisa, who *was* an experimentalist, was also not interested in competing for the position. She was still trying to recover from her depression which had resulted from their stay in London and the problems at the CBPF. She was at most willing to enter a competition for the position of lecturer, but even in that case she was not prepared to spend much time assembling her application materials.

However, when the news spread that Tiomno was going to take responsibility for the Chair, a vicious campaign against him arose: A theoretical physicist who was trying to appropriate the Chair of Experimental Physics! The affair spread to the press and reached the University of São Paulo, where Lattes had recently transferred. As a result, Tiomno refused to accept the temporary post, and wrote to Leite Lopes, at the time head of the Department of Physics at the UB,

> I sincerely regret that I must respond to your inquiry by affirming that I cannot accept my nomination to assume responsibility by limited agreement for the Chair of General and Experimental Physics at the FNFi. I hope that the Department of Physics, together with the Dean's Office of the Faculty, will be able to find a solution to this problem which is worthy of a university-level institution.[2]

In the end, in spite of her depression, it fell to Elisa to take over responsibility for the functioning of the Chair. On February 2nd, 1961, she was named as interim occupant, without additional salary, of the Chair of General and Experimental Physics, until the competition to fill the position had been brought to a conclusion. It had been agreed that this interim post would not last for more than one year. Her nomination for the post however evidently aroused Armando Dias Tavares' anger, causing friction in their relations and stress for Elisa. She also had difficulties in dealing with Eremildo Viana, the Dean of the *Faculdade Nacional de Filosofia*, a controlling and reactionary figure who caused all sorts of problems for her on a daily basis (see below).

Elisa asked several times to be discharged from responsibility for the Chair of General and Experimental Physics. But the competition was continually postponed, and she was again and again begged to continue for just a little longer. Finally, in March of 1964, she requested unconditionally to be released from the position.[3]

A New Particle

While at the Seattle Conference in 1956, Tiomno had become interested in the K mesons ('kaons')[4] which had been discovered by Rochester and Butler (they were originally called 'V particles'),[5] and which were later investigated using nuclear emulsions by the K⁻ collaboration including Burhop, Powell and Occhialini, along with many other authors[6] (in which Elisa had participated during their London stay in 1958/59). Tiomno was fascinated by the idea of symmetry principles. While in London, he had worked on the prediction of a new meson, which he later called the K'. When he returned from England in 1959, he proposed to his two assistants, Nicim Zagury and Luciano Videira, that they should work together on this topic. As Zagury later recalled:

> It was a study of the possibility of the existence of a new particle, the K'. So we spent more or less one year, Luciano Videira and I, examining the experimental data that were available at that time to see if they agreed well with the theory. Nobody knew how to explain those experiments, and Tiomno's idea was that they could be explained by the existence of a [*new*] particle with certain properties, which would be our K' meson. It was a particle that would decay into two others: a K meson and a π meson. We were trying to see if such a particle in fact existed, and if so, what its mass must be.[7]

In late August of 1960, Tiomno traveled again to Rochester, to attend that year's *International Conference on High-Energy Physics* (which had been moved from its previous time-slot in April). "This conference is not very interesting, and it has too many parallel sessions", he wrote to Elisa.[8] But that first impression was quickly modified. In the discussion following Gell-Mann's talk, in which he suggested the existence of a new particle with the opposite parity to that of the K meson, whose existence would make certain particle interactions more symmetrical, Tiomno presented the incomplete results of his own work with Zagury and Videira, showing that their calculations indeed indicated that such a particle could exist, and that they had estimated its probable mass. His remarks received considerable attention and were published in the Proceedings of the Conference ([JT33], 1960). When he returned to Brazil, Tiomno urged Videira and Zagury to finish the calculations as quickly as possible, and they published a more detailed article in the *Physical Review Letters* on February 1st, 1961, entitled '*Possible Existence of a New K' Meson*' ([JT35]). In parallel to their work, *Luis W. Alvarez* and his coworkers at the UC Lawrence Radiation Laboratory in Berkeley/CA (now the Lawrence Berkeley National Laboratory, LBNL) had hastened to verify Tiomno's prediction, using the Bevatron (the Berkeley synchrotron which could produce a proton beam of over 6 GeV particle energy), together with the recently-installed 72″ liquid-hydrogen bubble chamber as detector. On March 15th, 1961, Alvarez published his confirmation of the particle's existence—also in *Physical Review Letters*—, calling it the K* meson.[9a] The Brazilian group had calculated that its mass should be in the range of 1,300–1,800 electron masses, and that the new particle would decay into a K and a π meson with a decay lifetime of the order of 10^{-23} s. Alvarez found a mass corresponding to 1,732 electron masses in his experiments.[9b] (Alvarez received the 1968 Physics Nobel prize for his many particle discoveries using the Bevatron-bubble-chamber combination). Tiomno explained their theoretical results:

> These works [*JT33 and 35*] predicted the existence of a new (K') meson, analogous to the K meson, but of opposite parity so as to permit the existence of a K' – K,π interaction. This in turn was necessary to explain the strong asymmetry observed in the production of K mesons together with Λ particles following p–π collisions. The presentation of the preliminary results at the Rochester Conference in 1960 [JT33] coincided with Gell-Mann's suggestion of a meson which he also called K', having zero spin and the opposite parity to the K, and related to it by the structure of the weak interactions.[10]

The results obtained by Tiomno, Videira and Zagury were presented at a meeting of the *Academia Brasileira de Ciências* (ABC) in May of 1961, and were celebrated in the Brazilian press (Fig. 10.1). The magazine *Visão* noted in a feature article, "A new atomic particle, the K' meson, was discovered by the Brazilian nuclear physicist Jayme Tiomno (*Moinho Santista* Prize for

Fig. 10.1 A.L.L. Videira, J. Tiomno, and N. Zagury, at the announcement of the K'-meson paper, at the ABC in 1961. *Source Acervo* CBPF, original photographed at the ABC. Photographer unknown. First published in *Visao* magazine (Brazil), May 19th, 1961, p. 21. This magazine is no longer published

1957) and his two assistants in the Department of Theoretical Physics at the CBPF, Antônio Luciano Leite Videira and Nicim Zagury".[11]

The press generated some excitement over that discovery, by default through Tiomno. He made good use of one of the interviews to mention the difficult financial situation of the CBPF and what was needed there. The Brazilian President at the time, *Jânio Quadros*, read the interview with Jayme Tiomno and thereupon sent a letter to his technical advisor *Cândido Mendes*, saying that he wanted the necessary measures to be considered rapidly, and that a concrete suggestion was to be sent to him so that he could arrange for significant assistance for the *Centro* as soon as possible.

In July/August of 1961, Tiomno spent six weeks at the University of Wisconsin, Madison/WI, USA during the Physics Summer School. While there, he encountered one of his former fellow graduate students from Princeton, Wayne R. Gruner, who had joined the National Science Foundation (NSF) as an administrator for physics projects. They discussed the possibility that the NSF might support projects in Brazil, in particular a proposal by Roberto Salmeron, whom they hoped to attract to the CBPF, as well as the completion of a Cockroft-Walton accelerator at the CBPF.[12] Tiomno also used this visit to check on progress in studying the K' meson which he had predicted the previous year, visiting the accelerator center at Brookhaven, NY, and he conferred with C.N. Yang in nearby Stony Brook. In addition, he had an opportunity to work with Abdus Salam, and as a result, they published their article *'Isotopic Spin Relations in Hyperon Production'* in *Il Nuovo Cimento* ([JT34],

1961), on which they had begun work while Tiomno was in London, and they prepared their joint lecture on the '$\Delta S/\Delta Q = -1$ rule'.[13] It concerns a *selection rule*, indicating which particle reactions can occur and which are 'forbidden', and maintains that the change in the 'strangeness' quantum number S for the hadrons in a weak decay must equal the (negative) change of the electric charge Q in that same decay. This rule has been tested a number of times and remained important in high-energy physics for some years.

About their joint lecture ([JT37]), Tiomno[10] writes: "...the consequences of the experimental results of [*Ugo*] Camerini et al. are examined; they indicate evidence of the violation of the $\Delta S = -\Delta Q$ rule in leptonic decays of the K^0 [*meson*]". (The Camerini paper referred to dealt with the mass difference between the two types of neutral K mesons).

But, most importantly, he had the chance to relax and briefly forget the problems in Brazil: "I'm making good use of my time here, mainly because of the lack of worries about the *Centro* and the Faculty".[14]

But Tiomno could hardly imagine how quickly his serenity would come to an end. One or two days after his return to Brazil, the magazine '*O Cruzeiro*' published an interview with César Lattes, in which the latter claimed that Tiomno had stolen credit for the work of Murray Gell-Mann. Furthermore, Lattes attacked theoreticians in general, saying that *discoveries* are made only by the experimentalists—"The credit for a discovery should belong to those who obtain the experimental and objective evidence, about which there can be no doubts"; and he continued, "They are prophets, those grand theoretical physicists. They construct theories, and starting from those ideas, they predict results which can be verified objectively only by the experimentalist, interrogating Nature. Only then can one speak of a discovery. All other cases are simply a question of empty little promises...".[15]

However, the most seriously offensive items in Lattes' interview were the open insults directed at Tiomno personally: "He [*Tiomno*], in my opinion, until the contrary is proven, is a con man. There is a book in the files of the *Centro Brasileiro de Pesquisas Físicas*, in which was written, at the request of Prof. Leopoldo Nachbin, a declaration made by myself to the effect that Prof. Tiomno is not trustworthy. If he, or persons acting in his interest, try to take over the Chair of Experimental Physics in Rio, this information could be useful".[15] In fact, Lattes mixed criticism of the grand announcement of Tiomno's work on the K' meson with his own comments on the conflicts at the University (UB) related to the Chair of General and Experimental Physics, left vacant by the death of Costa Ribeiro; the result was rather confused.

Various Brazilian physicists expressed their solidarity with Tiomno. One group wrote that, "We judge that this magazine article, owing to the opinions expressed in it, is not worthy of serious consideration".[16] Another group of physicists from the University of *Rio Grande do Sul* reaffirmed Tiomno's competence, pointing out that, "The scientific reputation and capacities of Prof. Tiomno were once again made evident by the fact that he was the only physicist from Latin America who was invited to participate in the Seminar on Theoretical Physics held in July 1961 at the University of Wisconsin, whose invitees were a small, select group of physicists from all over the world".[17] Tiomno made no move to enter into a public discourse with Lattes. As he disliked this type of interviews by the press, he wrote a note to the newspapers, responding to Lattes' criticisms, and emphasizing that it would be his only statement on the subject.[18]

At the beginning of the following January, 1962, Tiomno once again departed for the USA, for a second visit of two months to the Physics Department in Madison, Wisconsin. He answered Robert Sachs' invitation to return there, writing,[19]

> … I have the best recollection of my stay there last summer and I am most thankful for this new opportunity of breaking the isolation of my work here… I did not answer your letter before because my wife has been ill and it was not clear if she would be able to go. She is much better now and we have decided that I should go alone…I should be arriving [*in*] Madison [*on one of*] the first days in January.

Tiomno had initially hoped that Elisa could accompany him to Madison, but her health was still not good, and he first postponed the stay, even considering canceling it altogether. In the end, he went to Madison alone and collaborated again with Abdus Salam, who was also on a visit there. The two of them gave a lecture together while at Madison, on the strangeness—charge selection rule (see above).[13]

During this period, Tiomno also completed several research projects at the CBPF, together with his graduate students. In 1962, returning to an uncompleted project begun with David Bohm while he was still at Princeton in 1949, Tiomno wrote a paper on '*Representations of Dirac Equations in General Relativity*',[20] published together with his colleague *Colbert Gonçalves de Oliveira*. It was described by Tiomno[10] as follows:

> After [*first*] having reformulated the equations of motion of a particle in a gravitational field in General Relativity in terms of tetrads, and obtaining the generalized Dirac equation in the Schrödinger representation, we thence obtained

the Foldy-Wouthuysen representation in the presence of gravitational fields. The most interesting result is that the gravitational 'gyromagnetic' factor is 1, and not 2, as is found for the electromagnetic case; this is in agreement with the equivalence principle for the inertial and gravitational masses. The red shifts of the atomic levels are obtained, coinciding with the usual predictions of General Relativity.

About the same time, Jayme Tiomno also presented a research paper entitled '*Octonions and Super-global Symmetry*'.[21] It was included in a volume called simply '*Theoretical Physics*' and published by the International Atomic Energy Agency in 1963. This article treated a rather different field of research, namely the application of symmetry principles to the classification of the elementary particles. It was authored by Tiomno alone; he says of this paper that in it, "... [*he*] examined the relation between Cayley's hypercomplex algebra and [*his own*] theory of isospin space in seven dimensions, to which [*Abraham*] Pais had called attention".[10]

Indeed, one of the more productive collaborations in which Jayme Tiomno participated, both for the resulting physics and in terms of promoting international organizations, was his long cooperation with the Pakistani physicist and Nobel prize winner *Abdus Salam.*

Mohammad Abdus Salam (1926–1996)[22] was born in the Punjab, in what is now Pakistan, on January 29th, 1926. He entered the University of the Punjab at age 14, with excellent marks on his entrance examinations, and later attended the University of Lahore on a merit scholarship, where he received his BA in mathematics in 1944 and his MA in 1946. He then went on a scholarship to St. John's College, Cambridge/UK, where he received a second BA in mathematics and physics with high honors in 1949, going on to complete his doctorate at the Cavendish Laboratory, Cambridge, in 1951 with a thesis on QED. He joined the University of Lahore as Professor of Mathematics and also served at the University of the Punjab from 1952 as Professor and Chair of the Mathematics Department. Because of opposition to his establishing a research institute there, and political unrest in Lahore, related to his affiliation with a minority branch of Islam, he returned to Cambridge as Professor of Mathematics in 1954, moving to the Imperial College in London in 1957, where he set up the Department of Theoretical Physics. He received an honorary doctorate from the University of the Punjab that same year, and established a scholarship program to bring talented Pakistani students to London. He spent a sabbatical at Princeton in 1959, working with Oppenheimer. He also became influential in science policy in Pakistan, and promoted theoretical physics there.

Abdus Salam's early work concentrated on QED and neutrino theory (where he introduced the term *chiral symmetry* early on; cf. Chap. 9, Salam's unpublished paper of 1956). Later, he worked mainly in particle physics (on the structure of the strong interaction, the Higgs mechanism, and on vector mesons), and in particular on the *electroweak unification*. It was during the period following the Conference on Theoretical Physics in Seattle in September 1956 that he began collaborating with Tiomno (see Chap. 9).

Later on, Abdus Salam collaborated with Sheldon Glashow and Steven Weinberg, beginning around 1964, on the Electromagnetic and Weak interactions and their unification, helping to formulate its mathematical structure. The three of them shared the Nobel prize in Physics in 1979 for that work, including their prediction of neutral weak currents. The electroweak unification forms another cornerstone of the Standard Model. The idea of neutral weak currents had been suggested in parallel by several authors. José Leite Lopes, in his history of physics in Brazil,[26] mentions his own contributions to the topic of neutral weak currents and the neutral heavy boson which mediates them (now known as the Z^0 *intermediate vector boson*), beginning in 1958.

Abdus Salam was the principal founder and the first director (until 1993) of the *International Centre for Theoretical Physics* (ICTP) in Trieste, Italy, which now bears his name. Salam was also a founder of the Third World Academy of Sciences (TWAS) [now 'The World Academy of Sciences'] (Fig. 10.2).

Especially during the 1960s, Salam made great efforts to support research in Pakistan, and he was instrumental in initiating the nuclear-power and space programs of his country. His role in the development of nuclear weapons by Pakistan was mixed; he initially supported the project but later distanced himself from it. He left Pakistan for good in 1974, but maintained his support of science there, bringing many young Pakistani scientists to the West for further education. He continued working in fundamental physics in later life, especially on the search for a 'grand unified theory', along with his many activities supporting science education and research, until his death at age 70 in Oxford, on November 21st, 1996.

The *International Centre for Theoretical Physics* (ICTP)

Jayme Tiomno also played an important role in the establishment of the ICTP, a major project initiated by Abdus Salam. In the early 1960s, he

Fig. 10.2 Abdus Salam in 1987. *Source* Dutch National Archives, Photo by Bart Molendijk. Used under Creative Commons License, published with permission. [Online at: https://commons.wikimedia.org/wiki/Category:Abdus_Salam#/media/File:Abdus_Salam_1987_(cropped).jpg]

took part as a discussion leader in the Seminar on Theoretical Physics held in August 1962 in Trieste, Italy, sponsored by the International Atomic Energy Agency (IAEA). In a letter to Tiomno, the Secretary of the Seminar, Haakon Storhaug, informs him of the plans for travel and accomodations and promises that the program for the Seminar would be sent to the participants soon.[27] Abdus Salam was one of the principal organizers of that seminar, which explored the idea of establishing an *International Centre for Theoretical Physics*; the original suggestion for the center had been made at a meeting on particle physics held in Trieste in 1960. The success of the 1962 seminar and of another one organized at almost the same time in the Lower Tatra, in Czechoslovakia, provided strong support for the proposal. Shortly thereafter, at the 71st Plenary meeting of the IAEA on September 26th, 1962, the decision to create the Centre was taken, having in mind that it would meet the needs of scientists from a large number of Member States, helping to support physics research and education in developing countries—in particular through fellowships to permit visits to those countries by prominent 'first-world' scientists, as well as visits by their own physicists to the Centre, where research would also be performed.

The IAEA sent information to all of its members about the intended founding of the Centre, and received many suggestions and offers to host it, the latter from Austria, Denmark, Italy, Pakistan, and Turkey, among others. In view of the number of offers made by Member States, the Board of Directors of the IAEA recommended to its Director General, Sigyard Eklund, that he should set up a small expert commission of three theoretical physicists to advise on possible scientific programs, the scientific staff to be recruited, the facilities needed for such a center and, taking into account all aspects of the offers, to make recommendations about its location. Those invited to serve on the commission were Jayme Tiomno from Brazil, Robert E. Marshak from the USA, and Léon van Hove from the Netherlands (representing Europe). They were sometimes referred to as 'the three wise men', not of course without irony.

The meeting of the expert commission took place in early April, 1963, in Vienna.[28] The commission considered that the new Centre should have as its main goals to help and encourage able theoretical physicists from still-developing countries to continue and expand their research work; to serve as a pilot project for future international research institutes; to promote the cross-fertilization of the various fields of activity within its scope; and finally to contribute to the furtherance of international contacts between widely separated areas of the world and between countries in different stages of scientific and technological development. The three chief fields of research were to be low-energy nuclear physics, solid-state physics, and elementary-particle physics. They also recommended that it be located in Trieste, whose university was willing to establish two full professorial chairs and four assistant-professor positions in theoretical physics for the Centre. "If outstanding theoretical physicists can be found to occupy these chairs, we feel that the University of Trieste might be able to provide the appropriate scientific climate for the Centre", they stated.[29]

There were problems in financing the ICTP, and the Italian government withdrew for a time from its original promise of support. There were a series of planning meetings in 1963 and 1964, and with the support of the IAEA and of UNESCO, the problems were resolved and the *Centre* began its operations in Trieste on October 1st, 1964, with Abdus Salam as its first president. Jayme Tiomno was one of its first Associates, with the privilege of visiting there at its expense for periods of one to three months up to three times during his three-year tenure. The Associates were elected by the Scientific Council of the ICTP each year.[30]

The ICTP, today called the *Abdus Salam International Centre for Theoretical Physics*, still maintains close relations with the CBPF in Brazil.[31]

Promoting Science Education

Jayme Tiomno had a special interest in teaching and in guiding the research of students from early on in his career,[32] and he shared it with his wife, Elisa Frota-Pessôa. Despite Tiomno's recognition of the importance of teaching, he was not a very well-organized professor himself; his blackboards were rather chaotic, and furthermore, he was shy and not the sort of professor who enjoys telling stories to make his lectures more interesting. Nevertheless, he was very clear about the limitations of the scientific-educational system in Brazil, especially regarding physics, both in the secondary schools and in the universities. Teaching was very formalized, and failed to inform the students about recent research results and new developments. Tiomno, in his own classes, emphasized problem-solving, giving exercises to the students, and introduced them to current publications and research. He also believed that it is very important for a physicist, in particular a theorist, to be able to carry out calculations rapidly and with a minimum of errors.[33]

The lack of suitable books in Portuguese was an additional problem for the students. Only a minority of them were proficient in English, in particular since the major external cultural influence on Brazil, dating from the time of the Empire, had been from France; but by the second half of the 20th century, English had become the 'language of science', no longer French, as in the 18th century, nor even German, which had predominated until the 2nd World War, as Tiomno had emphasized earlier.

When Jayme Tiomno went to the CBPF in 1952, he had taken on responsibility for setting up the Teaching Department and the undergraduate labs, and somewhat later, he established the Graduate School there. In 1953, following Feynman's 1952 talk on physics teaching in Brazil, Tiomno published a paper on '*Teaching Physics in Secondary-School Courses*', together with José Leite Lopes.[34] In it, they wrote:

> It is our opinion that the standard textbook to be adopted might well be that of Blackwood, Herron and Kelly, '*High-School Physics*', which could be translated and adapted to Brazilian conditions. Or else a book written especially for the purpose and in the spirit mentioned above [*could be used*]. It would be convenient to use the same book for students, also [*presumably in their early years of study as future secondary-school teachers*].

The two of them followed up on their own recommendation a few years later, translating the physics textbook mentioned from English to Portuguese, so that it could be used in Brazilian schools.[35] It was for many years the principal textbook for physics courses there. About that project, Leite Lopes had the following to say in an interview fifteen years later[36]:

> After that [*Feynman's 1952 talk*], Tiomno and I translated a physics textbook from the USA into Portuguese and adapted it for use in Brazil, and it was published. It must still exist. It was by Blackwood, [*Herron and Kelly*] … It was a book on physics written in an intuitive style, showing how things happen. This was in contrast to the empty formalism of those who tend to cultivate what is called 'traditional mechanics', writing formulas with the sole purpose of obtaining more formulas.

Tiomno had also been docent for the '*Final Course for secondary-school teachers of physics*', at *São José dos Campos*, in the State of São Paulo, in 1955. He wrote several articles on education for the science magazine of the SBPF, *Ciência e Cultura*, during the 1950s (see Appendix B).

During the early 1960s, he expanded these activities to a more national and international level. In September of 1960, after the Rochester Conference, he went on to Ottawa, Canada, where he served as the delegate from Brazil to the General Assembly of the International Union of Pure and Applied Physics (IUPAP), having been chosen by the Brazilian National IUPAP Committee.[37] After arriving at the Assembly, he met Oscar Sala, a colleague from USP/São Paulo, who claimed that *he* was the 'official' delegate to the Assembly, having been named by César Lattes (who was by then an interim professor at USP). Sala, after conferring with the Secretary General of IUPAP, Pierre Fleury, who emphasized that only the official Brazilian National IUPAP Committee could choose delegates, decided to leave Ottawa without attending the Assembly. This was an embarrassing situation for all concerned.

At the Assembly, the *Commission on Physics Education* was set up, and Tiomno was named as one of its members. He also nominated Sala and Lattes for two other commissions, attempting to moderate the conflict that had played out before the eyes of the Assembly, but they were not accepted. Somewhat earlier, it had been suggested that Brazil could host the next IUPAP congress, but since no commitment was made within the required time, it had been scheduled instead for Poland, a rival candidate. As a result, it was now proposed that Brazil should be considered as host nation for 1966, and could also host a meeting of the new Commission on Physics Education.

This suggestion was received well in Brazil, and in June/July of 1963, the second International Conference on Physics Education was held in Rio de Janeiro, sponsored by the International Union of Pure and Applied Physics (IUPAP) and other institutions, including the CBPF; its official title was 'First Inter-American Conference on Physics Education'. Jayme Tiomno served as a member of its Organizing Committee, Co-Chairman of the Conference (together with Sanborn C. Brown, president of the Commission), and co-editor of the Conference Proceedings volume, 'Why Teach Physics'.[38]

Tiomno's own lecture at the Conference, 'Science Education in the Contemporary World', was reprinted in edited form in the Proceedings volume.[39] In it, he gives a brief history of the sociology of scientific achievements, and discusses the social role of scientists, both in education and in the search for solutions to their countries' problems:

> Recognition of the need for drastic revision and reform in the methods and systems of scientific education and in physics, in particular, has instigated experiments in various countries, the results of which are worthwhile comparing. It is interesting to note that most of these experiments were made with the participation of leading scientists, who were the first to realize that education had to be revised. The attendance, at this and other Conferences on the teaching of physics, of outstanding researchers is a proof that the modern scientist no longer lives in his legendary ivory tower. On the contrary, he takes as active an interest as the average citizen, or an even more active one, in the problems of his national community and in problems of international significance. In the present case, the responsibility of scientists and educators is particularly heavy. They are the most able persons to point to solutions and to bring home to governments and to national and international institutions the need for planning and for taking urgent steps to ward off future disasters and the hazards of irreparable damage to the development of their countries.[38, 39]

Tiomno also summarizes the resolutions regarding the teaching of physics which were adopted by the Conference, and his paper is followed by another from a Brazilian physicist, Professor *Paulo Gomes de Paula Leite*, highlighting the special problems of a 'developing' country like Brazil in science teaching. That paper concludes with the following summary remarks[38]:

> Having to supplement the wholly inadequate foundation given in the secondary school, and having even to eliminate erroneous concepts from the students' minds, the [*university*] teachers fail again by insisting on an encyclopaedic coverage. The lack of experimental equipment, the excess of students, and the shortage of instructors make the situation worse. The pressure to restrict the teaching of basic sciences to the first year so that immediate

professional training can be started also has its ill effects. As the students have no time to *think* or to *understand*, they cannot enjoy the benefit of a solid foundation of basic science.

In fact, this general situation, especially in present times when an increasing proportion of the population is expected to attend colleges and universities, is still eminently relevant, and not only in 'developing' countries.

In the Conference Proceedings volume, edited by Tiomno together with Sanborn C. Brown and Norman Clarke,[38] the results of the conference are characterized[40]:

> The meeting dealt with physics as part of a liberal education. The serious prac-
> tical difficulties of teaching physics in a way that is appropriate to this purpose
> are now widely recognized in those countries that are highly developed scientif-
> ically, and many projects have been launched to solve them. Reports on some
> of these projects were given at the Conference, and still others are referred
> to in the book. In the comparatively underdeveloped countries, on the other
> hand, it is still necessary to establish the importance of including physics and
> other sciences in the curriculum. The book will be useful to all those who are
> concerned with science education. It will prove particularly useful to science
> educators working in Latin America, Africa, and Asia.

This conference coincided with Richard Feynman's fifth stay at the CBPF. He went there as lecturer at the Fifth *Escola Latino-Americana de Física* (ELAF), a cooperative physics school involving Brazil, Mexico, and Argentina, orga-nized by the *Centro Latino-Americano de Física* (CLAF). It had been founded in 1959 by Marcos Moshinsky from Mexico, Juan José Giambiagi from Argentina, and José Leite Lopes from Brazil.

Of course, Feynman also attended the conference on physics education, and in his keynote speech there, which he called '*The Problem of teaching physics in Latin America*',[41] he repeated and reinforced many of the points that he had made in his earlier talk in 1952 (on physics teaching in Brazil, quoted in Chap. 9). Feynman's 1963 talk on physics education has been ana-lyzed in detail by Castro Moreira,[42] who points out that it followed closely upon Feynman's own experience of giving the introductory physics courses at Caltech to beginning undergraduate students in the academic years 1961–63. These were the famous *Feynman lectures*, which were the source of the books of the same name.[43]

One must indeed note that, however excellent those books may be as refer-ence works for advanced students, Feynman's original lectures failed to meet their goal. Legend has it that the first- and second-year undergraduates gradually began to skip the lecture sessions, since they simply could not follow Feynman's pace and the depth of his lectures (not even *those* students, within that small, select group who had been admitted to the elite university *Caltech*). They were replaced by graduate students, who had heard about the brilliance of the lectures and, already having a foundation in the subjects covered, could appreciate the subtleties of Feynman's expositions. Feynman himself was skeptical about the effectiveness of 'frontal' teaching (i.e. in a lecture hall), and was especially bothered by the problem of putting enough advanced physics into his lectures to make them interesting to the best students, while still offering a clearly-defined 'minimal program' for the slower ones. And he was less optimistic than Jayme Tiomno about 'passing the torch' on to his own graduate students.[44]

Tiomno maintained his own interest in physics teaching and science education throughout his professional life (Fig. 10.3). Earlier, in 1960, he had also served as docent for the *Escola Latino-Americana de Física* (ELAF) in Rio de Janeiro. The list of docents for its Second Session, held in Rio in 1960, included in addition to Jayme Tiomno also *Chen Ning Yang*, Tiomno's old colleague from Princeton; *Giampetro Puppi*, his competitor as discoverer of the UFI, from Bologna; and *Gleb Wataghin*, who returned from Turin to give

Fig. 10.3 Jayme Tiomno at the blackboard, late 1950s. *Source* Private archive, [EFP]

lectures at the School. The Eighth Session was also planned to take place in Brazil (in 1966), but Leite Lopes was in France at the time, Tiomno was in Trieste, and the country was facing a difficult time, so that they were unable to organize it, and it was held elsewhere. In 1963, Jayme Tiomno was also responsible for structuring the curriculum in meteorology at the CBPF.

In early 1962, making good use of his stay in the USA (at Madison/WI), Tiomno also participated in a program of physics teaching. He had been invited by the American Association of Physics Teachers and the American Institute of Physics as part of a broad nationwide program to stimulate interest in physics, with the support of the National Science Foundation. It consisted of giving formal lectures, taking part in colloquia, meeting informally with the physics staff to discuss teaching and research in his own country, and visiting a variety of colleges and secondary schools to learn about the varied conditions under which physics is taught in the USA.[45]

Tiomno had a very intense schedule: He visited Carleton College, the University of Minnesota, the University of Chicago, the University of Illinois, Tufts University, Brandeis University, Williams College, Princeton University and the Institute for Advanced Studies. At each of those institutions, he gave at least one talk, generally on the fields of his research work, Elementary particle physics and Relativity. At the Colleges, he gave two or three more elementary talks for the students and more advanced lectures for the staff, among other activities, and spoke with the staff about teaching physics.[45]

Tiomno, as we have seen, had a lifelong commitment to improving physics education at all levels in Brazil, and public consciousness of science all over the world. He once expressed his attitude towards physics education in the broader sense as follows:

> ...When I was abroad, I was producing quite a lot; that would have been a part, perhaps an important one, within the machinery in which I, in reality, did not feel that I belonged [*i.e. the international research community*]. I would have been contributing to the universal scientific enterprise, that much would have been true, but I would not have been doing a thing which, for me, is very important: namely furthering the scientific development of my country. Thus, in Brazil, even though I accomplished much less than I could have done in research had I remained abroad, there was something which I did accomplish and which will continue even beyond my own death, which will be [*in fact*] my contribution to the development of scientific activity in Brazil. While I was retired [*by the military regime, in 1969*], I made a list of 50 physicists on whom I had had an influence that was at least significant, since it was recognized by them [*themselves*]. Currently, many of them are among the 80 or 100 best physicists in Brazil.

This gives me a very great satisfaction, the fact that I can see that those people are able to produce [*research results*] directly and through their own disciples; much more than what the few dozen or the hundred articles that I might have written had I remained abroad as a professor at this or that university would have contributed. That wouldn't have meant so much for Brazil, the output of such self-perpetuating work.[46]

Elisa shared Tiomno's ideas regarding physics education. But their personal characteristics were rather different: Perhaps due to his shyness, Tiomno remained distant from the students; he was closed up to them in personal terms. In contrast, Elisa was very open in her relations with her students, asking them about their private problems and inviting them to her home. In the words of their former student Mario Novello, "Elisa treated us like a mother".[33] Tiomno was very appreciative of Elisa's openness and warmth for their students, and helped her whenever possible. The students responded to Tiomno's and Elisa's dedication and enthusiasm, and the two of them had many loyal followers among their own current and former students.

Hard Times

In spite of his intensive and productive work, the beginning of the decade of the 1960s was a difficult time in the life and career of Jayme Tiomno. The Cuban revolution, led by Fidel Castro and Che Guevara, who overturned Fulgêncio Batista as Cuba's president on January 1st, 1959, intensified the cold war, with results felt in both Europe and Latin America. Fears that the communist sphere of influence might expand to include other countries in South America appeared as a danger signal to the US government, as well as to the elites in Brazil.

In January of 1961, Jânio Quadros took office as President of Brazil, with João Goulart (popularly known as '*Jango*') as Vice President. This represented an attempt to establish a coalition government: Jânio Quadros was supported by various conservative parties, while Jango belonged to the *Partido Trabalhista Brasileiro* (Brazilian Labor Party, PTB), rather left-leaning. However, in August 1961, just 8 months after taking office, Jânio Quadros suddenly announced his retirement from the presidency, alleging pressure from "occult forces". João Goulart was at the time on a trip to China to discuss possible commercial treaties. He was barely tolerated by the more conservative members of the military and the Congress, who wished to prevent him from assuming the presidency. This elicited a strong response from some of the state governors and some sectors of society, such as Leonel Brizola, in the state of *Rio Grande do Sul*, in the extreme south of Brazil, who threatened to take up arms to defend Jango's legal rights and the judicial system of the country. To moderate this conflict, the Brazilian Congress quickly passed a law reducing the President's powers and introducing a parliamentary system of government, with a Prime Minister. That system was however not very successful, and in January, 1963, a popular referendum was held on the question of whether the presidential system should be reinstated. The result was that over 80% of the voters favored a return to a presidential government.

In addition to the political conflicts, at the same time there was an economic crisis, with a very high inflation rate. The CBPF was facing a difficult period; salaries were low and there was a lack of funds for research. The support which had been promised by Jânio Quadros as a result of Tiomno's successful prediction of the K' meson was set aside when Quadros unexpectedly resigned. Of this difficult period, in the early to mid-1960s, Tiomno later said,

> Then, later, with the devaluation of the *cruzeiro*, and without new funds being approved, or an adjustment for inflation, the *Centro* began to fall into decadence. The salaries became lower and lower [*each year*]. There came a time when a tenured professor was earning barely more than 100 dollars [*per month*]. At that time, the *Centro* lost many professors, who went abroad. I had the opportunity to contact the President [*of Brazil, in 1961*], Jânio Quadros, who determined that a significantly larger amount of funding should be provided, in order to restore the original value of the support for the *Centro*. Unfortunately, this wasn't carried out, since he resigned a short time later[46]

During the period from 1960 until 1964, José Leite Lopes was Scientific Director of the CBPF.[36] He succeeded Jayme Tiomno in that office; Tiomno occupied it for a time in 1960. (Lattes had resigned as Scientific Director in 1955, before going to the USA for nearly two years). Leite Lopes, together with other scientists and scientific administrators, attempted to make use of Goulart's administrative reforms to establish a Ministry of Science and Technology, which would then absorb the CNPq and guarantee a consistent national science policy. He and others also proposed and supported the creation of the Latin-American Center for Physics (CLAF), to be financed by the Brazilian Government and UNESCO (along the lines of the ICTP in Trieste). It was in fact formally founded, but failed to receive sufficient support. The CBPF was also in need of renewed financial assistance, and that gave rise to various efforts to secure it during Goulart's tenure as President.

At the *Centro*, an attempt was made to take advantage of the influence of the anthropologist Darcy Ribeiro, who was close to Goulart's government. He had also been President of the CBPF during a brief period near the end of the 1950s. In August of 1962, he assumed the office of Minister of Education in the government of João Goulart, and a few months later, he became Goulart's Cabinet Minister, with direct access to the President; but the times were not easy for finding financial support.

In the midterm elections in 1962, when the National Congress and the regional parliaments were up for reelection, Mario Schenberg, who had already been a representative for a brief time in 1947, was again elected to the State parliament of São Paulo. However, since the Communist Party (PCB) was still illegal, he ran on the ticket of the Brazilian Labor Party, the party of Goulart. But he, along with three other representatives from the PCB, were prevented from taking their seats in the parliament, since they were considered to be "notorious communists".

After being reinforced in his office by the 1963 popular referendum, President Goulart began a program of reforms, including an agricultural reform, and maintained very independent international policies, provoking a reaction from the conservative forces, both military and civil. This eventually resulted in the takeover of the government in a *coup d'état* on March 31st, 1964. Evidently the US secret service (CIA) and the State Department supported the military takeover. (There is a strange story concerning the precise date of the takeover: In fact, the *coup d'état* took place on April 1st, the traditional 'April fools' day'. Afraid of appearing ridiculous and wanting to protect the credibility of their movement, the military forces claimed that the 'revolution', as they termed it, had actually begun on the previous night—March 31st).

Following the takeover, João Goulart fled into exile in Uruguay, and many other politicians had their rights to political action curtailed; thousands of people were arrested, including many members of the military who were suspected of having opposed the *coup*. At the same time, a general repression of all those areas held to be 'leftist' began, such as the *União Nacional dos Estudantes* (the National Students' Union), the *Ligas Camponesas* (an organization of agricultural workers and small farmers), the *Central Geral dos Trabalhadores* (a united workers' organization), and the *Juventude Universitária Católica* (the Catholic University Youth organization). The CLAF was also regarded with considerable suspicion by the military government due to its international (Latin American) character.

From the early 1960s, the CBPF was presided over by another Admiral, *Octacílio Cunha*, who had been the head of the Brazilian Nuclear Energy Agency. Cunha was elected by the senior scientists at the *Centro*, as was the custom owing to its democratic structure, but he was a supporter of the military regime and began to show authoritarian tendencies and to dismantle precisely that democratic structure. The universities, as well as various research institutes, were considered to be centers of dissemination of leftist doctrines, and became the victims of the so-called 'Operation Clean-up', which consisted of expelling all those who were held to be 'subversive'. A number of those institutions were invaded, their libraries destroyed and their personnel arrested. There were various bizarre occurrences caused by the eagerness to discover and destroy 'communist material', leading to the trashing of works such as Stendhal's 'The Red and the Black', or 'The Red Circle' by Arthur Conan Doyle, or any publication that had a red-colored cover.

The climate at the *Centro* had become very difficult. Leite Lopes resigned from the post of Scientific Director of the CBPF, permitting his old colleague who had come with him from Recife—*Hervásio de Carvalho*—to assume that position. Carvalho benefited from the military regime, occupying several positions in the government during the dictatorship. Tiomno, who was nominated for another term as a member of the Commission on Physics Education during the 11th IUPAP General Conference in Warsaw, September 18th–23rd, 1963, sent a letter to Sanborn C. Brown, resigning from the post.[47]

The increasingly authoritarian leadership, together with continuing financial problems, contributed to a decline of the *Centro* in the mid-1960s, compounded by the military regime and its accompanying general repression of freedom of expression in the years after 1964. This, in turn, led to the loss of much of the faculty and of many students from the *Centro*. Those faculty members who saw the 'handwriting on the wall' and who were able to do so sought positions elsewhere, often going abroad, where they were able to find more secure positions, even before the takeover that led to the military regime.

The situation in the Faculties was no better. Plinio Süssekind Rocha, at the Department of Physics of the FNFi/UB, and Mario Schenberg, at the Department of Physics of USP, were both arrested. The international scientific community began to mobilize a protest action directed at the Brazilian government. In August (1964), while Leite Lopes was attempting to obtain a passport so that he could go to France, he was also arrested. Tiomno's telegram to Wheeler was short and direct: "Leite Lopes arrested. Please contact Peter Bergmann".[48, 49] Wheeler answered that Bergmann was in Europe taking part in the *Congress for Cultural Freedom* in Paris. Oppenheimer sent a telegram to Bergmann, asking him to try to mobilize the participants at the Congress. At the same time, Wheeler insisted that Tiomno should accept his invitation, originally sent in April, to come to Princeton.[50] Soon after, Leite Lopes was released and left for France.

The Dean of the Faculty (FNFi) in Rio, *Eremildo Luiz Viana*,[51] a radical conservative, saw the 'Operation Clean-up' as an opportunity to rid himself in one fell swoop of all those whom he held to be communists, as well as those who were his potential rivals in internal academic politics or whom he simply disliked. Eremildo Viana had been Professor of Ancient and Medieval History at the FNFi/UB since 1946. He became Dean of the Faculty in 1957.

Viana was involved in an action[52] soon after the military *coup d'état* of March, 1964, during which he occupied a radio station operated by the Faculty—with the aid of military forces—claiming that it was the focus of a subversive conspiracy, and had its previous director, the history professor *Maria Yedda Linhares* removed from her post, taking over the station himself. He became a figure symbolic of the repression practiced by the dictatorship in the universities.

Even before the imposition of the dictatorship at the end of March 1964, Viana submitted a denunciation containing the names of 44 professors at the FNFi, with the title 'Communist professors at the *Faculdade Nacional de Filosofia*'. It listed all those who were supposedly members of a communist cell, allegedly called '*Anchieta*', which was claimed to operate mainly in the physical chemistry laboratories.

On April 19th, a few days after the *coup d'état*, the Minister of Education, Flavio Suplicy de Lacerda, ordered that every university was to establish an investigation commission to search for subversive and communist elements within the institution. At the UB, the 'Investigation Commission of the *Universidade do Brasil*', headed by General *Arcy da Rocha Nóbrega* was set up, and Viana's denunciation was sent to that Commission. He claimed that the alleged leader of the 'cell' was the occupant of the Chair of Physical Chemistry, *João Christóvão Cardoso*, a former president of the CNPq and a potential candidate for Dean of the Faculty, and thus a competitor of Viana's. The list included some of Cardoso's assistants, including Paulo Kobler Pinto Sampaio, Walter Faria, and Jayme Tiomno's youngest sister, Silvia Tiomno Tolmasquim. It also listed various 'malcontents', including Jayme Tiomno and Elisa Frota-Pessôa, José Leite Lopes as well as his wife, the mathematician Maria Laura Mousinho Leite Lopes, and Darcy Ribeiro, among others.

Nóbrega took no immediate action, since the denunciation contained no supporting evidence and seemed to him to be mainly a statement of personal animosities on the part of Viana.[53] In the end, the Commission to which the list had been submitted concluded that Eremildo Viana himself was the only member of the faculty who should be disciplined, under the charge of false accusations (perjury) and of misuse of public funds during his term as Dean of the Faculty. The inquest concluded that the professors on 'Eremildo's list', as it was now called, were free of any guilt and that the 'Anchieta Cell' had never existed. Legal action against Eremildo was however never taken, since he was considered to be a 'person of distinction' by the military regime.

Elisa later described her first meeting with Viana in the late 1950s, just before he was first nominated to be Dean of the Faculty (FNFi):

> ... Lattes asked me, 'Wouldn't you like to have a meeting with the personnel?' And I answered, 'I don't have time!' In truth, I find that in all the institutions that I know well, if they had only a tenth as many meetings, their problems would be discussed more effectively, since the meetings would be better organized. Then Leite [*Lopes*] said, 'I want you to at least come to lunch with Eremildo Viana, so that you can get to know him, since he will be the salvation of this Faculty. He is a good fellow, young, enthusiastic, and that is just what we need'. We went to lunch together, at the 'Philosophy Bar'.[54] After the lunch, I said to Leite, 'If that Eremildo becomes part of the Direction [*the administrative division of the FNFi*], that will be the end! That man is convinced of his own power and has a thirst for it'. I couldn't stand Eremildo – by the way, Leite Lopes later radically changed his own opinion of him.[55]

In 1965, the University of Brazil was reorganized and transformed into the '*Universidade Federal do Rio de Janeiro*' (UFRJ). Three years later, the FNFi was separated into different schools and institutes and spread across the city. The History Department was removed from the FNFi and integrated into the new '*Instituto de Filosofia e Ciências Sociais*' (IFCS), located in the center of Rio, and Eremildo Viana again became the director of that Institute, remaining in that position until the end of the dictatorship in the mid 1980s. At the same time, the Institute of Physics and the Institute of Chemistry were established, and both of them were transferred to a location far from the center of the city (the island of *Fundão*), where a new university campus was under construction.

Notes

[1] José Leite Lopes to Eremildo Viana (Dean of the FNFi), August 1st, 1960; [JLL].

[2] Jayme Tiomno to José Leite Lopes, October 14th, 1960; [JLL].

[3] Elisa Frota-Pessôa to Eremildo Viana (Dean of the FNFi), March 12th, 1964; [EFP].

[4] Jayme Tiomno to Elisa Frota-Pessôa on September 23rd, 1956; [EFP]. Tiomno reports in his later memoir Tiomno (2005) and [JT107] that he had an intuition already in 1956 that there would be another K-meson with positive parity, and it became the K' particle which he later predicted, with Videira and Zagury, in 1961 ([JT35]).

[5] Rochester and Butler (1947).

[6] Bhowmik et al. (1959).

[7] Interview of Nicim Zagury given to ATT on March 18th, 2019.

[8] Jayme Tiomno to Elisa Frota-Pessôa. August 27th, 1960; [EFP].

[9a] See Alston, Alvarez et al. (1961).

[9b] The resonance found by Alvarez et al. was at 885 ± 3 MeV, corresponding to 1731.9 electron masses. The particle is now called K*(892), 892 MeV being the current accepted value of its mass (1745.6 electron masses). See the Particle Data Group (PDG) site, http://pdg.lbl.gov/2013/listings/rpp2013-list-K-star-892.pdf.

[10] Tiomno (1966a).

[11] 'Mais uma do átomo' ('Another one of the atoms'), in Visão Magazine, May 19th, 1961, p. 21.

[12] Jayme Tiomno's Report of activities between July 1st and August 15th, 1961; [JT].

[13] [JT37].

[14] Jayme Tiomno to Elisa Frota-Pessôa, July 11th, 1961; [EFP].

[15] cf. 'O Cruzeiro', edition No. 45, August 19th, 1961, pp. 19–21.

[16] Copy of the petition quoted, without signatures, August 1961; [JT].

[17] Darcy Dillenburg, Gerhard Jacob and Th. Maris to José Leite Lopes, August 17th, 1961; [JT].

[18] Note to the press: Refutation by Prof. Jayme Tiomno of recent declarations made by Prof. César Lattes. August 23rd, 1961; [JT].

[19] Tiomno to Robert Sachs, Madison/WI, November 4th, 1961; [JT].

[20] [JT36].

[21] [JT38].

[22] cf. Fraser (2008) for a biography of Abdus Salam.

[23] See [JT29].

[24] [JT34].

[25] [JT37]; see also Chap. 9.

[26] Leite Lopes (2004); in this connection, see pp. 13–18.

[27] Haakon H. Storhaug to Jayme Tiomno, May 17th, 1962; [JT].

[28] P. N. Bhandari, Personnel Director, IAEA Vienna, to J. Tiomno, March 22nd, 1963; [JT].

[29] Jayme Tiomno to R. W. Lefler, June 20th, 1962; [JT].

[30] Abdus Salam to J. Tiomno, December 18th, 1964; [JT].

[31] The history and development of the ICTP in the decades following its founding have been summarized by Roederer (2001).

[32] Jayme Tiomno's youngest sister, Sylvia Tiomno Tolmasquim, reports that he began working with students early in his university career, giving individual lessons to those who were having difficulties with the material, and later taught physics courses at private secondary schools. However, after obtaining his *licenciatura*, he refused a well-paid position as secondary-school teacher at a public school in order to concentrate on his research. But his interest in secondary-school instruction continued, albeit in a latent form, and he took great interest in his sister's progress at the Colégio Pedro II, discussing her schoolwork and her teachers, many of whom were acquainted with him.

[33] Interviews with Mario Novello for this book by ATT, August/September 2019.

[34] Leite Lopes and Tiomno (1953).

[35] Blackwood et al. (1962).

[36] Leite Lopes (1977). See also [26].

[37] José Leite Lopes (as Scientific Director of the CBPF): Report on Tiomno's attendance at the 10th General Assembly of IUPAP, Ottawa, Canada, Sept. 1960, dated 1.12.1960. JT Archive, MAST; [JT]; Jayme Tiomno, Report on the 10th General Assembly of IUPAP, Ottawa, Sept. 1960, to the President of the CBPF, dated 12.11.1960. JT Archive, MAST; [JT].

[38] Brown et al. (1964).

[39] For a pdf of the lecture, cf. http://mitp-content-server.mit.edu:18180/books/content/sectbyfn?collid=books_pres_0&id=7236&fn=9780262523769_sch_0001.pdf. See also [JT39] and [JT40].

[40] cf. https://books.google.de/books/about/Why_Teach_Physics.html?id=tv7LQgAACAAJ&redir_esc=y.

[41] Feynman's 1963 talk was published separately as: R.P. Feynman, 'The Problem of teaching Physics in Latin America ', in Engineering and Science, Vol. 27, p. 21 (1963); cf. Feynman (1963).

[42] cf. Castro Moreira (2018).

[43] Feynman, Leighton and Sands (1964–2006).

[44] See Mehra (1994), Chaps. 22 and 25.

[45] News item from the American Institute of Physics, February 26th, 1962; [JT].

[46] Tiomno (1977).

[47] Jayme Tiomno to Sanborn C. Brown (President of the Commission on Physics Education, MIT), April 13th, 1964; and Sanborn C. Brown to Jayme Tiomno, May 19th, 1964; [JT].

[48] Telegram from Jayme Tiomno to John Wheeler, August 1964.

[49] Bergmann was an assistant to Einstein at the IAS in Princeton from 1936 to 1941. By 1964, he was a professor at Syracuse University. Tiomno evidently hoped that he would have some political influence to aid in securing the release of Leite Lopes, whom Bergmann knew well from the former's visits to Princeton.

[50] Telegram from John Wheeler to Jayme Tiomno, August 12th, 1964.

[51] Viana's family name is often found in the literature as 'Vianna'. Similarly, his middle name, Luiz, is often spelled as 'Luis'. We have used the simpler form, 'Eremildo Luiz Viana' consistently.

[52] See Moraes Ferreira (2014).

[53] Gama Pereira (2008).

[54] This bar still exists today, and has kept its name, although its namesake, the FNFi, ceased to exist in 1968.

[55] Frota-Pessôa (1990).

11

A Utopian Academy

A New University

The colonization of Brazil, like that in many other places, began at its coasts, starting in the east and continuing north and south along more than 7,300 km of coastline. After the end of the Empire in 1889, the new republican government had plans to move the capital from Rio de Janeiro, on the southern coast, to a location in the interior of the country. This would encourage a movement of population to the central parts of the country, which were very sparsely inhabited, and would represent a further break with the royalist tradition that was omnipresent in Rio de Janeiro. As early as 1892, a commission of astronomers from the National Observatory cartographed a region called the *Planalto Central* ('high central plain'), more than 1,100 km from Rio, delineating the territory where the future capital was to be erected. However, the move there was accomplished only much later, under the presidency of Juscelino Kubitschek, elected in 1955. The construction of the new capital – *Brasília* – became the trademark of his administration, anchored to the motto of 'developmentism'. He invited the urban planner *Lucio Costa* to work out the plans for the new city, and architect *Oscar Niemeyer* to design the main structures, such as the Congress, the Supreme Court, the seat of government and the ministries, as well as a cathedral, a museum, and a university.

Coordinating the project of establishing the future university was entrusted to the educator *Anísio Teixeira*, who had been responsible for founding the *Universidade do Distrito Federal* (UDF) in Rio de Janeiro during the 1930s. Created as a Foundation on December 15th, 1961, the University itself was

© Springer Nature Switzerland AG 2020
W. D. Brewer and A. T. Tolmasquim, *Jayme Tiomno*, Springer Biographies,
https://doi.org/10.1007/978-3-030-41011-7_11

initiated on April 21st, 1962, two years after the new capital had been officially inaugurated. The plan was supported in particular by the ethnologist/anthropologist *Darcy Ribeiro* (Fig. 11.1).

He shared the vision of a modern university with the plan's principal author (and his mentor), Anísio Texeira. Its conception and planning had already begun in 1958, and many scientists and intellectuals had been called upon to participate in the planning. A detailed outline of its academic and administrative structure (the '*Plano Orientador*', Plan of Orientation) was prepared. It was a new project in Brazilian higher education, with similar goals but a much grander scope than the short-lived UDF.

In contrast to the other existing universities in Brazil at that time, whose principal function was the training of professionals, the *Universidade de Brasília* (UnB) had as its primary objective "the education of citizens capable of searching for democratic solutions to the problems confronting the Brazilian people in their struggle for economic and social development". In second place came the goal of the education of research scientists, and only after that the training of professionals.

Fig. 11.1 The Inaugural Meeting of the. UnB, 1962. The speaker (*standing, with microphone*) is Darcy Ribeiro. Seated are the Minister of Education and Culture, the Mayor of the Federal District, and the Cabinet Chief of the President. *Source Archivo Central, UnB, AtoM UnB*. Used under Creative Commons license 4.0 International.

In order to attain its objectives, the University was to be governed "by the principles of freedom of investigation, of freedom of teaching and of expression" and should maintain itself "true to the requirements of the scientific method and always remain open, with the objective of increasing knowledge, to all the currents of thought, without participating in political or partisan groups or movements". [1] Its functional structure was also rather innovative: It was based on eight Central Institutes (Arts, Letters, Humanities, Biology, Geosciences, Mathematics, Physics and Chemistry) as well as various faculties.

The central institutes were to be spaces oriented towards research and the education of future academics, and would provide the basic courses for all the students in their first two years of study. Those who wanted to specialize in academic teaching or in research would remain in the central institutes, while the others could obtain their professional education from the faculties. In this way, research could be combined with teaching right from the beginning. Furthermore, each student would be assigned on admission to a tutor who would help him or her in finding the appropriate curriculum and choosing suitable courses. There were some courses of basic education which would be obligatory for all the students, and others which could (and must) be chosen individually, depending on the interests and goals of the student. In the first two years, the students would be required to register for at least one of the latter during each semester. In addition, it was planned that the graduate courses would be initiated soon after the opening of the University, so that teaching assistants could be drawn from its own ranks and not taken from other universities. All of the central institutes would be combined within the Central Institute of Science (ICC), in a building designed by Niemeyer as its first (and still largest) major edifice on the original campus, often called the 'Minhocão' ('giant worm') (Fig. 11.2).

The Universidade de Brasília would not be organized with a professorial system, like the other Brazilian universities at the time, where each full professor remained the single authority responsible for his/her area (Chair) as long as she/he was active. Instead, the UnB was to be governed by commissions and councils, in which all of its members would be represented. As in the model used for the CBPF, there would be a Directorate Council above the Rector, to be occupied by well-known intellectuals and entrepreneurs, for the most part from outside the University. (This is much like the governance of the University of California, for example, where the Board of Regents and its President have authority over the Chancellor of each campus). In administrative terms, the University was established with the express intention of

Fig. 11.2. The building for the Central Institutes (ICC) at UnB, designed by Oscar Niemeyerand built in the years 1963–1971. Courtesy of the *Universidade de Brasília, Arquivo Central*, AtoM UnB, Fotográfico—2A58-00,729-08. Reproduced with permission *Source* Arquivo Central. AtoM UnB. https://onsomething.tumblr.com/post/134878055996

avoiding the bureaucratic interference and incompetence which were problematic in other state-supported, directly-administered institutions.

For physics, the CBPF group was called upon to join the project, and a planning commission was established, consisting of *Guido Beck*, *Gabriel de Almeida Fialho*, *Jayme Tiomno* and *José Leite Lopes* (chairman),[2] as well as external consultants, among them *Roberto Salmeron*. It played a significant role in the early conception and planning of the physics institute. The CBPF was already facing a deep financial crisis, and they considered the possibility of transferring the Center to the new Physics Institute at UnB, moving the equipment, the whole library and all the researchers to the new university. Many years later, in an interview, Tiomno commented that in Darcy Ribeiro's plan, one person would initially go from the CBPF to the UnB during the founding process. "Thereafter, successively, in the following years, the classes would increase in size, and a number of physicists from the CBPF would go [*to Brasília*], setting up research groups there, so that within five years, the Institute of Physics would be operating at full capacity".[3]

Darcy Ribeiro was the founding Rector of the UnB, and Anísio Texeira presided over its Foundation and the Directorate Council; he then took over as Rector 6 months after the official founding of the UnB, when Darcy Ribeiro became Minister of Education (and then also assumed the duties of President of the Directorate Council; i.e. they exchanged their positions). After the founding of the UnB, the task of filling it with life—with teachers, researchers, and students—was begun. It commenced operations in April, 1962, with initially only the Humanities as well as Arts & Letters, and this continued throughout the following year. In 1964, various other curricula were initiated, including Chemistry, Mathematics, and Biology. Thus, in its third year of operation, the University was already offering 19 possible majors.[4]

The Physics Institute

In late 1963, the Brazilian physicist Roberto Salmeron arrived from Geneva with his family to become Director of the *Instituto Central de Física Pura e Aplicada* (ICFPA), the new Physics Institute of the UnB. He had been thinking of returning to Brazil, and this provided an excellent opportunity. Salmeron began his work in January 1964, and shortly thereafter, he was asked to take on the additional function of General Coordinator of the Central Institutes, as a sort of Dean of the Central Institutes (ICC).

Roberto Aureliano Salmeron is a physicist and engineer from the same generation as Jayme Tiomno and Elisa Frota-Pessôa. He was born on June 16th, 1922 in São Paulo, and studied electrical engineering at the Polytechnic School there before continuing with physics at the FNFi/UB in Rio de Janeiro. He completed his undergraduate studies in 1947. That same year, he returned to São Paulo to serve as instructor at the Polytechnic School, and as research assistant in Gleb Wataghin's department at USP, studying cosmic rays for three years. In 1950, after Wataghin's departure, he joined the recently-founded CBPF in Rio, so that he also numbers among the founding scientists there. He worked at the *Centro* for another three years before going in 1953 to the University of Manchester/UK to write his doctoral dissertation under the guidance of Patrick Blackett, who had in the meantime gone there from Cambridge. After receiving his doctorate at Manchester in 1955, he became one of the very first researchers at CERN, the European Cooperation for Nuclear Research, in Geneva.[5] Roberto Salmeron had a permanent contract as a senior researcher at CERN by 1963, when he was invited to return to Brazil and contribute to the project of establishing the *Universidade de Brasília*, for which he had been an external consultant in the field of physics. Salmeron had a similar idealism and devotion to the goal of enabling scientific research and education at a high level in Brazil as was manifested by the other founders of the CBPF.

However, in March of 1964, three months after Salmeron arrived in Brasília, and while the University was still in its initial stages of development, the military took over the Brazilian government in a *coup d'état*, as we have seen. The *Universidade de Brasília* was thought by the armed forces to be a center of subversion. Its founding principles of freedom of education and research were viewed as a private license for disseminating Communist ideas. Troops occupied the University campus, its Rector Anísio Teixeira was removed from office, all the members of the Directorate Council were dismissed, and 15 professors were arrested. Anísio Teixeira went to Columbia University and later to the University of California, returning to Brazil in 1966. In 1971, in one of the most violent periods during the military regime, after having gone missing for two days he was found dead in the elevator shaft of the building where a friend lived, whom he had planned to visit. Many years later, it was discovered that he had been arrested by the military forces before being found dead.[6] Darcy Ribeiro, for his part, went into exile in Venezuela, Chile, and Peru, returning to Brazil only after the end of the military dictatorship.

An 'intervenor' was named as the new Rector of the UnB—*Zeferino Vaz*, a professor of veterinary medicine from USP. In addition, a military tribunal was set up to investigate the UnB (*Inquérito Policial Militar*, IPM, a 'Military Police Inquest'). Nevertheless, the inquest was unable to determine the existence of any subversive groups, and was retired after a time. In spite of the tensions, the rector promised that there would be no new invasions of the campus, and the new courses were initiated as planned. Thus, Salmeron invited Jayme Tiomno and Elisa Frota-Pessôa to begin organizing and teaching the physics courses, and Tiomno was also named Coordinator for establishing the Physics Institute. They believed that the worst was over, and expected the situation to normalize more and more as time passed.

In fact, other scientists from the CBPF were expected to join the small group of professors at the Physics Institute of UnB. But owing to the *coup d'état* in 1964, and the political instability of the country, various researchers at the CBPF were still thinking over whether or not they should go to Brasília, and others, like Leite Lopes, were in exile abroad. Tiomno and Elisa had considered the situation before leaving Rio and had decided to go, but in addition that it would have to be " 'all or nothing'; either we would start up all the courses in the following year, from the first through the fourth undergraduate years, plus the graduate courses, or else we wouldn't go there at all. We realized that the situation would be so unstable that if we didn't succeed in starting up the department immediately, it would later have difficulties in growing at a normal pace".[3]

Despite the difficult political situation that the country was facing, they still cherished the dream of the University of Brasília as it had been planned: a modern university for teaching and research. Tiomno would later call it a 'university for development', or a 'super-university'. They were very enthusiastic about its conception and shared with their students the dream of a new university with research as one of its principal activities, providing graduate courses. Students who had already completed their undergraduate education also joined them; they could begin graduate work at the UnB and be contracted as teaching assistants to Salmeron, Tiomno and Elisa, while those who were still undergraduates were engaged to give lessons at various high schools in the city, but also participated in teaching the still-younger students. They all wanted to be a part of this ambitious project.[7] Many of the students had given up or refused jobs paying the order of a million *cruzeiros* per month to work as teaching assistants and attend their own graduate courses, with salaries of around 200 thousand *cruzeiros* monthly. "We never talked about our employment", said Mario Novello, one of those students, about their motives for following Tiomno and Elisa to Brasília, explaining, "It was simply the possibility of doing fundamental physics. Tiomno exuded fundamental physics. 'I am preparing you to do fundamental physics', he would say. That was an honor for us".[8]

Many years later, Elisa also recalled their move to Brasília:

> We left [*Rio*], on leave from the Faculty [*FNFi*] in 1965, and went to [Brasília], taking more than 30 students with us; they left in the middle of their courses and went along with us.
>
> We took more than half of the students in my class, all those who agreed to leave Rio and go with us to Brasília; among them were: *Carlos Alberto da Silva Lima, Mário Novello, Sérgio Joffily, José Carlos Valladão de Matos, Marcelo Gomes, Maria Helena Poppe de Figueiredo* and *Miguel Armony*. Once there, we managed to make arrangements to support them: Since there were many government schools that were lacking in teachers, the UnB made an agreement for them to teach in the public schools and to earn their living that way. The pupils in the *ginásio* [*high school*] in Brasília received lessons from those students. It was a lovely time.[9]

The students that Elisa mentions later on started a, 'fan group', called GOELTI-64.[10] She modestly fails to mention her own daughter, Sonia, who was the eighth member of that 'Group of Eight'. One of the other members later said of this time, "The enthusiasm and the will to support the transformation of our society moved some of us to follow our mentor-teachers to the desert of *Brasília* in 1965, already completely dominated by the military forces that had beset our country the year before, and which would persist for the following twenty years".[11]

And the situation was far from being settled. There was a lack of laboratory equipment and technicians. The budget of the University was some 4.2 billion *cruzeiros*, roughly 1/20 of that of USP, which was around 70 billion *cruzeiros* at the time. The central library was very small and the students had no study halls. Some of the laboratories were improvised, including the physics teaching labs. The lack of infrastructure at the University was exacerbated by the shortage of housing in Brasília. The professors and their families had to make do with hotel rooms or rented private rooms. Often, the financial resources were not sufficient to pay the operating expenses of the university. Furthermore, Brasília, in spite of its official inauguration as capital, was still a city under construction, with dirt roads and many building sites still unfinished. "…Many times we thought of you, burdened with the difficult task of bringing science to the heart of Brazil and teaching those hopeful students who had arrived from every corner of this immense country", wrote the Argentine physicist/chemists Myriam and Mario Giambiagi to Tiomno.[12]

All of these problems nevertheless failed to diminish the spirit and the vision of the dimensions and the importance of the project that they were carrying out at the *Universidade de Brasília*. The memories kept by Elisa of those first months in Brasília were dominated by her enthusiasm for what they were constructing:

> In Brasília, I got to know people working in plastic arts, in music and in many other areas. We all had the same spirit, we all felt like part of a family. At the FNFi, I never felt like that. Everyone at the UnB had the same enthusiasm for helping the younger people to make progress—the opposite of what happened here at the FNFi. At the *Universidade de Brasília*, all of these people were gathered who simply wanted to do something different in Brazil. And, in fact, we were managing to do it there. It was a year of dreams at that University.[9]

Indeed, the University was in a state of effervescence. It was experiencing what Salmeron called "contagious enthusiasm". He had already worked at a number of different institutions and universities in a variety of countries, but

he maintained that he had never seen such enthusiasm as what he encountered at the *Universidade de Brasília*.[2] It offered courses, lectures, films and
concerts to the local population. Claudio Santoro, Director of the Music
Department, established a complete (although small) symphony orchestra.
Another professsor of music, Yulo Brandão, set up a baroque ensemble.

There was likewise intense activity at the Physics Institute. The faculty
of that Institute was composed of Tiomno, Elisa and Salmeron. Salmeron
indeed was not only professor of physics, but at the same time general coordinator for all the central institutes.

In spite of being only three professors, they resolved to offer courses at
all undergraduate and graduate levels to accomodate all of the students, who
were at different stages in their studies. Besides those students who had come
from Rio, they had other students from many cities within Brazil and Latin
America. They developed collaborations with the CBPF and with the *Université de Paris*, and briefly began a partnership with the University of California.
Furthermore, it was planned to organize the next Latin-American School of
Physics (ELAF) at the UnB the following year; it would last for two months
and was organized in turn in Mexico, Argentina and Brazil. Salmeron recalls
this time:

> ... I received inquiries from physicists who had university or industry positions
> in the USA, others who worked in England. Ten recently-graduated engineers
> from the ITA [*Institute of Technological Aeronautics*] in São José dos Campos
> wanted to all come together and set up a group specializing in systems research.
> In my function as Coordinator, I was contacted practically every week by some-
> one interested in working at the University [*UnB*].[13]

The workload was very heavy. Entering the first year course, the majority
of students had come from different parts of the country to matriculate at
the UnB; the first-year class had 100 students. In the second year, there were
students from Brasília and others who had come with Tiomno and Elisa from
Rio. The third- and fourth-year classes were composed almost exclusively of
students from Rio. All together, there were around 200 students majoring in
physics. The lectures were two hours long and each of them (Tiomno, Elisa
and Salmeron) gave courses for four different years at the undergraduate level.
Apart from that, it was the instructors, the aides and the assistants who set up
the problem sessions, developed and corrected the exercises and monitored
the lab courses. Elisa had also set up the teaching labs as well as research
labs for experimental physics. "We had no other choice, because this was the
price we were paying for having a structure that would permit various other

physicists to return from abroad [*and join the UnB*], the next year or in later years",[3] explained Tiomno.

A teacher from the Engineering School in Pará (later part of the *Universidade do Pará*, in the north of Brazil), *José Maria Filardo Bassalo*, was one of the students invited to the UnB by Jayme Tiomno to complete his own studies in physics, and he was also hired as a teaching assistant; he well remembered the intense daily routine there:

> Our life as students at the ICFPA/UnB was rather hard. For example, in addition to solving about 40 exercises per week in the disciplines of the courses I was taking, I had the duty as instructor (together with *Carlos Lima* and *Miguel Armony*), [*other teaching assistents*] for Prof. Salmeron in the courses 'Physics I and II', of correcting in turn the 600 exercises of his roughly 200 students.[14]

As Tiomno later recalled,

> We carried out something which I really believe to be unique in the history of Brazilian universities, or of North American or any other countries'. Three tenured professors, Salmeron, Elisa Frota-Pessôa and I, set out to put into operation [*within one year*] a complete physics curriculum, from the first through the fourth year, as well as post-graduate courses.[3]

The great majority of the professors who were invited to go to Brasília were well-known names in their areas of research, both nationally and internationally. And they were full of the same energy and will to construct something new in their country, just like Tiomno, Salmeron and Elisa. The results were impressive. After three years of operation, the *Universidade de Brasília* already offered 19 undergraduate majors and had 1,600 students. Among them were 156 graduate students, quite possibly the highest number at any of the Brazilian universities of that era. It had granted 31 Masters degrees and there were 6 doctoral theses waiting to be evaluated. Various international foundations were beginning to pay attention to the new university. It had begun to receive financial aid from the Ford Foundation, the Rockefeller Foundation, UNESCO, the National fund of the United Nations, The International Atomic Energy Agency, the French Technical Cooperation, the Atomic Energy Commission of the USA, and the UN/FAO, among others.

The Beginning of the End

In spite of (or perhaps because of) the good results that were being obtained at the *Universidade de Brasília*, it remained under threat. The country was ruled by a dictatorial government, which had as one of its main preoccupations to 'throw overboard' (i.e. to eliminate) all persons and activities which it considered to be its potential enemies, or who could represent some sort of danger to the regime—the so-called 'subversives'. In May of 1965, the Catholic professor *Ernani Maria Fiori* was contracted to give lecture courses in philosophy. There was pressure from the security services for the Rector to cancel his contract, since he was held to be a communist sympathizer. A student was also expelled from the University by orders from outside. The faculty and student body were disturbed by this external interference. Admissions and hiring, as well as dismissals, should be carried out only according to academic criteria; and furthermore, it was essential that the Rector reconstitute those collegial functions that had been dissolved in April 1964. The students began a strike, but it was ended by the promise that the demands of the students and the faculty would be heard. At this point, pressure from the military authorities became very strong, and Rector Zeferino Vaz made use of the vacation period in July to dismiss the professor in question. A renewed strike by the students caused him to resign from his post as Rector. He had put himself in a very delicate position *vis-á-vis* the security services by not consulting them before offering a contract to Fiori, causing the whole difficult situation.

Although there was a climate of instability and uncertainty, no-one could have imagined the series of events that were to occur following the installation of the new Rector, *Laerte Ramos Carvalho* (a professor of education from USP) in early September. In the middle of that month, Tiomno traveled to the UK to participate in the *Conference on Elementary Particles* in Oxford. He took advantage of the trip to speak with various physicists about the project which was underway at the *Universidade de Brasília*, and invited some of them to spend some time at the new university. From Oxford, he continued on to Paris and Geneva.[15]

Within a short time, the new Rector decided to dismiss two professors, *Edna Soter de Oliveira* and *Roberto Décio de Las Casas* from the Humanities Institute, in order to "eradicate the subversive focal points". Once again, professors and students demanded that proposed dismissals be justified only by applying academic criteria. The Coordinators of the central institutes and faculties tried to arbitrate, and asked the students and professors to give a vote of confidence to the new Rector. At first view, this was in agreement with the demands brought to the Coordinators. But in the following sequence of events, probably due to external pressure, the rector dismissed Las Casas and reproached the Coordinators for trying to put him under pressure. As a result, the Coordinators felt unable to carry out their duties and asked to be released from them. They meanwhile compromised themselves by continuing to respond to the institutes on a day-to-day basis, to avoid paralyzing the activities of the University. At the same time, the students decided to stage a general strike and the faculty planned a work stoppage for 24 h (the military government had promulgated a decree rescinding the right of employees to strike).

On the 10th of October, by request of the Rector, armed forces invaded the University (Fig. 11.3). Anyone moving on the campus was stopped and searched; members of parliament and representatives of international organizations were prevented from entering the campus. The bus service that operated within the campus was suspended. Many professors and students lived on the campus; their children were not allowed to go to school, and people who needed to travel had to do so on foot. The French physicist Michel Paty and the Indian architect Shyam Janveja were arrested. Four biology professors who had gotten out of their car were taken to the precinct station and submitted to interrogations concerning their 'subversive activities'. The Rector and the military forces were convinced that the UnB was a 'focus of subversion' and claimed that they were 'eradicating' the causes of the supposed subversion.

Elisa later recalled of this time,

> I remember a Colonel Lázaro ... he had been ordered to close the University. And he said, 'I am closing this University with tears in my eyes, because my son, before this University came here, before you all arrived, was in a group that was on a very bad path. Nowadays, he is studying [*here*] and is enjoying his studies; but I can't tolerate subversion'.[9]

The Collapse

On the 18th of October, the Rector summarily dismissed an additional 15 professors. This produced a situation which the rest of the faculty could no longer tolerate. It was impossible to carry out the project of setting up a university with the uncertainty that at any moment, a professor could be dismissed or arrested simply because the security services of the military government considered that he or she might have subversive attitudes or contacts. As a result, 210 professors, more than 80% of the faculty, resigned from the

Fig. 11.3 The occupation of the UnB in October 1965 by the military. Courtesy of the *Universidade de Brasília, Arquivo Central*, AtoM UnB, Fotográfico—1A2-00,007-01. Reproduced with permission Source: Arquivo Central. AtoM UnB.

University. Jayme Tiomno explained the reasons for their collective resignation:

> A purge within the University, carried out by its own Rector; that is not in any way acceptable. If there had been a purge from outside, like the first one, there might still have been some hope that the intervention would be limited, even admitting that there would be losses at the scientific level and all the rest. If it hadn't continued, we might have stayed on, taking the long view; for Brazil, it was much more important for us to stay on than to leave. But a purge from within, carried out by our own Rector, which wiped out all the institutions for collective decision-making and left [*that Rector*] in reality as the only one to make decisions, that left us with no option of remaining there.[3]

A Parliamentary Commission was set up to investigate the crisis that was occurring at the University. Scientists and intellectuals from various locations in Brazil and abroad began sending messages of solidarity to the professors who had resigned. Tiomno received tens of messages sent from physics departments at the Universities of Rome, Milan, Turin; and from the Sorbonne, Buenos Aires, Mexico, the UK, California, Columbia, and CERN, among others. Various physicists, including Feynman, Gell Mann, S. Shale, S. Lang, Abdus Salam and many others offered support and solidarity.

The professors who had resigned still remained in Brasília during the next month, waiting for the facts to be laid out. An emergency fund was set up to assist them, since they were no longer receiving their salaries. Finally, on the 30th of November, 1965, those professors who had remained in Brasília trying to correct the situation and to resolve the crisis released a note informing that they were seeking closure: "We consider that our tasks at the UnB have been permanently terminated". And they predicted that "The *Universidade de Brasília* represents an irreversible experience: The idea was able to blossom for a day, in spite of the determination with which its destruction was carried out".[16]

That mass resignation of course meant the end of the idealistic project of the *Universidade de Brasília*, for the time being at least. And it signalled the failure of the ambitions, plans, and all the hard work of Jayme Tiomno, Elisa Frota-Pessôa, Roberto Salmeron and the other faculty and students who had set out so optimistically to create a new and better University for Brazil. For Jayme and Elisa, that project had also envisioned the inclusion of their long-held dream of an independent but associated Center for excellent physics research, the CBPF, into an appropriate university environment.

Many of the professors went into exile. They knew that they would be persecuted if they were to remain in Brazil. Roberto Salmeron and his family initially intended to stay in the country; he tried to find employment at another Brazilian university, or even taking up his original engineering profession in Brazil, but he was 'tainted' by the affair at UnB and by fear of reprisals from the military regime. He received several offers to return to CERN, and after speaking to Victor Weisskopf, its director at the time, he finally decided to go back to Geneva, which for him was psychologically admitting defeat, much like what Tiomno and Elisa experienced on returning to the CBPF. After a year and a half at CERN, he was invited to apply for a faculty position at the *École Polytechnique* in France (at *Paliseau*, near Paris). He accepted the position, in particular to give his children the chance to be educated in Paris, and remained there until his retirement. He is currently *emeritus* Research Director of the CNRS in France. In 2005, he was presented with an honorary doctorate by the (in the meantime fully functional) University of Brasília.[17]

Tiomno and Elisa had an additional problem: their students. They returned to Rio de Janeiro and occupied themselves with finding places for their students from Brasília, who were left abruptly without a university. They took all the students who had accompanied them to Brasília back with them, and saw to it that they were all able to continue their studies without significant loss of time. As Elisa later said,

> We were left with the problem of taking all those students back and putting them into other universities. We didn't let even one of them lose a year. As I was always very decisive, I didn't ask anyone [*in the FNFi*], I just did what I wanted and in a general way, it all worked out. So it goes.[9]

It was a bitter period for them, not only because their hopes of establishing a new, innovative university, and of finally anchoring the CBPF within an academic environment had been dashed and their year's work wiped out, but also because they had left Rio on an optimistic, idealistic mission and were forced to return in defeat. But Tiomno consoled himself later with the delayed fruits of their labors:

> We left [*Brasília*], and the truth is that all that work was wiped out. Now, it is impressive to see that the great majority of those youthful students whom we had collected there are today [*1977*] carrying on successful work in physics. They are physicists, many of international renown, who have completed their doctorates in Brazil or abroad, and are producing scientific results. What remains—to our satisfaction—is that even though our pioneering work was interrupted and destroyed, it still came to fruition.[3]

It was clearly time for Jayme Tiomno and Elisa Frota-Pessôa to find some other place to work. The CBPF was at its low point; many researchers and students had already left it, its financing was still uncertain, and its President, Admiral Otacilío Cunha, was moving towards an authoritarian management style. And the military government regarded it with suspicion, as a 'focal point of subversion', a 'den of Jews and Communists', much as they had treated the UnB. The new scientific director of the CBPF, Hervásio de Carvalho, who had taken that position after Leite Lopes' dismissal, began boycotting Tiomno.[18] The climate at the FNFi was also not conducive to their work; their confrontations with its Dean, Eremildo Viana, made their return there all the more difficult. "The situation here in Rio was very unfavorable for our work, and that derives from the general atmosphere that currently exists in Brazil", Tiomno would later say.

Already during the period of crisis in Brasília, in November, 1965, Abdus Salam had written to Tiomno, reminding him that he could go to Trieste whenever he wanted.[19] C.N. Yang also invited him to spend the academic year from September 1966 until June 1967 with him at the State University of New York (at Stony Brook, near Brookhaven).[20] But Tiomno and Elisa preferred at that point to remain in Brazil; and thus the option to return to São Paulo became a real possibility. Tiomno tried to arrange a cooperation agreement between the FNFi and USP which would permit him to work in São Paulo and go to Rio weekly to give his lectures. But that didn't work out, so in the end, he requested leave to move to USP.[21] As Tiomno explained: "We came to São Paulo in order not to have to emigrate from Brasil, since after the tragic experiences of the destruction of the *Universidade de Brasília* and of the Physics Department at the FNFi, along with the exhaustion of the CBPF, that seemed to me to be the last possibility for remaining in the country".[22]

Notes

[1] '*Estatutos da Universidade de Brasília*'—'The Statutes of the University of Brasília'; [JT].

[2] cf. Salmeron (2013). A complete version may be found in the archive [JT]. See also Salmeron (1999–2012).

[3] Tiomno (1977).

[4] The early history of the UnB, up to the military *coup d'état* in March, 1964 (after which he went into exile in Venezuela, Chile and Peru) is described in the book by Darcy Ribeiro, '*UnB—Invenção e Descaminho*' ('UnB—Invention and Derailment'), written during his exile; cf. Darcy Ribeiro (1978).

[5] The fascinating story of how Roberto Salmeron started working at CERN right from its outset is told in an interview given by Salmeron in 2005 to the magazine *Passages de Paris*; cf. Salmeron (2005).

[6] Rocha, João Augusto de Lima, '*Desmontada a versão da ditadura de 1964 sobre a morte de Anísio Teixeira*', In: Nassif, Lourdes, *Jornal GGN,* April 2nd, 2018. Online at: https://jornalggn.com.br/ditadura/desmontada-a-versao-da-ditadura-de-1964-sobre-a-morte-de-anisio-teixeira-por-joao-augusto-de-lima-rocha/.

[7] What happened in the following months is recounted in Salmeron's book, '*A Universidade Interrompida—Brasília 1964–1965*', Salmeron (1999–2012), and in more abbreviated form in his lecture given at the UnB in 2005, Salmeron (2013), as well as in his testimony before the Parliamentary Inquest Commission on the University of Brasília, given in November, 1965: Salmeron 1965. See also the interviews of Jayme Tiomno, Tiomno (1977), and Elisa Frota-Pessôa, Frota-Pessôa (1990), as well as the articles by J.M.F. Bassalo, Bassalo (1998) and Bassalo (2012).

[8] Interviews with Mario Novello, carried out by ATT for this book, August/September 2019.

[9] Frota-Pessôa (1990).

[10] cf. Chapter 15, 'Gravitation, Field Theory, and Cosmology', and Chapter 17, 'Looking Back'.

[11] Valladão de Mattos (2015), p. 408.

[12] Myriam and Mario Giambiagi to Jayme Tiomno, October 22nd, 1965; [JT]. Mario was a chemist and the brother of Juan José Giambiagi. They worked at the CBPF for many years; see the memoir by Myriam, Giambiagi (2001).

[13] Salmeron (2013).

[14] Bassalo (1998); see also Bassalo (2012).

[15] Jayme Tiomno to Elisa Frota Pessôa, September 21st, 1965; [EFP], and Jayme Tiomno to José Leite Lopes, September 20th, 1965; [JLL].

[16] Press release from the ex-professors at the *Universidade de Brasília*, November 30th, 1965; [JT].

[17] Salmeron (2005); Salmeron (2013).

[18] Statements that Prof. Jayme Tiomno was prevented by the President of the CBPF from taking part in its *Assembleia Geral* on December 12th, 1967. Available in the archives [JLL] and [JT].

[17] Abdus Salam to Jayme Tiomno, November 8th, 1965; [JT].

[18] C.N. Yang to Jayme Tiomno, July 8th, 1966; [JT].

[19] Jayme Tiomno to Plínio Süssekind Rocha, head of the Physics Department of FNFi, April 29th, 1966; [JT], and Elisa Frota-Pessôa to Marcos Moshinsky, June 24th, 1966; [EFP].

[20] Quoted from Tiomno's interview in the newspaper *Folha de São Paulo*, August 5th, 1968.

12

The Military Regime. The *'Leaden Years'*. Repression and Resistance

A Nobel Prize for the Samba School

In early 1966—just at this sad time for the CBPF, for the University of Brasília and for scientific research in general in Brazil, soon after Tiomno and Elisa had returned from Brasília—the Secretary for Tourism in Rio de Janeiro, *João Paulo do Rio Branco*, was busily organizing the Carnival, which would take place in February. He planned to invite prominent names from popular culture, such as Ingrid Bergman, Sean Connery, Gina Lollobrigida, and even Salvador Dalí. The list of official invitees also included the name of the physicist Richard Feynman, who had shared in the Nobel prize for Physics the previous year. Feynman had already spent five sojourns in Brazil as a lecturer and visiting researcher at the FNFi and the CBPF (cf. Chap. 9). During those diverse visits, in particular his second one in 1951/52, he had become fascinated with the samba and had learned to play some percussion instruments; and he had helped to establish and played with a Samba School in the Carnival of 1952—the group was called '*Farsantes de Copacabana*'.

In the end, the great names of the arts and culture were all unable to attend, so that only Feynman accepted the invitation, bringing his English wife, Gweneth Feynman (he remarried in 1960). The Brazilian newspapers had a field day repeating the story of how his previous wife had asked for a divorce because she couldn't stand Feynman's practicing carnival music all night long. This time, however, in contrast to his previous sojourns in Rio, the reason for this visit was not at all scientific. He didn't come as a relatively unknown physicist, as he had before, but rather as a celebrity, a Nobel prize winner. It had originally been planned that he would at least give a lecture

© Springer Nature Switzerland AG 2020
W. D. Brewer and A. T. Tolmasquim, *Jayme Tiomno*, Springer Biographies,
https://doi.org/10.1007/978-3-030-41011-7_12

at the CBPF, but in the event, that was substituted by a luncheon at the restuarant '*Sol e Mar*'.

The intention of the Secretary had been to bring some of the members of the Samba School, and even its percussion section, to meet Feynman at the airport. In fact, due to restrictions by the airport security police, the planned grand reception was replaced by something much more modest. He attended the performances of the samba schools, participated in the ball at the Hotel Copacabana Palace, and marched in Copacabana with the group '*Farsantes de Copacabana*', which he had helped to found 14 years earlier (Fig. 12.1).

Fig. 12.1 Richard Feynman (*below, second from left*) with the Samba School *Farsantes de Copacabana* in 1952 at the Carnival Ball in the *Teatro Municipal*, Rio de Janeiro. Published in: *O Cruzeiro*, March 15th, 1952

It was, in other words, a public-relations stunt. This would be Feynman's last visit to the country. Tiomno later described it:

> ...He [*Feynman*] returned several times to Brazil, until, the last time that he was invited to come, already after 1964, he became aware of the situation at the *Centro* and he never came back again, neither to the *Centro* nor to Brazil.
>
> "On that last journey to Brazil, he was invited not by the CBPF, nor by any other scientific institution, but instead by the *Prefeitura* [*Mayor's Office*] of the city of Rio de Janeiro—which at that time was called *Guanabara*—to participate in the Carnival of Rio de Janeiro. He was met at the airport by the samba school that he belonged to, or had belonged to when he was in Brazil earlier, with a large sign reading, '*Salve Richard Feynman, Primeiro Prêmio Nobel da Escola de Samba ...*' [*Hail Richard Feynman, first Nobel prizewinner of the Samba School* 'Farsantes de Copacabana']... He had played the *frigideira* [*a small frying pan beaten to produce the samba rhythm*] with that samba school and was very proud of that, since quite truly the *frigideira* is an authentically Brazilian instrument, and he had ended up being the premier *frigideira* player of the samba school. He overtook all the others and became the premier player. He always went out with the samba school, including, many times, he played with the samba school at the houses of millionaires, for parties, to earn money to pay for their costumes. He was very proud of that.
>
> ... One mustn't forget that this was connected with the period of decadence of the *Centro*. The most that happened was that the *Centro* invited Feynman to lunch.
>
> But the interesting point is that it was not that [*Feynman's*] Nobel prize was being celebrated as the first Nobel prize for the *Centro Brasíleiro de Pesquisas Físicas*, but as the first Nobel prize of that Samba School.[1]

This somewhat sad and demeaning story of Feynman's last visit to Brazil is a crowning irony of the year 1966, which must have been a low point not only for the CBPF, but also for Jayme Tiomno's hopes of establishing excellent research and education for Brazilian physics. Whether Tiomno himself met with Feynman during his 1966 visit is unknown; they however maintained friendly relations for many years afterward, and Tiomno visited Feynman later in Pasadena and met him at other places in subsequent years—but never again in Brazil.

Italy, 1966/67

After the collapse of the UnB project in Brasília in late 1965, the options of Tiomno and Elisa in Brazil were becoming more and more limited. The

CBPF was at the height of its financial crisis; its personnel had been decimated and it was struggling with the authoritarian administration of Admiral Otacílio Cunha. The FNFi, for its part, was suffering a smothering atmosphere, mainly due to the false accusations raised by Eremildo Viana. Thus, when the possibility of returning to São Paulo appeared, their interest was aroused. USP had announced a public competition for the professorship in Advanced Physics (*cadeira de Física Superior*), which was open for applications. Tiomno went to São Paulo to prepare himself to apply for it.

Thus in May, 1966, he was in São Paulo, writing and submitting his resumé[2a] and his thesis[2b] to the *concurso* (competition) for the professorship at USP/São Paulo. As soon as that was accomplished, in July 1966, he and Elisa departed for Italy. They hoped that this would provide an interlude of 'cooling off' following the UnB collapse, and perhaps would spare them rebuffs of the kind experienced by Roberto Salmeron during that period, while enabling them to maintain a low profile with regard to the military regime. At the time, they were in need of some respite from the political and scientific situation in Brazil, and they had resolved to spend some time abroad—Tiomno at the *International Centre for Theoretical Physics* (ICTP), which had been founded in 1964 on the initiative of Abdus Salam (who was its first president), and with the support of the International Atomic Energy Agency and the help of many scientists, including Tiomno himself. They planned to make use of Tiomno's privileged status as an Associate of the ICTP, allowing him to visit there at its expense.

Elisa, as an experimental physicist, worked at the *Italian National Institute of Nuclear Physics* in Trieste at the same time. "I had my observations and experimental results and he [*Tiomno*] had his pencil, his paper and his work on field theory".[3]

Thus, they were able to work in their respective fields while traveling together, as they had done during their stay in London 8 years earlier. When Tiomno's tenure as Associate at the ICTP ended in late 1966, he spent another 8 months there as a Visiting Professor of the Centre.

While in Trieste, Tiomno worked with two Italo-Argentinians, *Carlos Guido Bollini* and *Juan José Giambiagi*, who were also Associates of the ICTP at the same time as Tiomno. Giambiagi was just four years younger than Jayme Tiomno. He was born in Buenos Aires, and he had worked for a time at the CBPF in the mid-1950's, where he published a paper together with Tiomno in 1954.[4] He later became head of the Physics Department at the

University of Buenos Aires (UBA; from 1959–66), and developed its program very successfully. He was also one of the founders of the CLAF (Latin-American Center for Physics), which sponsored the Latin-American Physics Schools (ELAF) in the 1960s (Fig. 12.2).

Bollini was two years younger than Giambiagi, and had completed his education in mathematical physics in La Plata, Argentina. He spent two years (1958–60) at the Imperial College in London, and had met Tiomno then. Bollini and Giambiagi themselves met at the Argentine National Commission on Atomic Energy (CNEA), where they both worked after 1956, and they continued collaborating for the next 30 years.

Fig. 12.2 Juan José Giambiagi, Jayme Tiomno, and Carlos Guido Bollini at the ICTP Trieste, during the Winter of 1966/67. *Source* Private, [EFP]

During their stay at the ICTP in 1966, a military dictatorship took control of the Argentine government. It had even more drastically negative effects on scientific and academic life there than did the dictatorship in Brazil, not to mention its brutality in pursuing and eliminating 'subversives'. Back in Argentina, Bollini and Giambiagi left the University in Buenos Aires, working privately for a time, until they were able to obtain positions at the University of La Plata, where they did their most important work in the early 1970s.[5]

Later on, as the persecution from the dictatorship in Argentina grew worse, first Giambiagi (in 1976), and then Bollini, a year later, escaped with their families to Rio de Janeiro, where they were able to find work at the CBPF (ironically, at a time when three of its founders, Jayme Tiomno, José Leite Lopes, and Elisa Frota-Pessôa were themselves banned from working there). Tiomno continued to collaborate with them during that period, writing six more joint articles between 1970 and 1981, as well as another with Bollini alone in 1980. Giambiagi became the head of the Department of High-Energy Physics at the CBPF from 1978 to 1985 and again in 1994–96; he continued working there at the *Centro* until his death in 1996 (cf. Giambiagi (2001)). Bollini returned to the University of La Plata in 1985, when the dictatorships in both Argentina and Brazil had ended, and worked there until his retirement in 1995; he passed away in 2009.

Their first joint article with Tiomno[6] resulting from the stay in Trieste was published in *Il Nuovo Cimento* in 1967 and bore the title, '*On the Covariance of Equal-time Commutators and Sum Rules*'. It deals with details of relativistic calculations and the properties of sum rules in field theories. This is an indication of Tiomno's revival of interest in field theory and gravitation.[2a]

In an interview, Elisa makes the following remarks describing her own view of their sojourn in Trieste (Fig. 12.3):

...we went to work in Italy, in Trieste, for a year. Jayme [*worked*] at the *Instituto de Física Teórica de Trieste* [*the ICTP*], and I at the National Institute of Nuclear Physics. … I had my work and he had his. Often, we would discuss the work of one or the other. There was never anything that made me feel at a disadvantage – or an advantage – from being married to him. We were great friends even before we became a couple, and we had many mutual friends.[3]

Back to São Paulo

The dictatorship in Brazil seemed by that time to be becoming less stringent, in spite of its paranoia regarding 'subversion'. A campaign was initiated by the Brazilian Society for the Progress of Science (SBPC) to encourage the

Fig. 12.3 Jayme Tiomno and Elisa Frota-Pessôa in Trieste, 1966/67. *Source* Private, [EFP]

return of scientists who had left Brazil, called '*Operação Retorno*' ['Operation Return']. In the first semester of 1967, Leite Lopes came back from France, reassuming his functions at the CBPF and at UFRJ (the *Universidade do Brasil* had acquired a new name and was now called the *Universidade Federal do Rio de Janeiro*). He had been invited by the Rector *Raymundo Muniz de Aragão* to become head of the Physics Institute. In addition, the military government, following a policy of national development, had begun to invest more in science. The CNPq, which had dispensed 553 scholarships in 1963, increased that number to 1,300 by 1967. The situation seemed to be improving.

After their year in Trieste, it was time for Tiomno and Elisa to return to Brazil. They arrived back in São Paulo in August of 1967. The competition for the professorship in Advanced Physics at USP, which Tiomno had entered the previous year, was approaching its final stages.

Tiomno's thesis for the competition at USP[2b] was entitled '*Contribuições à Física das Partículas Elementares*' ('Contributions to the Physics of Elementary Particles'). It summarizes his work in elementary-particle physics following his doctoral thesis, over the subsequent 15 years. He had written a considerable quantity of articles during that period, many of them influential within the international physics community. The deliberations of the Examination Committee and the referees took place while he and Elisa were in Italy, and he passed his thesis defense and public lecture on returning to Brazil. He thus obtained the position officially in November 1967, and began his new duties in São Paulo at the beginning of the academic year 1968. Elisa was invited as a visiting scholar with a fellowship from the CNPq by *Ernst W. Hamburger*, Director of the Physics Department at USP, to reorganize the Nuclear Spectroscopy Laboratory there, which made use of nuclear emulsions, her specialty.

As is customary, when Tiomno assumed his new professorship at USP in early 1968, he delivered an 'Inaugural Lecture' (on March 4th, 1968). It bore the title '*Ciência, Universidade e Desenvolvimento*' ('Science, the University, and Development').[7] Its theme, as indicated by the title, was the role of a nation's universities in advancing scientific education and research, and their increasing importance in the modern world. He began by quoting a still earlier lecture, his acceptance talk for the '*Premio Moino Santista*', the prize that he was awarded in 1957, still relevant in 1968. He made use of the opportunity to express his revulsion and sadness at the destruction of universities and research centers in Brazil:

> In spite of the significant development of scientific research in Brazil during this period [*since the 1950's*], the situation is now still more grave, because in the more advanced countries, the development has continued much more rapidly, and serious measures to create a solid infrastructure in Brazil have not been undertaken. One of the reasons for these omissions by the government is the lack of awareness of the Scientific Revolution on the part of public opinion, politicians and those in the governments of the less-developed countries...
>
> ... By means of what *Alceu Amoroso Lima* [*a Brazilian intellectual of the time*] has called 'cultural terrorism', the mediocrity implanted in various sectors of Brazilian culture has managed, with or without the help of the Castelo Branco government [*the military dictatorship*], to destroy or weaken some highly-developed sectors in Brazilian science and in the universities. Among them are the *Instituto Oswaldo Cruz*, the *Centro Brasileiro de Pesquisas Físicas* and the Departments of Physics and Mathematics of the *Faculdade de Filosofia*

of the *Universidade do Brasil* [*now the UFRJ*] in Rio de Janeiro, the *Instituto Technológico de Aeronáutica de São José dos Campos* and others. Even the *Universidade de São Paulo* has not been immune.[7]

He could not fail to mention what had happened in Brasília in 1965: "But the most terrible of all this, the crime of treason against the fatherland, whose judgment is yet to be made by History, was the destruction in 1965 of the first 'university for development' in Latin America—that was the *Universidade de Brasília*, which had already suffered a severe purge in 1964". He pulls no punches in his lecture, and goes on to say, "For me, my participation during six months in 1965 in the activities and the construction of the UnB, in particular of the *Instituto Central de Física*, was the most stimulating experience of my university career. My resignation during the crisis in Brasília is to me the most valiant act in connection with the Chair of Advanced Physics of that Faculty". And he concludes with a promise: "I will not hesitate or temporize in my work and with my responsibilities in *this* Faculty, nor in the struggle for the progress of science in Brazil".[7]

Times of Turmoil

As things turned out, Tiomno did not have an easy time at USP, beginning with the *concurso* itself. When César Lattes left the CBPF in 1960, he went to the Physics Department of USP, where he was contracted as interim professor. There, he set up the Nuclear Spectroscopy Laboratory, making use of large, layered photographic plates with nuclear emulsions to study cosmic ray events. Beginning in 1962, he actively promoted a Brazilian-Japanese Cooperation (CBJ) on cosmic-ray research, which had been proposed by Hideki Yukawa. Yukawa had suggested that they make use of the Laboratory for Cosmic-Ray Physics that Lattes had set up in 1951 on Mount *Chacaltaya*, in Bolivia, with the help of a team from the CBPF. The project required considerable effort, because many plates with nuclear emulsions and many absorber plates had to be transported via the Bolivian capital La Paz to the *Chacaltaya*. It was such a great quantity of material that they joked that they were "exporting lead to Bolívia".

Mario Schenberg, head of the Physics Department at USP, had expected that Lattes would continue to occupy the position, becoming the tenured holder of the Chair of Advanced Physics of the Department. However, when the competition was announced in 1966, Lattes was very annoyed with Schenberg for its timing, due to the CBJ activities. That project was just entering its critical phase of moving from instrumentation development and testing to actual physical measurements. In addition, he felt that the procedure of the *concurso* was too bureaucratic and outmoded, and let everybody know that he would not take part in it.

At that point Tiomno, as well as the physicist Ernst W. Hambuger, already at USP, decided to apply for the position themselves. The assistants in Lattes' group tried to persuade him to enter the *concurso*, and they helped him to gather the documents, prepare his resumé, and even aided in writing the thesis. By that time, Tiomno was already working on his own thesis and resumé, and he decided to continue. Hamburger however withdrew his candidacy.

Clearly, it would have made a difficult choice for the Examination Committee if Tiomno and Lattes had both been competing. While Lattes was very well known and revered for his early scientific contributions, he had not published so much science in the years since founding the CBPF; he had been more occupied with administration and scientific planning and organization than with research.

In fact, the Examination Committee was spared that difficult situation: Lattes did not enter the competition. Tiomno was approved as the unique candidate and that precipitated a crisis at the Physics Department of USP. The Department, in particular Mario Schenberg, had advertised the position only to satisfy the formal requirements, and clearly intended to hire Lattes on a permanent basis to the position that he had occupied for over six years as interim professor.

At the same time, in Campinas, less than 100 km from São Paulo, the founding of the *Universidade Estadual de Campinas* (Unicamp) was in progress. It was intended to be a kind of showcase for the military government, to demonstrate their dedication to science, and salaries and equipment budgets were more generous than at USP. To coordinate the establishment of that university, Zeferino Vaz, the rector who had withdrawn from the *Universidade de Brasília* in the middle of the crisis there, had been employed in Campinas. Marcelo Damy, an experimental physicist who had spent his whole career at USP, had transferred to Campinas to set up the Physics Institute of the new university. Knowing of Lattes' dissatisfaction with Schenberg and with the Physics Department at USP, Damy invited Lattes to also transfer to Unicamp and contribute to the establishment of the new Institute. Lattes set up a Department of Cosmic Rays and a Laboratory for Nuclear Emulsions at Unicamp, where he continued the CBJ—the Brazilian-Japanese Cooperation that he had initiated while at USP. One of Lattes' assistants later wrote of this period.

Unfortunately, just as had happened during the previous competition for the Chair of Nuclear Physics at the *Faculdade Nacional de Filosofia* in 1962, he [*Lattes*] also refused to participate in this competition, for the Chair of Advanced Physics at the FFCL/ USP. It was a great frustration for everyone: the group had worked hard to help Lattes to prepare his thesis, which he never considered to be complete, and I myself took the documents and the thesis to enter the competition at the FFCL/USP. Then a great crisis began, which subsided only when he went to Unicamp, at the invitation of Prof. Marcelo Damy de Souza Santos.[8]

Lattes' thesis, written for the *concurso* at USP but never formally submitted, was forgotten until his 80th birthday (in 2004), when a retired physicist and former Scientific Director of the CBPF, *Alfredo Marques*[9] located a copy and asked Lattes for permission to publish it. A limited edition was issued by the CBPF in 2005 (in time for the International Year of Physics).

The incident remains only a footnote to the lives and careers of Jayme Tiomno and César Lattes; however, it was evidently a sign of the end of their friendship, which had been quite close in earlier years, but had never recovered from the shock of the 'scandal' at the CBPF in 1954. And it apparently also left some strains between Tiomno and some of the physics faculty at USP, in particular with Mario Schenberg, as later became apparent.

Disgusted by the crisis, Schenberg asked to be dismissed from his function as Director of the Department. Hamburger took on the position, with the condition that there be a climate of tranquility in the Department. However, some sources of friction remained, perhaps inevitably, since Tiomno's occupying the Chair of Advanced Physics—as a theoretician—had inescapably changed its research emphasis from what it had been during the (interim) professorship of César Lattes, an experimentalist who specialized in particle observations using nuclear emulsions.

In August, 1968, some six months after Jayme Tiomno took over the Chair (and Elisa Frota-Pessôa, at the invitation of the Department's new director, Ernst W. Hamburger, had begun refurbishing the Nuclear Spectroscopy Laboratory, from which Lattes had presumably taken most of the apparatus along to Campinas, where he planned to continue the Brazilian-Japanese Cooperation (CBJ) that he had founded six years earlier),—another conflict erupted.

The other experimental physicist in the Department, Oscar Sala, asked Tiomno to let him use the Laboratory of Nuclear Spectroscopy that had been set up by Lattes, for his own use and for the work of Elisa Frota-Pessôa. This was the fuse that set Schenberg off—he accused Sala and the Technical Administration Council of the Department, and asked to be transferred to a different department. As a result, Sala submitted his resignation to the University, followed by two other important researchers. Hamburger, for his part, also asked to be released from his duties as head of the Department. The students accused Tiomno of being reactionary and authoritarian.

Mario Schenberg was perhaps disturbed that his original plan to keep Lattes as permanent occupant of the Chair, with the CBJ project firmly rooted at USP, had been derailed, and the Chair was now occupied by Jayme Tiomno, a fellow theoretician. He began accusing Tiomno of damaging the Nuclear Spectroscopy Laboratory. His accusations were made indirectly, in letters to colleagues and to the press, while he failed to confront Tiomno directly or challenge him openly in meetings where both were present. Tiomno turned the matter over to an inquest commission to be set up by the Department adminstration and left any actions regarding his Chair in the hands of the Director of the Department.[1] The affair resonated for a week or two in the press in Rio and São Paulo, and was finally resolved without lasting damage. Hamburger declared to the press that having independent, all-powerful Chairs within a departmental structure was not feasible, and should be replaced by a more modern structure—Schenberg recalled this period in an interview ten years later.[10] Little by little, the tension was reduced. Sala ceded to the pressure of the great majority of the academic faculty, including Tiomno, urging him to remain at USP, and Schenberg also stayed in the Department.

At the same time, the idea of a more sweeping university reform was gaining momentum. Since 1945, when Brazil had no more than 5 universities, it had increased that number within a period of 20 years to 25 public universities, both federal- and state-supported. The number of students in higher education had more than quadrupled, increasing from 30,000 to 142,000. A consensus had been reached that the universities, for the most part created by combining traditional faculties of medicine, engineering, and law, were born defective and were inefficient. Put together in the pattern of the old *Universidade do Brasil,* as collections of independent schools and faculties, mainly in the traditional professions, they had kept the old professorial system (*catedrática,* Chairs) that permitted few interactions among their various subunits. The demands for a university reform reflected the changes that Brazil had experienced during the 1950s and 60s, when ideas of modernity and development had become predominant. In that period, the country had seen a significant economic growth, with industrial expansion, an increase in population, and the rise of a considerable middle class. The new universities founded during that time were molded to an archaic pattern and were controlled by traditional groups.

In spite of its collapse, which occurred for political reasons, the model proposed for the *Universidade de Brasília* of a combination of teaching and research, with a stronger integration of the various faculties and the end of the professorial system, among other aspects, had gained more and more support, even within the government. Furthermore, the proposed reform had a justification which was eagerly accepted by the governing forces: increasing the efficiency of the universities.

In 1968, the discussion about the planned university reform became a major topic within the universities throughout the country. Committees were formed to discuss what measures should be taken. The atmosphere was agitated, in particular because of student activists. In addition, the students' uprising and the general strike in France in May, 1968 spread to various cities in Europe and North America and soon arrived in Brazil.

The climate of conflict within the Department and the agitation throughout the whole University notwithstanding, Tiomno continued setting up his research group within the Physics Department at USP, working in particle physics and gravitation, and it later became the original nucleus of the current Department of Mathematical Physics. As the holder of the Chair of Advanced Physics, he established a research program involving both those theorists and experimental physicists who were already in the Department, and initiated collaborations with other groups. He organized new courses for undegraduates and graduate students, and started a cooperation specializing in elementary particle physics which involved various theoreticians. It quickly became the largest theory group in the country, having 9 permanent members, 6 of whom had their doctorates, in addition to 8 graduate students and a number of visitors and undergraduate students with scholarships, involving all together 26 members in its activities. He also was able to invite 4 visiting professors for shorter periods. The results were beginning to appear: 10 articles published in the specialized literature, 6 Masters, doctoral and *livre-docência* theses, participation in national and international conferences. The group grew so rapidly that in 1968, of the ca. 15 graduate-school courses given by the Physics Department, 9 were directed by members of the Elementary-Particle Group. However, due to his involvement with administrative duties and with establishing and coordinating research, Tiomno himself published nothing during this period.

Near the end of 1968, there were still more changes. On the 28th of November, the Federal Law No. 5540 was approved, which initiated the university reform in Brazil. It was inspired to a great extent by the model of the UnB, eliminating the professorial system (Chairs), strengthening the power of the central university administration *vis á vis* that of the various faculties and departments, reinforcing the connection between teaching and research, and making extension programs for the wider community obligatory. It formalized the graduate schools and established standardized entrance examinations for university studies. The professors at UnB who had prophesied at the end of 1965 that the ideal of the *Universidade de Brasília* would prevail in spite of its collapse could not have imagined that this would occur so rapidly, within only three years. The reform was indeed inspired by the model of the UnB, but it lacked the liberal premises of that institution; it was instead based mainly on the ideas of modernity and development held by the military government.

Thus, having been instituted in an authoritarian manner, the university reforms ended up producing a great deal of chaos. Without the professorial Chairs. the universities lost their reference points for taking decisions. For the new *Instituto de Física* at USP, the professors suggested the creation of no fewer than 10 individual departments. However, without the Chairs, there was no longer a mechanism for reasonable academic decision-making. Lacking the capacity for taking decisions, the University administration asked Tiomno to propose a form of organization for the new Physics Institute which would include only three departments. During the internal squabbles, Tiomno felt it to be most reasonable (in the positive sense of being most considerate of his colleagues) to organize the departments by affinities rather than by topics. But that in the end was not carried out. Furthermore, the modernization was accompanied by repression.

Dark Times

Unfortunately, the last month of 1968 brought still another unpleasant surprise: On December 13th, 1968, the military dictatorship in the person of its President-Marshal *Artur da Costa e Silva* issued the infamous *AI-5*, the 'Institutional Act Number Five'. It was a decree which overrode the Constitution, with no judicial recourse. It gave the government permission to suspend human rights and due process, and allowed arbitrary interventions into local and state governments and individual rights. It permitted censorship, the suspension of legislative power, immediate dismissal of employees without justification, and the suppression of meetings and demonstrations, among other things. The AI-5 provided the basis for a number of specific decrees, for example wiping out the central student organization, the '*União Nacional dos Estudantes*' ('National Students' Union'). The legislative bodies in several cities were shut down, and numerous people were arrested, dismissed from their positions or forcibly retired. It affected over a thousand persons within the first few months following the decree, who were considered to be 'subversive' and therefore subject to removal from their functions, whether they be military, parliamentary, administrative or teaching personnel.

The AI-5 decree was the opening act of the '*leaden years*' in Brazil, when no-one was safe from arbitrary repression, punishment or even torture. There was a notable difference with respect to 1964, when the military regime first took power: Now, there were no longer inquest commissions. People were simply arrested without explanation or justification.

Even during such a difficult time, Tiomno tried to continue with his scientific work, and in January of 1969, he departed for the USA to take part in the Coral Gables *Conference on Fundamental Interactions at High Energy*, which was scheduled to begin on January 22nd at the recently-founded 'Center for Theoretical Studies' of the University of Miami, Florida.

In Brazil, one of the principal targets of the dictatorial regime were researchers and university professors, considered to be fomenters of unrest and inciters of student protest. Thus, on the 28th of April, 1969, a list of professors to be forcibly retired was published. They were essentially all professors from USP and from the former *Faculdade de Filosofia* (FNFi) of the *Universidade do Brasil*, now known as UFRJ (where Tiomno was included). Of the latter, 12 of the 44 academics who had been denounced by Eremildo Viana in 1964 were now blacklisted, including Tiomno, Elisa and Leite Lopes. In fact, it had already been more than 18 months since Tiomno had left the UFRJ (Federal University) and had transferred to USP (State University)—but the files of the Security Services had not been updated. Little by little, other lists were being released, including Mario Schenberg along with many other scientists, and now retiring Tiomno from the correct institution. About 200 recognized scientists were forcibly retired. While in 1964, most of the accused professors were young and active, the regime now wanted to attack the academic leaders.

The lists of retirements and dismissals included many people of scientific and intellectual prominence, which might have made them potentially dangerous to the government; or their blacklisting might have been due simply to vindictiveness from within their organizations. In the case of many of those from Rio de Janeiro who were dismissed, that could mean in particular that they had opposed *Eremildo Viana*, former Dean of the FNFi (who had been driven from that post by a series of student strikes), who was now director of the new Institute at UFRJ. Viana (and other faculty members who denounced their colleagues) may have actually believed that the persons on the list were subversive and potentially dangerous to the University, but the denunciations were also acts of personal vengeance and wounded pride because they had questioned Viana's authoritarian management of the Faculty—or simply a venal attempt on their part to 'eliminate' potential rivals. It is however another good example of the misuse of administrative authority coupled with the power of a dictatorial regime.

The Department of Physical Chemistry of the FNFi offers an example of this. *João Christóvão Cardoso*, who held the Chair of Physical Chemistry, was a potential competitor of Viana as Dean of the FNFi. He was denounced by 'Eremildo's list' in 1964, and forcibly retired by the AI-5 in 1969. Sometime later, Jayme Tiomno's youngest sister, Silvia, who at the time was a professor of physical chemistry in Cardoso's department at UFRJ, 40 years old and mother of three children, was held under house arrest by security agents without anyone's being able to understand the reason for her arrest, since she had maintained a certain distance from the student movements. There was a concerted effort by her colleagues and friends to have her freed, which was finally successful—although no-one offered an explanation as to why she had been arrested in the first place.[11] That remained an open question for fifty years, and the answer was discovered only recently during research in the files of the Security Service (DOPS) for this book.[12]—After Cardoso's forced retirement, his position was 'up for grabs', and a new competition was to be held to name his successor. Silvia, who was qualified as a university teacher (she had obtained her *livre-docência*), was a potential candidate for that position (although she in fact was not interested in it). One of her 'colleagues' took advantage of the situation to denounce her as a "member of the '*Grupo de Ação Partisan*' ['Group of Partisan Action']", supposed to be a radical cell, thus presumably blackening her reputation and eliminating her as a potential rival in the competition for Cardoso's position. This incident provides just another example, if one is needed, of how authoritarian regimes spread corruption and bring out the very worst in people, even those not directly connected with the regime.[13] The denunciation proved groundless, and she was finally released after some days of house arrest.

In the same way, the Security Services kept a file on Elisa, in which she was characterized as a "militant Communist, a technician for nuclear emulsions"; she "had been involved in incidents that had occurred at the *Universidade de Brasília*"; she was an agitator, subversive, connected to the '*Comando dos Trabalhadores Intelectuais*' (the 'Commando of Intellectual Workers', thought to be a Communist front organization). She was "…dangerous. Chronically ill". Among the activities which had contributed to that characterization, they described how "she and her companion, Jayme Tiomno, were responsible for taking diverse militant Communist students from the *Faculdade Nacional de Filosofia* to study in Brasília". After arranging the transfer of these students to the *Universidade de Brasília*, she and her companion "also went there, becoming members of the teaching faculty of that educational institution". She was declared to be a Communist since "while a professor at the FNFi, she systematically gave poor grades, out of pure spite, to the democratic students to punish them. The physics students detested her".[14]

The dismissals and compulsory retirements affected universities and research centers throughout the country. In the case of the *Instituto Oswaldo Cruz*, 10 researchers were retired, amounting to roughly 10% of the research team there. The most serious aspect was that—as likewise at UFRJ and USP—the group leaders were the targets, completely disrupting the research groups. In physics, those forcibly retired included *José Leite Lopes, Elisa Frota Pessôa, Plinio Süssekind Rocha, Sarah de Castro Barbosa, Jayme Tiomno* and *Mario Schenberg*.[13] All of them were originally from Rio, with the exception of Schenberg, who was the only avowed Communist in the group (and was hardly 'subversive', in spite of his political leanings).

Again, similarly to what happened at the *Universidade de Brasília*, Tiomno's only consolation was that he had managed to plant the seeds of progress in physics for the country within the short time that he was able to work at USP:

> So, once again, during one year – they gave me only one year – I had the opportunity to set up another group. Alone, because it was a single professorship, it had only the weight of that single position; there was no external support. Within that position, I set up a research group which has managed to survive, by itself, after my forced retirement. At present [*1977*], it is the best research group in theoretical physics at the University [*USP*]. That is to say, it is the most productive. [...] It consists of those whom I chose to receive scholarships at that time; they were just beginning their education, and they finished it and went abroad. They obtained their doctorates abroad and came back. Others stayed here to do their doctorates. At present, the group is rather active and quite productive.[1]

On the same day on which the decree forcibly retiring Tiomno, Leite Lopes and Elisa was published, *Roland Koberle*, an assistant in Physics at USP, wrote to Wheeler:

> I have the duty to inform you that Profs. J. Tiomno, Leite Lopes and other university professors have been forcibly retired by presidential decree. It is possible that other retirements will follow. This will probably mean the end of scientific research and in particular of theoretical physics in Brazil. Direct appeals of influential people from abroad to President Costa e Silva could be of extreme importance at this moment.[15]

Wheeler began to mobilize the international community of physicists. At the same time, various other physicists started sending letters and telegrams to the President of Brazil, asking that he rescind the decree: Robert Marshak, H. Gove, J. B. French, Elliot Montreal, C. N. Yang, and the presidents of the American Physical Society and the Federation of American Scientists, of the *Sociedade Mexicana de Física*, as well as Nobel prize winning scientists from diverse fields, such as Fritz Lipman, Robert Holley, Linus Pauling, Marshall Nirenberg, John Bardeen, and Harold Urey. In France, 240 professors signed a petition asking for the revocation of the forced retirements. Major newspapers in the USA, Europe and Latin America wrote of the repression of scientists and intellectuals in Brazil.

Wheeler, nevertheless, could not understand why Tiomno, who had never been politically active, had been forcibly retired and dismissed from the University. It made no sense to him that the country should dispense with researchers of his caliber for no apparent reason. In his correspondence with several Brazilians, he asked them to explain what was really happening, so that he could intervene in a more effective way in favor of Tiomno and the other physicists. The explanation came from *José Goldemberg*, Executive Secretary of the *Sociedade Brasileira de Física*:

> The reason for such decisions in the government are beyond our knowledge; the same government has been trying to give better support to science and technology for the last few years, and my own impression is that the retirement of professors has been made because of the blindness of government officers in realizing the serious consequences that will follow. There is much of personal vendettas in this, and also the influence of extremists in the government who want to eradicate so-called "left-wing tendencies" at any cost, regardless of the means and consequences. The three components: ignorance, hate, and extremism, can be seen in the list of 67 'victims' [*in the following months, there would be many more*]; they cover a broad range of political ideas and many of them have had no political activity at all for many years. To label them 'subversive' is an extremely rough simplification.[16]

Several members of the Institute of Advanced Studies in Princeton, as well as Marvin ('Murph') Goldberger, head of the Physics Department at Princeton University, invited Tiomno to spend the academic year 1969/70 there. They were still half stunned by the situation. Tiomno felt that it would be better to take advantage of invitations that he had received to give seminars and lectures outside Brazil, in order to sound out the situation and hope that the landscape would become clearer. He was particularly worried about Elisa, whose depression had returned with a vengeance. Whatever solution might be found, it was important to consider a place where Elisa could work, even if she would not be paid. It was not good for her to remain at full stop, far from her students and her laboratory. He departed on the 14th of July for New York, continuing on to Trieste to participate in a major conference that would last two weeks.

Wheeler was especially concerned that he had not received a reply from Tiomno regarding his invitation to Princeton. He had heard of many cases of people who were arrested in an arbitrary manner, as well as censorship of the mail. He worried that a repeated invitation sent to him directly might even worsen Tiomno's situation. For that reason, he began sending letters to various colleagues of Tiomno, asking them to pass on the invitation.

Having gotten the message, Tiomno traveled from Trieste to Switzerland, where he met with Wheeler in the small city of *Interlaken*, and then continued on to Paris to confer with Salmeron. Tiomno's trip also included returning via New York to take part in a conference at Columbia University and to give seminars in Rochester and Princeton. He made contact with *Behram Kurşunoğlu*, from the University of Miami, whom he had met at the January meeting, and who wanted to develop a cooperation with the USP group. Tiomno could make no promises regarding the group in São Paulo, but he was able to go to Miami for a few days to give a seminar on '*Linear, non-local theories of gravitation*', and to discuss the possibility of his and other members of the group spending some time in Miami in the near future, perhaps laying the ground for a real collaboration when the political situation in Brazil had become more settled. Kurşunoğlu welcomed Tiomno's suggestion and organized his visit and lecture. Tiomno then traveled to Los Angeles, where he met Sonia, Elisa's daughter, with her husband Zé Carlos and their daughters; both were in graduate school at USC. They visited Feynman at his Pasadena home on that occasion. Finally, he arranged with Marcos Moshinsky to return to Brazil via Mexico, where they could discuss future plans and he could give a seminar lecture. All in all, he spent three months traveling, making contacts and trying to keep in touch with current research. At the meeting in Switzerland, Tiomno agreed with Wheeler that he would go to Princeton in the spring of the following year. He also received an invitation from Philippe Mayer, from the *Département de Physique Théorique*, Faculté des Sciences, in Orsay/France to work there for a year. The idea of going to France rather appealed to him.

Tiomno wanted to use the intervening months to help to relocate members of his team, both students and assistants, and he planned then to return to the CBPF.[17] Concerning his theory group, which continued working at USP, he later made the following comments:

They invited some visiting professors, I myself made a great effort. At that time, no-one, none of the physicists from abroad who were invited to come to Brazil, no-one wanted to come, especially to São Paulo, to the place where a number of professors had been forced to retire.[18] But, still, I made an effort, as a physicist, showing that, in truth, the *Universidade de São Paulo* was free of guilt in that process; or rather, the Department of Physics was not at all guilty, since the list [*of those to be forcibly retired*] had in fact been submitted by the University. It was the visiting professors from abroad who helped out during that phase of bare survival.[1]

Ernst Hamburger, head of the Physics Department at USP, wanted to keep Elisa on as a visiting researcher at the Department, in spite of her forced retirement from UFRJ, which included her in the blacklist of the Security Services. The CNPq, which was paying for her fellowship, was willing to conceal the fact that she was receiving a fellowship from that institution and to make the payments via a third party. But Elisa refused to accept that—either she would receive the fellowship officially, or she would not receive it at all, so that the agency would have to make the payments in her name. Tiomno had the payments that he was receiving from the CNPq for research projects transferred to André Swieca and other researchers.[19]

Jayme Tiomno and Elisa Frota-Pessôa had tenured positions at the CBPF, which was not a public institution, but rather a foundation. They planned to return there. Tiomno felt that their difficult situation would be temporary and would soon improve. In the meantime, he and Elisa could continue developing their activities at the CBPF. However, they were becoming more and more circumscribed within a shrinking circle of repression: On the 21st of October, 1969, the government released a new decree stating that whoever had been subject to compulsory retirement could not be hired by any public or private educational institution, or any institution that received government support.

In various institutions, the administrations found ways to subvert the dismissals, through repeated consultations with government agencies, by establishing commissions to study the problem and postponing any action, or simply by ignoring them and waiting to act until they were officially notified. However, the President of the CBPF, Admiral Otacílio da Cunha (also chairman of the Brazilian Nuclear Energy Agency), a supporter of the military *Junta* that was now ruling the country, saw an opportunity to remove some of the more disaffected members from the *Centro*, and he decided (without consulting the Scientific-Technical Council) to dismiss Tiomno, Leite Lopes and Elisa from there as well, with the argument that the *Centro* was supported by public funds, although it was not a public institution. Even Mario

Schenberg, who had never been a member of the CBPF, was also dismissed, to thwart a plan for him to give a course at the *Centro*.

Leite Lopes referred to this decision as double jeopardy,[20] since those members of the CBPF had already been punished by forced retirement from their university positions. The dismissal from the CBPF was a severe blow, perhaps more so than the compulsory retirements of Elisa and Leite Lopes from UFRJ and of Tiomno from USP. The CBPF was their 'home', the institution that they themselves had created, and now it also was rejecting them. Some scientists who were abroad doing graduate work or in post-doctoral positions, such as Samuel MacDowell, Fernando Souza Barros, Moysés Nussenzveig and his wife Micheline, resigned from the CBPF in solidarity with Leite Lopes, Tiomno and Elisa. In a visionary fashion, Leite Lopes had educated the young researchers *Mario Novello* and *Sergio Joffily* so that they could soon go abroad to do graduate work. Novello still remembers the words of Leite Lopes: "You will be more important when the normal structure of the CBPF is re-established. Right now, it has a totally abnormal structure. It has suffered an invasion from outside, and the scientists have lost control of the CBPF. So it would be very difficult for you to continue your education here during this period. It is most important that you go abroad to obtain your qualifications at the highest level and then come back to help in restructuring the CBPF"; and he added, "It was a generation of talented scientists who returned [*after that difficult period*] to help the CBPF".[21]

Jayme Tiomno himself suffered a period of depression, and was unable to take any initiatives towards organizing their stay at Princeton, as he had agreed with Wheeler. For the two physicists, around 50 years old, who had dedicated their lives to teaching and research, being prevented from working was a major shock. Tiomno began using his home address for correspondence and scientific publications.

After the Dismissal

The situation of the other physicists who had been dismissed was also not easy. *José Leite Lopes* had been a full professor at the FNFi/UB since 1948 (by now, that University had been reorganized as UFRJ) (Fig. 12.4). In the late 1960's, he had been leading a project to set up an accelerator center there; but after learning that even if the cost of the accelerator itself could be covered, there would be no funds for the necessary infrastructure, he resigned as director of that project, in January, 1969.[20] Shortly thereafter, in April, he was retired from his position and forbidden to even enter the University.

Fig. 12.4 José Leite Lopes with his wife, the mathematician Maria Laura, and their son in Rio, ca. 1955. (Compare Fig. 16.2 in Chap. 16). *Source CanalCiencia – Ibict BiblVirtual.* Used under the Creative Commons license 3.0 with attribution. Online at: https://www.canalciencia.ibict.br/ciencia-brasileira-3/notaveis/292-jose-leite-lopes

Concerning his dismissal from the CBPF by the decision of its President, Admiral Otacílio da Cunha, Leite Lopes said of that period, "[*I*] reaffirm what I have always done, have always said, have always written, and continue to affirm, to write and to say: My fundamental problem was to fight for the development of science and technology in Brazil and in the interest of Brazil. Not necessarily in the interest of other countries or interests foreign to Brazil".[20] He was vocal about his loyalty to Brazil, not realizing that he could not remain there for long.

At that point (in mid-1969), he also received several invitations from foreign institutions, and he accepted the offer to go as Visiting Professor to the Carnegie-Mellon University, in Pittsburgh/PA, USA in the near future. However, the American Consulate, which maintained a close contact to the Security Services in Brazil, took note that Leite Lopes was in grave danger there—his name appeared on a list of people who were to be 'liquidated'. They wanted him to leave the same day, but then they decided to allow him and his family one day to organize their affairs; they departed the following day on a PanAm flight.[20], [22] Leite Lopes liked to joke that he had traveled to Pittsburgh with his family as paid passengers. He left Brazil on the 11th of September.

He remained in Pittsburgh for a year, but because of the US support for the *coup d'état* in Brazil, he preferred not to stay there longer, and thus accepted another invitation to go to the *Université Louis Pasteur* in Strasbourg/France, where he became Professor and Director of Research of the CNRS.

He stayed abroad until after the dictatorship in Brazil had ended, returning to Rio definitively only in 1985, when he became Director of the CBPF (i.e. he was in charge of the *Centro*).[23]

Mario Schenberg was another of those dismissed by the AI-5 in 1969. Shortly before the forced retirements were announced, Schenberg had traveled abroad in France, Switzerland, and Mexico, and had received invitations to go to each place to work. However, when he was retired by the AI-5 in April of that year, his passport was also confiscated, so that the option of exile was closed to him.[24] He lived a very secluded life in the following ten years, as he wrote to the author Clarice Lispector in the 1970s.[25] He regained his passport in 1972, but for private reasons, he was no longer in a position to leave Brazil. He reduced his work in physics, dedicating his efforts mainly to the philosophy of science and to his work as an art critic, which was very well received.

He was, however, able to continue his work in theoretical physics to some extent, in spite of being barred from the USP campus, and published several papers during those 10 years. Schenberg returned to USP in 1979, after the lifting of AI-5, and gave graduate courses there (one of which even led to a book, *'Pensando a Física'* ['Thinking of Physics'], published posthumously in 2001). He died in São Paulo in November, 1990, aged 76 (Fig. 12.5).

One of the saddest cases is that of *Plinio Süssekind Rocha* (1911–1972), Elisa's first mentor from her high school and later professor at the FNFi/UB. He had continued teaching at the Faculty, interrupted by a stay in Paris in the late 1930s, and he occupied the Chair of Classical Mechanics, Celestial Mechanics, and Mathematical Physics at the FNFi/UB (later UFRJ) from 1942 to 1969. Plínio was not a researcher in the usual sense, but he was an excellent teacher who inspired whole generations of physicists. He had no connections with other institutions, and no strong ties to universities or institutes outside Brazil. He had apparently no family and very limited resources.

Fig. 12.5 Mario Schenberg working, 1937. *Source Acervo, USP* (Public domain)

After his forced retirement, Plínio gave up his apartment in Copacabana, which had been conveniently close to the University but was now too expensive; so he moved to a modest apartment in *Lapa*, nearer the city center, where rents were lower. Some of his former students, many of whom apparently genuinely revered him, took up a collection for his support, which he accepted only reluctantly.[26] He lived a secluded life there, visited by the more loyal of his former students, and gave up physics entirely, maintaining his interests in the cinema and in philosophy, which had been important to him for many years. He died just over three years after his retirement, at age 61. Whether his death was hastened by his unjustified dismissal is uncertain, but that compulsory retirement successfully derailed his life, in any case (Fig. 12.6).

Antonio Luciano L. Videira, who was a former student of both Plínio Süssekind Rocha and of Jayme Tiomno, and a collaborator with the latter, wrote in a memoir of Süssekind Rocha[26]:

> Having been one of the most critical figures whose acquaintance I have had the privilege to make, Plínio was, quite consistently, his own most severe critic. Plínio simply did not publish. Since there is practically nothing of his in print, it is imperative for those who had the good fortune of knowing him personally to leave a record, even though it may be only an incomplete outline, even though it can only lighten the shadows of the remembrance, of the memory of Plínio Süssekind Rocha.

Fig. 12.6 Plinio Süssekind Rocha, photographed in the 1950s. *Photo* Saulo Pereira de Mello, *Archivo Mário Peixoto*. Used with permission

An obituary/tribute to Plínio had been written already in November, 1972, after his early death in August of that year, by two of his former students, Videira and *Jorge André Swieca*, both accomplished and well-known physicists. Most likely because of the dictatorship, it was not published at that time, but remained sequestered as a manuscript until many years later. Swieca himself died before the end of the dictatorship, in 1980, at only 44 years of age. The manuscript was finally published in 2013[27] by Videira (who himself passed away in October, 2018 in Portugal). It is called '*O Tempo de Plínio*' ('Plínio's Time'), and ends with the words, " – not one of the great epistemological problems escaped his analysis. His interests covered all of them, although one stands out as the most prominent: *time*. Time, preserved in thousands of annotations and notes. Time which ran down, drained off, and which, finally, came to an end".[27]

In the late 1950s, the Physics Institute at the Catholic University in Rio de Janeiro (*Pontifica Universidade Católica do Rio de Janeiro*, PUC/Rio) had been established, and many of Tiomno's students moved there in the late 1960s and early 70s. They included *Nicim Zagury* and *Erasmo Madureira Ferreira*, from the CBPF, and later on *Jorge André Swieca* and *Antonio Luciano*

Leite Videira from USP. Tiomno continued to practice physics as a '*privatier*', and they invited him to use their department as a safe base, even though he was not formally employed there—by government decree, he could not be contracted even by that university, a private entity.

Elisa said later of that time,

> For us, Jayme and me, it was always a question of educating researchers, and inspiring their enthusiasm for going ahead. Shortly after our forced retirements, quite soon afterwards, many of our former students went to PUC/Rio to build up the Department of Physics there.[3]

Still much later, in the year 2000 during the celebration for Tiomno's 80th birthday, one of his former students, Sergio Joffily, made a parallel between the Tiomno triangle and the drastic situation of Tiomno's life during the dictatorship, to show his capacity for resilience and his permanent search for a place where he could simply do physics:

> There was thus a different type of 'Tiomno triangle; that is the triangle of Rio – Brasília – São Paulo; a decay in Rio determined a capture in Brasília; a decay in Brasília determined a capture in São Paulo; a decay in São Paulo determined a capture in Rio, and *vice-versa*. All of them were caused by a single interaction: the military government.[28]

At the same time, Tiomno's North American colleagues, in particular John Wheeler, maintained their concern about his situation in Brazil; in addition to Wheeler's invitation letter of June 24th, 1969, the following year (September 18th, 1970) he received a letter of invitation from Freeman Dyson (at that time professor at the IAS) to come to Princeton for the calendar year 1971:

> We at the Institute have been discussing with our colleagues at Princeton University how we could invite you to Princeton. We have agreed that I should make you a definite offer to come here for the calendar year 1971. You would be supported by Princeton University from January to June, and by the Institute [*IAS*] from July to December. You would of course be free to take part in the scientific life at both institutions as much as you like. Since the Institute is at the moment very short of office space, while the University is not, you will probably be given an office in Jadwin Hall, the new physics building at the University. If these dates are not convenient for you, please let us know what you would consider to be a more suitable arrangement. Please also let us know soon if you are willing to accept this invitation, so that I can ask our Director to write you a formal letter of appointment.[29]

Little by little, Tiomno and Elisa were recovering from their depression. In July, 1970, Tiomno went to La Plata at the invitation of Giambiagi, and he later visited Chile and Uruguay. On his return to Rio, the new invitation from Dyson arrived. Tiomno finally agreed to this new appeal, replying to Dyson on September 24th that he would be honored to accept the invitation.

Sonia, Elisa's daughter, who had obtained her first degree in physics at the *Universidade de Brasília* in 1965, returned to Brazil at the end of 1970 after taking her MS at the University of Southern California/USA—and, in an irony of fate, in November of 1971, while Tiomno and Elisa were in Princeton, she was contracted as professor by the Physics Department at USP, where Tiomno had been forcibly retired not long before. She continued working on her doctorate at Unicamp in the following years, obtaining her Ph.D. there under the guidance of Roberto Luzzi in 1975.

Notes

[1] Tiomno (1977).

[2a] Tiomno (1966a)—his resumé for the Competition at USP; [JT].

[2b] Tiomno (1966b)—'Thesis submitted to the Competition for the Chair of Advanced Physics at the FFCL/USP, São Paulo, July 1966'; cf. [JT41] and Archive [JT].

[3] Frota-Pessôa (1990).

[4] [JT18]. J. J. Giambiagi spent a postdoctoral year at Manchester/UK in 1952/53. During that year, the University of Buenos Aires experienced a political purge, and so on returning to South America, he went instead to the CBPF at the invitation of Leite Lopes. He remained there as an exile from 1953–56 (not for the last time), and collaborated with Tiomno on this article. Biographical information on Giambiagi can be found at e.g. https://www.df.uba.ar/es/actividades-y-servicios/hemeroteca/6613-juan-jose-giambiagi (in Spanish) and at https://portal.ictp.it/icts/facilities/lecture_rooms/AGH/giambiagi/giambiagi_biography.

[5] See the article Bietenholz and Prado (2014) in *Physics Today*. Also, by the same authors, '40 Years of Calculus in $4 + \varepsilon$ Dimensions', arXiv:1211.1741v1 [physics.hist-ph] 8 Nov 2012; online at: https://arxiv.org/pdf/1211.1741.pdf.

[6] [JT42].

[7] [JT75]: CBPF, Ciência e Sociedade Document No. CBPF–CS–011/85; cf. [JT].

[8] Igor Pacca, '*O "boy wonder" volta à USP em 1960*', in: *Jornal da Unicamp*, March 30th–April 3rd, 2005—p. 8.

[9] Marques (2011); see also Marques (2005).

[10] Schenberg (1978); see also [JT83]: Jayme Tiomno, 'Experiências de uma vida na Universidade Brasileira', ['Experiences of a lifetime in Brazilian universities'], in: CBPF, Ciência e Sociedade, Document CBPF–CS–005 (1986) ([JT83]). This is a Semester Inaugural Lecture given at the UFPA (Universidade Federal do Pará) in March, 1986; supplementary documents are also included in the file.

[11] Without her knowing it, Silvia had been the target of a denunciation. She had imagined for 50 years that the motive for her arrest had been that University rules required the elections for the Student Academic Council to be observed by a representative of the Congregação ['Assembly'], as the Department Council was called, in order to guarantee that they were carried out correctly. Silvia had therefore signed the official election results as "representative of the Congregação". She assumed that since the security agents didn't understand the meaning of the term, the word 'Congregação' appeared highly suspicious to them and they supposed it to be the name of a subversive organization.

[12] File on Silvia Tiomno Tolmasquim, from the Brazilian Security Service (DOPS, 'Department of Political and Social Order'; Inform. No. 9, May 14th, 1969. Series BR9, Vol. 80) during the dictatorship; Archive [APP].

[13] See e.g. Knobel & Vieira (2016); this is the 50th Anniversary Memorial Volume issued by the Sociedade Brasileira de Física (SBF), containing a chapter on the persecution of physicists by the military regime.

[14] File on Elisa Frota-Pessôa, from the Brazilian Security Service (DOPS) during the dictatorship; Polpol. BR23 of February 4th, 1969, page 12; [APP].

[15] Roland Koberle to John Wheeler, April 28th, 1969; [JWA].

[16] José Goldemberg to John Wheeler, May 19th, 1969; [JWA].

[17] Note from Wheeler to Goldberger, September 1969; [JWA].

[18] In this connection, cf. also the section on *Enrico Predazzi*, Chap. 15. He was the first and most influential of the visiting professors to Tiomno's group at USP, and he also motivated others to follow him there.

[19] Manoel da Frota Moreira (Director of the CNPq) to Jayme Tiomno, July 2nd, 1969; [JT].

[20] Leite Lopes (1977).

[21] Quoted from an interview with *Mario Novello*, conducted by ATT for this book, on August 29th, 2019.

[22] Personal communication from Bert Schroer, 2019.

[23] There had been an organizational restructuring of the CBPF after its original Foundation expired in 1974, and it became an institute attached to the CNPq. Its leader now no longer held the office of 'President', as at the time of its founding, but instead was its 'Director'.

[24] Schenberg (1978).

[25] Quoted by Amélia I. Hamburger, August 3rd, 2009, in '*Obras científicas de Mario Schénberg*'; cf. https://www.scielo.br/scielo.php?script=sci_arttext&pid=S0103-40142002000100012.

[26] Videira (1997) and Videira (2017): These are memoirs of Plínio Süssekind Rocha published in 1997 and 2017 in the series CBPF: Ciência e Sociedade by Antonio Luciano L. Videira. Available online.

[27] Swieca and Videira (2013): This is the belated publication of the obituary for Süssekind Rocha, originally written in 1972.

[28] Sergio Joffily, *'Homenagem aos 80 anos do Professor Jayme Tiomno'* ('Homage on the occasion of the 80th birthday of Prof. Jayme Tiomno'); Document CBPF-CS-016/03, December 2003.

[29] The invitations from Wheeler and from Dyson can be found in the Tiomno Archive at MAST, Rio de Janeiro; [JT].

13

Exile—Princeton Once More: Gravitation and Field Theory

Princeton *Redux*

Jayme Tiomno and Elisa Frota-Pessôa arrived in Princeton at the beginning of 1971. Things had changed quite a lot there since Tiomno had departed just over 20 years before. The Old Guard—Einstein, Weyl, von Neumann, and Oppenheimer—had passed on, and Gödel was retired and withdrawn; but others with whom Tiomno had worked before—such as Abraham Pais, Eugene Wigner, and of course John Wheeler—were still at work.[1]

Twenty years earlier, Jayme Tiomno had been a not-so-young student, 'footloose and fancy free', and had lived in graduate-student housing at the University, working long hours on his research. He had dreamed then that Elisa might join him so that they could share the experience, but that never came to pass. Back then, he was eager to return to Brazil; certainly for reasons of principle, because he had decided to devote himself to establishing excellent scientific education and research in his native country—but also because he wanted to rejoin Elisa, and because it is difficult for a true *Carioca* to live far from Brazil (and from Rio de Janeiro) for all too long.

This time, he was returning as a 'Visiting Senior Research Physicist'. Elisa accompanied him, under very different circumstances. They had lived together for most of the past 20 years. Her children, Sonia and Roberto, were now grown, and both were married and finishing their own studies. At other institutions where Tiomno and Elisa had gone together as visiting or tenured scientists (in London, Brasília, Trieste, São Paulo…), she was also offered an opportunity to work, but in the USA there were strict rules against 'nepotism'

© Springer Nature Switzerland AG 2020
W. D. Brewer and A. T. Tolmasquim, *Jayme Tiomno*, Springer Biographies,
https://doi.org/10.1007/978-3-030-41011-7_13

which prevented that (those have since been relaxed, as educational and scientific institutions have come to realize that it is to everyone's benefit if both members of a professional couple can work, together or in separate functions) (Fig. 13.1).

But there were two additional problems. She was still very depressed by the situation that they had experienced in Brazil in the previous two years. "The worst is that I feel an immense emptiness because of not working and not having the strength, even the physical strength, to start something new",[2] she wrote to her friends Caden and Haity Moussatché; and furthermore, Princeton had just begun accepting women as graduate students and researchers, and it would not be easy for her to find something to do in physics there.

She was very annoyed at the situation of the women in the University community. They were mainly wives, and were known by their husbands' names—she was 'Mrs. Jayme Tiomno'. "The women here in Princeton have no names!", she exclaimed. It was typical that there were 'wives' programs',

Fig. 13.1 Jayme Tiomno and Elisa Frota-Pessôa at Princeton, in 1971. *Source* Private archive, [EFP]

organized breakfasts or lunches for the wives of faculty and staff members. "Today is Sunday, and I woke up with a yearning to have breakfast with Caden and Haity [*Moussatché*]" (Fig. 13.2). But that was unfortunately not a possibility where she was. "What exists here is an invitation to the ladies' coffee on the 10th and the 30th! Are you getting the sense of tragedy? I can't even complain, since everything is being done to please us…". More interesting were the mixed social gatherings with both the men and their wives, where she was able to find interesting people to talk to. She spent most of her time reading, taking English lessons, attending lectures at a museum, doing fitness exercises at home and, of course, taking care of the housework.[3] They were living in an apartment on Von Neumann Drive, an apartment with a pleasant varanda and a lovely view of the gardens outside.

Nevertheless, after the preceding two hectic and threatening years in Brazil, they could live a calm life in the shady suburban streets of Princeton. "It is sad to find that, in spite of all this, and in spite of missing all of you, I find it splendid to have gotten out of there".[3] Her daughter Sonia had the following to say about this period in her mother's life[4]:

Elisa was 48 years old [*at the time of their forced retirement*]. They were very active, engaged with various projects, and their retirement was an enormous setback. Jayme spent hours in his room, rather depressed, and Elisa started to devote all her attention to him. Jayme was well known, and soon, invitations began to arrive from abroad. He accepted a position at Princeton, one of the

Fig. 13.2. The two old friends Haity Moussatché and Jayme Tiomno, 1995. *Source* Private archive, [EFP]

most famous universities in the USA, and *mamãe* accompanied him, as his wife. It was a rather different experience for Elisa. Jayme worked much of the time and with great success, and Elisa was always an active supporter, a participant, giving him any help that she could. She was an excellent cook and hostess, joined in the local life, invited colleagues and brought young relatives from Brazil to visit their house. At least three nephews spent several months with them.

Elisa herself, in later interviews, doesn't mention their time in Princeton at all. It seems for her to have been a not-unpleasant but also an unimportant period, perhaps because she was simply a 'housewife' while there, a role that she could readily fulfill but which she found uninteresting.

For Jayme Tiomno, in spite of experiencing a stimulating time scientifically, a certain background sadness must have permeated his thoughts—the homeland which he clearly loved and for which he had made such great efforts had not only rejected those efforts, it had revoked the very possibility for him to work in his chosen profession. In some ways, going to Princeton must have been like a 'coming home' for him, since he had spent nearly three years there before, at a formative time in his life. But for him—and for Elisa—it could never be what Brazil was for them. In that sense, he was more attached to his native land than were his colleagues and friends Roberto Salmeron and José Leite Lopes, both of whom spent large portions of their adult lives away from Brazil (Leite Lopes in part under threat of death). Of the 22 years between 1964 and 1986, both spent only three in Brazil, and Salmeron was a genuine 'French-Brazilian', although at times reluctantly.

But it was indeed a very productive time for Tiomno's science, and that was a consolation to Elisa: "Here, everything is in order. Jayme is working very hard, and he has already nearly finished two articles that are to be sent for publication in the coming days. I believe that Princeton is ideal for the work he is doing, and that he is happy with that".[3] As soon as he arrived there, he started catching up on what other people were doing scientifically and finding out what the significant problems were that they were trying to solve. "Jayme is a new man!", Elisa wrote.

Wheeler's New School: Gravitation and Relativistic Astrophysics

Scientifically, Princeton in 1971 also presented a very different aspect from what it had in 1950, when Tiomno had last worked there. His old mentor John Wheeler had made his first 'sea change' soon after Tiomno's departure in late 1950—he had ended his *'everything is particles'* period, during which he had worked in nuclear and particle physics for over 20 years, and entered the '*everything is fields*' phase, where he concentrated on relativity, gravitation and field theory. He had made some tentative moves in that direction already while Tiomno was completing his Master's, in 1948/49, but as Tiomno wrote at the time to Elisa, Wheeler himself was not yet very sure-footed in that field.

Wheeler dated his transition to the academic year 1952/53.[5] He had announced a lecture/seminar course on Relativity for that year, and planned to make it the basis for a book. This was his method of making the transition from particles to fields—giving the course would force him to learn the subject thoroughly himself. The book,[6] entitled simply '*Gravitation*', finally appeared in 1973, coauthored with his former students *Charles W. Misner* and *Kip S. Thorne*.[7] Wheeler was probably working on finishing it while Tiomno was in Princeton in 1971/72.

One must remember that the field of General Relativity had been something of a 'Sleeping Beauty' for over 25 years by 1952. After Einstein's publication of his seminal papers in the years 1915–1920, and some important work by other authors in that period, interest in it had subsided. This early period included the solution to Einstein's field equations found by *Karl Schwarzschild* (1873–1916). He showed that a 'gravitational singularity' could arise if a certain high mass density were to occur in a limited region of space; it would collapse to a region of infinite density, where General Relativity would no longer hold. Inside a given radius surrounding that singular region, now known as the 'Schwarzschild radius', even light would no longer be able to escape the strong gravitational field. This defines a surface around the singularity, now termed the 'event horizon'. Schwarzschild worked out his solution while serving as a soldier on the Russian Front during the First World War, in 1915. He died of an autoimmune disorder contracted at the front, at age 42 in 1916.

Several other important consequences of General Relativity (GR) were identified in that early period, such as the possibility of *gravitational waves*, propagating through spacetime at the speed of light, and the '*gravitational lensing*' effect, as well as the solution implying an *expanding universe* (Friedmann's equations, 1922; observations beginning in the late 1920s, especially by Edmond Hubble). Einstein himself was however skeptical of the physical relevance of these mathematical solutions. A few authors continued investigating GR throughout the period 1925–1950, but its show was in the main stolen by quantum physics in those years, occupying most theoreticians during that time, while Einstein himself spent his last 30 years in the (vain) search for a 'Unified Field Theory' which would combine gravitation and electromagnetism (and eventually include quantum mechanics). It has still not been formulated today.

One of the scientists who was still interested in GR and gravitational theory in the 1940s was Mario Schenberg, as we have seen from Tiomno's early work with him in São Paulo. But it was still quite an empty field when Wheeler turned to it in 1952.

Another author who had considered the implications of gravitational collapse in the real world, in the late 1930s, was *J. Robert Oppenheimer*. With his students *Hartland Snyder* and *George M. Volkoff*, he treated the collapse of a 'burnt-out' star, or stellar remnant, leading either to a neutron star (when the pressure within the stellar remnant becomes so high as to force the electrons and protons in its plasma to combine to yield neutrons and neutrinos, in a 'reverse beta decay'). The resulting object has the density of nuclear matter and a mass comparable to that of the Sun within a sphere of diameter ca. 20 km[8a]; or else, when its mass is even greater, the collapse of the stellar remnant will continue unchecked, leading to a *gravitational singularity* (today termed a 'black hole') as postulated by Schwarzschild.[8b] Wheeler re-read those articles in 1952 and they set him to thinking about the implications of gravitational collapse for astrophysics. This led in the following 20 years to many important contributions, a number of highly successful students from his research school, and the revival of the field, which has become a major branch of theoretical (and, more recently, observational) physics: *Relativistic Astrophysics*.

As we have seen, Wheeler popularized the name 'black hole' for a collapsed gravitational singularity in 1967/68, a term which has captured the imagination of both (astro-) physicists and the general public. Many of his works, and those of his students in the 1950s through the mid-1970s and continuing onward, dealt with various aspects of black holes. Wheeler made a strong contribution to the 'reanimation' of General Relativity and of gravitational theory in the wider sense,[9] and to the founding of relativistic astrophysics. Some of Wheeler's students, in particular *Joseph Weber* and *Kip Thorne*, also contributed essentially to the experimental observation of gravitational waves[10] and thus to the founding of a new branch of observational astronomy, for which the theoretical foundation was laid to a considerable extent also by Wheeler's school of relativistic astrophysics.[We note, however, that the roots of relativistic astrophysics go back to the early 1930s, and several other groups besides Wheeler's, in particular in the UK, also pioneered the field in the 1960s.[11]

New Collaborations

By the time that Jayme Tiomno arrived in Princeton in early 1971, he had himself already written a handful of papers related to gravitation and field theory, and their connections to quantum mechanics and particle physics. These include five of his earlier publications,[12] spanning the years 1955–1970. Those areas had, to be sure, not been his major research focus in the years before his return to Princeton; however, he was certainly ready to apply himself in the general direction of field theory when he arrived there. During the subsequent 1–1/2 years, he did not work directly with Wheeler, as he had done over 20 years before—very likely, Wheeler himself was by then simply too busy with his many projects. But there was a group of younger academics, graduate students, visitors and postdocs, both at the IAS and at Princeton University, who had been attracted by Wheeler's school of relativistic astrophysics and were eager to work in the rapidly-developing area of gravitation and field theory, especially the electrodynamics of black holes; and Tiomno, who had a strong background in electromagnetic theory, collaborated with a number of them in his usual open and cooperative manner during his stay in Princeton, culminating in over a dozen publications in the next few years. His stay was thus extremely productive in the scientific sense, and might be termed a renewal of his life and work as a theoretician. In addition to the collaborative works with younger physicists at Princeton, he published several papers in this new field on his own during the years up to 1975, and continued working in the field with his own students back in Brazil for years afterward.

Tiomno's more senior collaborators in Princeton were *Remo Ruffini* and *Leonard E. Parker*. Both were already well established when Tiomno arrived in Princeton: Ruffini was Assistant Professor at Princeton University, and Parker was Associate Professor at the University of Wisconsin, working as a visitor in Wheeler's group at Princeton (and later at the IAS).

Remo Ruffini, born May 17th, 1942 in *Briga Marittima*, Italy—now *La Brigue*, France,[13] studied physics and mathematics at the University of Rome '*La Sapienza*', completing his doctorate there in 1966. He then went as a postdoctoral fellow to the University of Hamburg, Germany, where he worked in 1967 with Pascual Jordan, one of the pioneers of quantum physics. In 1968, he moved to the Palmer Physical Laboratory at Princeton University, working with John Wheeler, and continued working in Princeton at the Institute for Advanced Study (IAS) through 1970, when he became Instructor and then Assistant Professor at Princeton University. He held that position while Tiomno was there. Later on, in 1974–76, he was again a Fellow at the

IAS, and was on leave in 1975 for guest stays at the University of Western Australia (Nedlands) and the University of Kyoto in Japan. The following year, he accepted a professorship at the University of Catania in Sicily, and in 1978 he became Professor of Physics at his *alma mater, La Sapienza* in Rome, a position which he held until his retirement in 2012. He was Chairman of the Department of Theoretical Physics there.

Ruffini, in addition to his research on gravitational theory and relativistic astrophysics, has been very active in scientific organization and advising. He was an advisor to NASA on the scientific potential of space stations in 1986–88. He co-founded the *Marcel Grossmann Meetings* on relativistic astrophysics, together with Abdus Salam, in 1975; they are held every three years at varying locations around the world, and continue with great success today. He helped found the *International Center for Relativistic Astrophysics* (ICRA), headquartered in Pescara/Italy, and was elected to be its President in 1985, an office which he still holds. In 2003, he founded its international network organization, ICRANet, and has been its Director since then. In 1987, he became co-chairman of the *Italian-Korean Meetings on Relativistic Astrophysics*, and in the years 1989–93, he was President of the Scientific Committee of the Italian Space Agency. He edits several series of conference proceedings and is on the editorial board of various journals. He has co-authored a number of books on neutron stars, black holes,[14] gravitational waves, relativistic astrophysics and cosmology.

Ruffini and Jayme Tiomno collaborated on four scientific articles[15] during the latter's stay in Princeton.

Leonard Emanuel Parker (originally Pearlman), born in Brooklyn/NY in May, 1938,[16] obtained his BA in physics at Rochester University in 1960, going on to graduate school at Harvard, where he completed his MA in 1962 and his Ph.D. in 1967, with a thesis on '*The creation of particles in an expanding universe*' under the guidance of Sydney Coleman. He joined the physics faculty of the University of North Carolina (Chapel Hill) in 1966 as Instructor (following in the footsteps of John Wheeler, 30 years earlier). He then became Assistant Professor (1968), Associate Professor (1970), and Full Professor (1975) at the University of Wisconsin (Milwaukee). In 1971/72, he spent a year as guest scientist in the group of John Wheeler at Princeton University, and returned several times to the IAS in Princeton during the 1970's. He and Tiomno collaborated on two articles during Tiomno's stay at Princeton.[17]

Leonard Parker's research has concentrated on QED in curved spaces and on relativistic astrophysics and cosmology. His work on particle-antiparticle pair production in changing gravitational fields was seminal for later developments, including the radiation from black holes (Hawking radiation) and the fluctuations in the cosmic microwave background (CMB) radiation. He has dealt with rapidly-rotating neutron stars, the effects of curved spacetime on atomic spectra, and inflation models of the early universe.

Several more junior colleagues were also co-authors of publications with Tiomno during his Princeton stay. They included *C. V. Vishveshwara*,[18] *Marc Davis*,[19] *Jeffrey M. Cohen*,[20] *Frank J. Zerilli*,[21] *Robert M. Wald*,[22] and *Reinhard A. Breuer*.[23]

Work on Relativistic Astrophysics

Jayme Tiomno's stay in Princeton in 1971/72 resulted in no fewer than thirteen publications ([JT45–57]). Thus it is clear that Tiomno rapidly became proficient in gravitational theory and was able to write articles alone or with relatively inexperienced co-authors, as an independent researcher in the field, as well as collaborating with well-established scientists.

The first seven of these articles were already published in 1972, while Tiomno was still in Princeton or soon after he returned to Brazil. The others appeared during the following year, 1973, when he was back in Brazil, but were no doubt formulated in Princeton and published with the usual delay time of 6–8 months by the journals (for the most part, the *Physical Review*). The final article dating from the Princeton stay, [JT57], with Breuer and Vishveshwara as co-authors, was a continuation of work begun earlier ([JT51]). It appeared only in 1975 in *Il Nuovo Cimento*, and was apparently delayed by some problem of formulation or editing. Both of Tiomno's co-authors were working elsewhere by that time.

The first of those articles from Tiomno's Princeton stay, published by Ruffini, Tiomno, and Vishveshwara, is entitled '*Electromagnetic Field of a Particle Moving in a Spherically Symmetric Black-hole Background*', and it appeared as a Letter to *Nuovo Cimento* ([JT45], submitted in November 1971). It deals with the possibility of detecting black holes by means of radiation emitted from charged particles moving in their strong gravitational fields. It is a continuation of work by Ruffini and other researchers at Princeton.

[This publication and other related papers have gained special significance today, and as of this writing: On April 10, 2019, the first direct detection of a black hole was announced, making use of the high-frequency radio waves emitted by its accretion disk, the swirl of matter orbiting, and being drawn into, the central black hole. The black hole in question is a supermassive one

(6.5 thousand million solar masses) at the center of the large galaxy M87, some 55 million light years from Earth. Its event horizon has a diameter considerably greater than that of our solar system. The detection was accomplished by 8 radiotelescope arrays working in synchrony over a large area of the globe, and shows a bright ring of radio emissions surrounding a black disk—the shadow of the enormous black hole. The detection and imaging were accomplished by the *EHT Collaboration*[24] and represent the beginning of a new field of observational astrophysics.]

Their next article, [JT46] was published in the prestigious journal *Physical Review Letters* by Davis, Ruffini, Tiomno and Zerilli (submitted in March, 1972), and it asks the question, '*Can Synchrotron Gravitational Radiation Exist?*'. *Electromagnetic* synchrotron radiation is produced by charged particles moving at relativistic velocities on a curved path, and therefore subject to a radial acceleration (for example electrons in a snychrotron accelerator ring or storage ring). If the motion of a massive particle in the gravitational field of a black hole exhibits similar properties, the resulting *gravitational* waves would be strongly focused (or 'beamed', in the terminology of the article), making them easier to detect. The authors use the Regge-Wheeler formalism developed previously in Wheeler's group to answer the question for scalar, vector and tensor gravitational radiation, depending on the masses of the particles and of the black hole, and on the orbital parameters of the particles. The result is that the concentration of energy within a plane, as observed with electromagnetic synchrotron radiation, is not so effective for the gravitational waves emitted by particles in orbit around a black hole. The authors suggest that a more detailed analysis of the properties of the gravitational radiation should be undertaken.

In June, 1972, Parker and Tiomno submitted a short note on '*Pulsars and Pair Production in Electric Fields*' ([JT47]) to *Nature Physical Science*. *Pulsars* are compact objects which radiate electromagnetic waves (typically in the radio-frequency region) in the form of pulses at a nearly constant repetition rate. They were discovered in 1967 by Anthony Hewish and Jocelyn Bell (the first pulsar is denoted as PSR B1919+21). Hewish received the Nobel prize in 1974 for their discovery. Their nature was initially mysterious; it was later explained in terms of a rotating neutron star which sends out a beam of radio waves like a searchlight—the beam appears as a pulse as it passes the observer. In 1974, a binary system of two neutron stars, one of them a pulsar (PSR 1913+16) was discovered by Russell Hulse and Joseph H. Taylor. This system radiates energy in the form of gravitational waves, causing the pulse period of the pulsar to increase gradually, and it thus permitted the first *indirect* observation of gravitational radiation. Hulse and Taylor received the Nobel prize in 1993 for their discovery.

Parker and Tiomno, in their 1972 paper, written before the neutron-star explanation for pulsars was generally accepted, propose a model consisting of a strong electric field from a central object, in which pair production (i.e. of electron–positron pairs) is occurring. The central object producing the field is surrounded by an oscillating plasma containing the pairs, which emits pulses of radiation. They calculate the ratio of emitted pulse widths to the pulse period, and find good agreement with observed values from pulsars. Other calculated quantities, such as the change in pulse period over time, and the emitted power, also agree with observations. They speculate about the nature of the central object and suggest that it might be a black hole; they however admit that their model is not likely to apply to real pulsars, but may be relevant to other observable processes. A longer, more detailed article was also published in the *Astrophysical Journal* ([JT50]; received in June, 1972).

In December 1971, Davis, Ruffini and Tiomno submitted [JT48] to the *Physical Review D*. The resulting publication is entitled '*Pulses of Gravitational Radiation of a Particle Falling Radially into a Schwarzschild Black Hole*', a title which describes its contents rather completely. The shape of the energy pulses and the angular distribution of the gravitational waves emitted by a particle falling into a black hole are calculated using methods developed by Regge, Wheeler and Zerilli, continuing work by Davis, Ruffini and other members of the Princeton group. In this article, the authors note that there is a strong burst of energy directed *inward*, which may affect the trajectory of the falling particle and thus indirectly the *outwardly-directed* (in principle observable) radiation. This article, along with the first joint paper by Ruffini, Tiomno and Vishveshwara ([JT45]), was selected for special mention in a later book summarizing then-current research in relativistic astrophysics.[25]

Subsequently, Tiomno published a Letter in *Nuovo Cimento* as sole author (received in June 1972), dealing with the '*Maxwell Equations in a Spherically-Symmetric Black-Hole Background and Radiation by a Radially-Moving Charge*' ([JT49]). It again treats the possibility of observing a black hole directly, by means of gravitational or electromagnetic radiation emitted from matter moving nearby. The model used in this paper is exact, an idealized model of the rather complicated real situation. Power spectra for ingoing and outgoing waves are computed.

At nearly the same time, Breuer, Tiomno and Vishveshwara submitted a Letter to *Nuovo Cimento*, [JT51], entitled '*Polarization of Gravitational Synchrotron Radiation*'. It also considers the possibility that the gravitational waves emitted by particles in orbit around a black hole might have similar properties to those of (electromagnetic) synchrotron radiation (this was originally suggested by Charles Misner). In this paper, an extension of the work by Ruffini, Davis and Tiomno, the emphasis is on the polarization of the emitted radiation, both gravitational and electromagnetic, as a basis for future observations. The authors calculate the Stokes parameters which characterize polarization, and the energy fluxes of the emitted waves. Detailed derivations and numerical results are promised for a later paper, which was published by the same authors in *Nuovo Cimento* (1975, received in April 1974, [JT57]). By then, Tiomno was back in Brazil, Breuer was finishing his doctorate in Oxford, and Vishveshwara was at the University of Pittsburgh. This followup article was entitled '*Polarization of Geodesic Synchrotron Radiation*' and indeed gives a detailed account of the calculation of the Stokes parameters for gravitational radiation and their comparison with the electromagnetic case.

Tiomno, as sole author, published a paper called '*Balancing of Electromagnetic and Gravitational Forces and Torques Between Spinning Particles at Rest*' in the *Physical Review D* (received in May, 1972; [JT52]). This article refers to earlier work by Tiomno in 1955 and by Tiomno and Oliveira in 1962[26], and it treats the precession of a spinning object in a gravitational field with pseudo-magnetic components, discussing the force on the object, calculated from the Dirac equation and the correspondence principle (the 'classical limit' of quantum calculations). The results are compared to contemporary work by Robert Wald, also at Princeton, and by Ruffini and Wilkins. It is based on the equivalence principle (of inertial and gravitational forces), which was found in the 1962 paper [JT36][26] to imply that the gravitational 'gyromagnetic ratio' is exactly unity (instead of two as found in the electromagnetic case for the electron). This article is essentially a continuation of Tiomno's 1962 work, which he was able to pursue while he had the time in Princeton, but it is also relevant to more recent research by others on the solutions of the Maxwell-Einstein equations for spinning charged masses, as well as to work on spinning black holes.[27]

A further article published by Jeffrey M. Cohen, Jayme Tiomno, and Robert M. Wald in the *Physical Review D* ([JT53], received July 10th, 1972) is entitled '*Gyromagnetic Ratio of a Massive Body*'. The *gyromagnetic ratio* or '*g*-factor' is the ratio between the magnetic moment and the angular momentum of a rotating, charged body (or a ring current); classically, it has the value *one*. But if the body is very massive (like a black hole), it will curve the surrounding space and the *g*-factor will no longer be exactly one. In this paper, the *g*-factor of a slowly rotating, massive charged shell with uniform charge density is calculated. When its radius is large compared to the Schwarzschild radius, $g = 1$ is still found, but increasing the mass of the shell increases *g*.

As the shell approaches the Schwarzschild radius (with increasing mass), its g-factor approaches 2 (the relativistic value found from the Dirac equation for electrons). This paper is the result of Tiomno's earlier work on spinning massive particles ([JT36], 1962) and the more recent work of Wald on the motion of spinning test bodies in strong gravitational fields.[28]

Another paper by Tiomno alone was published in the *Physical Review D* ([JT54], received June 2nd, 1972). It deals with the '*Electromagnetic Field of Rotating Charged Bodies*'. This is a very old topic, going back to the work of Ampère in the early nineteenth century, but is meant here in the modern sense of a body rotating in a strong gravitational field, such as a charged black hole. The Kerr solution of the Einstein field equations describes a rotating (non-charged) black hole, and the Kerr-Newman solution extends this to the Einstein-Maxwell equations describing a charged, rotating black hole (i.e. with electric charge as well as angular momentum or 'spin'). Following a proof by Werner Israel in 1967 for stationary, non-charged black holes (Schwarzschild metric), Wheeler and his student Jacob Bekenstein conjectured that black holes could exhibit only the three properties *mass*, *charge*, and *spin*; this is called the '*no-hair conjecture*' ("black holes have no hair", i.e. no other observable properties; they are like (enormous) elementary particles). Tiomno here makes use of the analogy with a classical system: a rotating charged ellipsoid with infinite electrical conductivity and infinite (or unit) magnetic susceptibility. He calculates the properties of such a rotating ellipsoid and applies them to understanding the Kerr-Newman solutions for a charged, rotating black hole (in particular its g-factor of 2). This paper is thus closely related to [JT53], the *Gyromagnetic Ratio of a Massive Body*.

The article '*Vector and Tensor Radiation from Schwarzschild Relativistic Circular Geodesics*' ([JT55]) by Breuer, Ruffini, Tiomno and Vishveshwara was submitted for publication to the *Physical Review D* on June 8th, 1972. It deals with a topic originally suggested by Misner, and gives a detailed treatment of the gravitational and electromagnetic radiation emitted by a massive charged or uncharged particle in a relativistic orbit around a Schwarzschild black hole (no charge, no spin) of mass M. The authors calculate the power spectrum of the radiation and compare the cases of scalar, vector and tensor fields [spin $s = 0$ (scalar), 1 (electromagnetic), 2 (gravitational)] for higher multipoles. It is an elaboration of the work published earlier as a Letter (*Can Synchrotron Gravitational Radiation Exist ?* [JT46]), giving more details of the calculations and their results. The authors give a simple formula for the power spectrum of the radiation, they discuss the implications for the observation of gravitational-wave radiation, and they describe the enhancement due to 'beaming' or orbital-plane focusing of the higher multipoles considered in this paper.

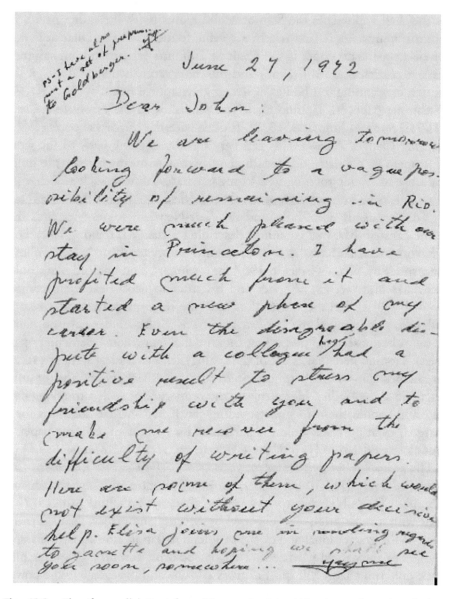

Fig. 13.3. The 'farewell letter' from Tiomno to John Wheeler, written just before Tiomno and Elisa's departure from Princeton in June, 1972. *Source* Private archive, [JT] (owner); used with permission. Also in the John Wheeler Archive [JWA], Princeton University/APhS

This article was however the subject of a certain disagreement among its authors—after it had been submitted to the *Physical Review*, in early June, 1972, Remo Ruffini, on his own initiative, contacted the journal editor

Simon Pasternack, asking that its publication be blocked. Breuer, Tiomno, and Vishveshwara then planned to publish it without his name, maintaining that it had been sufficiently discussed within the group. Wheeler was evidently involved in the discussion, and in the end, the article appeared with all four authors' names.[29]

Finally, one further article was published by Tiomno alone in the *Acta Physica Austriaca* in 1973 ([JT56]). Its title is '*On Gravitation-Induced Electromagnetic Fields*'; it is however not readily available today (the journal is not digitally archived). The article is presumably an elaboration of [JT46] and of Ruffini's 1972 paper about the bremsstrahlung (electromagnetic 'braking' radiation) produced by a charged particle falling in a strong gravitational field (e.g. the field of a black hole). Both are cited in a paper by Stephen Boughn of Stanford[30] on '*Electromagnetic radiation induced by a gravitational wave*'.

The number of these articles submitted between March and July of 1972 is no less than astonishing: there were 9 (of which 3 were published by Tiomno alone). Even taking into account that he was of course helped by his co-authors, Jayme Tiomno exerted a tremendous effort to finish and submit all of those articles in his last few months before leaving Princeton in the summer of 1972. His daily routine was hectic and filled with work, beginning at 8:30 am and continuing until 2 pm, when he took an hour's break for lunch at home, then again working uninterruptedly from 3 to 8 pm. At 8 pm he went home for dinner, then worked at home until midnight. On Saturdays, he worked 'only' until 2 p.m.[31] His motives and feelings at the time are reflected in his farewell note to Wheeler, written on June 27th, 1972[32]:

> We are leaving tomorrow, looking forward to a vague possibility of remaining in Rio. We were much pleased with our stay in Princeton. I have profited much from it and started a new phase of my career. Even the disagreeable dispute with a colleague has had a positive result to stress my friendship with you and to make me recover from the difficulty of writing papers. Here are some of them, which would not exist without your decisive help. Elisa joins me in sending regards to Janette and hoping we shall see you soon, somewhere . . . Jayme.
> PS. I have also sent a set of preprints to Goldberger.

The handwritten letter is shown above. The 'dispute with a colleague' mentioned in it refers to the article that Tiomno had written together with coauthors Breuer, Vishveshwara and Ruffini ([JT55]; see above). The discussion about its content and authorship took place just two weeks before Tiomno's departure from Princeton (Fig. 13.3).

Their initial idea had been to spend a year in Princeton and then 6 months more in Mexico, but the IAS offered to extend their stay in Princeton by another half-year, through June of 1972. Jayme Tiomno and Elisa were evidently eager to return to Brazil, although they couldn't know what conditions there would be like on their return. Perhaps they were hoping that the dictatorship would soon come to an end, after more than 8 years. But if so, that hope was disappointed, and some difficult years still lay before them.

Notes

[1] Wigner retired in 1971, and Wheeler in 1976, although both remained active in physics for some years afterward; Pais had moved to Rockefeller University in 1963.

[2] Elisa Frota-Pessôa to Caden and Haity Moussatché, March 21st, 1971; [HM]. Haity Moussatché was born in *Izmir* (*Smyrna*), Turkey in 1910. He emigrated to Brazil and studied medicine at the UB, then became a research assistant at the *Instituto Oswaldo Cruz* in Rio. He was later director of its Pharmacodynamics Department and was well known as a biologist and physiologist. He participated in the founding of the UnB in 1965, and was active in the SBPC. He was dismissed during the dictatorship and went into exile in Venezuela, from whence he returned in 1985. He was a co-founder of the International Society of Toxicology and of the Brazilian Society for Biology, and a member of several scientific academies. He and his wife Caden were close friends of Jayme Tiomno and Elisa Frota-Pessôa for many years, until his death in 1998. See the photo (Fig. 13.2).

[3] Elisa Frota-Pessôa to Caden and Haity Moussatché, May 19th, 1971; [HM].

[4] Personal communication from Sonia Frota-Pessôa, 2019; [EFP].

[5] See Wheeler and Ford (1998).

[6] Misner, Thorne and Wheeler (1973).

[7] Misner has had a distinguished career in gravitation physics at Princeton and at the University of Maryland. Thorne occupied Feynman's old chair at Caltech until his retirement in 2009, and he was co-recipient of the Nobel prize for physics—for his contributions to the detection of gravitational waves—in 2017.

[8a] Oppenheimer and Volkoff (1939).

[8b] Oppenheimer and Snyder (1939).

[9] This is described for example in the lecture '*John Wheeler and the Recertification of General Relativity as True Physics*' given by his former student Charles Misner at the 'J. A. Wheeler School', Erice/Italy in 2006, in honor of Wheeler's 95th birthday, and included as an article in a *Festschrift* volume; cf. Misner (2010). Misner holds that this was indeed Wheeler's most important contribution to science.

[10] See the website of the LIGO/Virgo Collaboration, at https://www.ligo.org/.

[11] The term 'relativistic astrophysics' was first used in connection with the 'Texas Symposia on Relativistic Astrophysics', which take place every two years, beginning in Dallas/TX in 1963, and are still active—the latest (30th) meeting was held in Portsmouth/UK in December, 2019; see https://texas2019.org/. The 'Marcel Grossmann Meetings', originated by Remo Ruffini and Abdus Salam in 1975, also continue today (see section on Remo Ruffini in this chapter, and Chap. 16), and additional conference series also exist, reflecting the activity of many research groups in this field around the world.

[12] [JT22, 31, 36, 42, and 43].

[13] See the web page of Remo Ruffini, at https://www.icra.it/People/Ruffini.htm.

[14] Ruffini also contributed to the popularization of the term 'black hole'; cf. Ruffini and Wheeler (1971).

[15] [JT45, 46, 48, and 55].

[16] See the web page of Leonard Parker, at https://uwm.edu/physics/people/parker-leonard/.

[17] [JT47], [JT50].

[18] Vishveshwara (1938–2017), often known as 'Vishu', was an Indian physicist who had done graduate work at Columbia University and the University of Maryland, where he received his Ph.D. in 1968, working under Wheeler's former student Charles W. Misner. Vishveshwara was officially a postdoc at Boston University and at New York University during Tiomno's stay in Princeton, but he had evidently been working with Ruffini after receiving his doctorate at Maryland, and continued the collaboration. He later moved to the University of Pittsburgh, and then returned to Bangalore in 1976, where he became professor at the Raman Institute and then at the Indian Institute of Astrophysics, retiring from the latter in 2005. He was called "the black hole man of India".

[19] Davis (b. 1947) did his graduate work at Princeton University and worked initially with Ruffini, and was a co-author of four articles with him, two of which were also co-authored by Tiomno; they deal with gravitational radiation. He then switched to experimental physics, and was later professor at UC Berkeley.

[20] Cohen (1940–2003) had received his Ph.D. from Yale in 1966 and was a postdoc at the IAS from 1969–1971. He joined the faculty at the University of Pennsylvania in Philadelphia in 1971, and remained there until his early death in 2003.

[21] Zerilli (b. 1942) was an earlier Ph.D. student in the Wheeler group; he is known for the *Wheeler-Zerilli equation*, describing the emission of radiation from black holes. By the early 1970s, he had moved to the University of North Carolina (Chapel Hill), but was still collaborating with Ruffini and others at Princeton.

[22] Wald (b. 1947) was also a doctoral student at Princeton during Tiomno's visit, and received his Ph.D. there in 1972. After postdoctoral positions at the University of Maryland and the University of Chicago, he joined the faculty of the latter, where he is today Distingished Professor at the *Enrico Fermi Institute.*

[23] Breuer (b. 1946) is a German physicist who studied in Würzburg, at the University of Michigan and at the University of Maryland, where he was located when he worked with Ruffini and Vishveshwara, and with Jayme Tiomno, in the early 1970s. He obtained his DPhil from Oxford in 1974 and then spent a postdoc tenure at the *Max-Planck-Institut für Astrophysik* in Munich, after which he turned to science journalism. He retired in 2010 as Editor-in-Chief of *Spektrum der Wissenschaft*, the German version of *Scientific American* magazine.

[24] Event Horizon Telescope Collaboration: cf. https://eventhorizontelescope.org/.

[25] Rees, Ruffini and Wheeler (1974).

[26] [JT22]: J. Tiomno, '*Relativistic Theory of Spinning Point Particles*' (1955); [JT36]: C.G. de Oliveira and J. Tiomno, '*Representations of Dirac Equation in General Relativity*' (1962).

[27] Kerr (1963).

[28] Robert M. Wald, in the *Physical Review D*6, 406 (1972).

[29] Reinhard A. Breuer, Jayme Tiomno and C.V. Vishveshwara to Remo Ruffini, June 12th, 1972. Also enclosed in this file: Reinhard A. Breuer, Jayme Tiomno and C.V. Vishveshwara to Simon Pasternack (Editor of the *Physical Review*), undated; [JWA].

[30] Boughn (1975).

[31] Elisa Frota-Pessôa and Jayme Tiomno to Annita Tiomno and Jayme's sister Riva, May, 23rd, 1971; [STT].

[32] Handwritten letter from Tiomno to Wheeler, dated June 27th, 1972; [JWA]. See also Bassalo and Freire (2003), Footnote [29] there.

14

PUC—The Catholic University of Rio de Janeiro

Jayme Tiomno and Elisa Frota-Pessôa arrived back in Rio de Janeiro from Princeton in mid-1972 to find the dictatorship still in full force. 'Intelligence units' had been set up in practically all of the public universities, and most of the universities in Brazil were public. They kept track of everything that was happening within the university, noting even the topics of masters and doctoral theses in the social sciences. They formed a communicating network, which could be consulted before authorizing the hiring of a new faculty member, giving a grant for a research project, or even for renewing a passport or granting a visa for a visit abroad. This procedure of double-checking and controlling every action of faculty members and researchers by the government was called '*cassação branca*' ('white cassation', disbarment without due process). It included all of those who had been dismissed or forcibly retired from their positions, or who had participated in any form of protest or disagreement with the government, such as signing a petition in support of a professor who had suffered some punishment by the regime, or even simply carrying out research on the role of blacks in Brazilian society, or on previous revolts and anti-government movements in Brazilian history, or whatever else might possibly be construed to oppose the powers of the dictatorial government. Nothing was officially publicized or justified; only the contract or the grant was not authorized, the visa for going abroad or the funding for travel to a conference was not granted.

There was a climate of fear and distrust, since any colleague at the university was potentially a spy or an agent of the government. Nevertheless, there were various groups who tried to organize and practice resistance and the mobilization of a popular movement against the repression. One of the

© Springer Nature Switzerland AG 2020
W. D. Brewer and A. T. Tolmasquim, *Jayme Tiomno*, Springer Biographies,
https://doi.org/10.1007/978-3-030-41011-7_14

most important of them was the *Sociedade Brasileira para o Progresso da Ciência* (SBPC), whose annual meetings gathered thousands of participants and were a focal point for discussions of the national problems. In 1977, the government of the State of São Paulo prohibited having the annual meeting of the SBPC at USP, as planned. A great effort was made at the last minute, and the meeting was successfully transferred to PUC [*the Catholic University*] in São Paulo. Indeed, the Catholic universities were among the few institutions of high-quality higher education that were not instruments for governmental manipulation, and they often became a sort of shelter, a refuge from arbitrariness and authoritarianism, in spite of some cases of collaboration with the repressive forces.

Many of Tiomno's and Elisa's former students had moved to the Catholic University of Rio de Janeiro (PUC/Rio), which became a kind of 'safe haven' during the most difficult period of the dictatorship.

PUC/Rio

The *Pontifical Catholic University of Rio de Janeiro*, PUC/Rio, was established in 1940 by the Jesuits (the Society of Jesus, SJ), to "emphasize humanistic values in the pursuit of knowledge". Its campus is located to the southwest of the city center, in the relatively quiet suburb of *Gávea*. Its Physics Department was founded (as the *Instituto de Física*) in 1957 by Father *Francisco Xavier Roser*, SJ, an expert on natural radioactivity, who died in 1967.[1] Some of their former students had tried before, in 1969, to find Tiomno a position at the Catholic University, but that had been frustrated by the AI-5 and its accompanying laws. By 1972, however, there was perhaps more leeway for getting around the occupational ban than there had been soon after its original enforcement in 1969/70.

This time, they had the support of the institution, in particular of the Physics Department's head, Father Collins; but that was not sufficient to overcome the opposition of the government to their employment, in spite of the fact that PUC was and is a private institution, supported mainly by the Catholic Church. In some manner, the university administration asked the Department of Political and Social Order (DOPS), the state Security Service of the dictatorial regime, for information on Tiomno so that his contract could be finalized, and that was not authorized.[2] It was nevertheless made possible finally through the intervention of Pope Paul VI,[3] combined with a positive recommendation by PUC. Most likely this was due to the influence of Carlos Chagas Filho, Tiomno's former professor and first mentor at the Medical School of the old *Universidade do Brasil.* Chagas Filho was a researcher who had great prestige on the national and international level, and he had been invited to serve as President of the Papal Academy of Sciences at the Vatican, where he had some access to the Pope. However, no concrete written evidence for his intervention has been found. Finally, at the end of 1973, Tiomno was allowed to join the faculty at PUC: "Jayme's contract with PUC was signed. The Rector decided to offer him the contract *without consultation* [*emphasis in the original*]. We are hoping that the situation will remain stable. Can you believe it?", Elisa wrote to her friends Caden and Haity Moussatché[4] (not knowing of the intervention that had made the offer possible).

This was a fortunate intervention for Tiomno and Elisa, as well as for PUC/Rio. Jayme Tiomno's presence there, as an integrating personality and a prominent figure in Brazilian physics, was without doubt an advantage for the success and the reputation of its Physics Department. However, there was a restriction: Tiomno was not officially allowed to be a thesis advisor for students at PUC. This was a rule that was invented especially for him, a 'lex Tiomno'. In any case, his position at PUC gave Tiomno and Elisa an academic home during a period when the only alternatives would have been isolation or a long-term exile, something that neither of them wanted or would have accepted without personal damage. Elisa might have adapted, as she did in Princeton, but for Tiomno, that would have been a kind of personal defeat from which he would have not soon recovered.

In 1973, when Tiomno joined its faculty, the Physics Department at PUC was small and homogeneous, but with a growing positive reputation, not least because of its theoretical physics group, composed in the main of his own students. Jayme Tiomno says of this group in an interview in 1977,

> ... But many of them are here at PUC, for example professors Jorge André Swieca, [*Antonio*] Luciano Leite Videira, Alceu Pinho, Erasmo Ferreira, Nicim Zagury, and others. Practically all of the infrastructure at PUC was set up by personnel educated under these conditions, that is to say that group of students from the *Faculdade de Filosofia* [*UB*], who afterwards had their specialized [*education*] at the *Centro* and who, with the adverse conditions at the *Centro*, found refuge here [*at PUC/Rio*]. Even at the *Centro* – practically all those who are still there – the majority of the older ones were our students.[5]

[PUC/Rio was recently (2016) ranked as the "fifth-best university in Latin America" by the *Times Higher Education* survey, and was ranked first among all the private institutions of higher education in Brazil by the Brazilian Ministry of Education. It currently has over 10,000 undergraduate and

7,500 graduate students (including 4,000 external (extension) students), and around 800 faculty members. It was of course much smaller and less well-known in the early 1970s, when Jayme Tiomno and Elisa Frota-Pessôa went there to work. We note that they were not the only victims of the AI-5 who were able to continue working in physics at PUC/Rio. Their colleague from the UB/UFRJ, Sarah de Castro Barbosa,[6] who was also forcibly retired in 1969, was likewise able to continue her career at PUC.]

The 'PUC Quartet'

Many scientists and academics in Brazil felt the restrictions levied by the military dictatorship, even if they were not themselves arrested, terrorized or blacklisted; and that pressure increased generally after the imposition of the 'AI-5' in late 1968. A private university like PUC/Rio offered reduced visibility and less direct interference from government agents than large, state-supported institutions such as USP or UFRJ. Furthermore, it was able to promote faculty members and make adjustments to the salaries of its employees more easily, correcting for the omnipresent inflation. The public universities, in contrast, were dependent on government authorizations. Several younger researchers, among them *Jorge André Swieca*, *Nicim Zagury*, *Antonio Luciano Leite Videira*, and *Erasmo Madureira Ferreira*, all in their mid-thirties and all former students of Tiomno and Elisa, either as undergraduates at the FNFi/UB and/or as graduate students at the CBPF or USP, had obtained faculty positions at PUC/Rio in the period 1967–1970. The Physics Department there at the time was relatively small. These four, whom we might call the 'PUC Quartet', had career stories that were intimately bound up with Tiomno's and Elisa's efforts and plans to establish physics research and education in Brazil—they belonged to the first generation of students who profited from those efforts, and who were inculcated with the ideals and values of the founders' generation, which included Tiomno and Elisa.

Jorge André Swieca was born in Warsaw, Poland in December 1936, as *Jerzy Andrzej Swieca* (Fig. 14.1). His family, of Jewish origin, decided to escape from Poland at the beginning of the 2nd World War in September, 1939, when that country was invaded from the west by Nazi Germany and from the east by the Soviet Union. They were able to travel by rail to Vladivostok and from there to reach Japan, hiding the 3-year-old Andrzej (for whom they had no identity papers) as necessary. After two years spent in Japan and Argentina, they finally settled in Rio de Janeiro, shortly before Brazil entered the war. Jorge André, as he was now called, attended

Fig. 14.1   Jorge André Swieca, 1970s. *Source* Private collection of J. A. Swieca. Reproduced with permission [This photo was also published online at https://jaswieca.if.uff. br/reminiscencias/reminiscencias_moyses.pdf. A very similar, but distinct photo can be found in the *Emilio Segré Archives* of the AIP.]

school in Rio and obtained his Bachelor's degree in physics and mathematics at the FNFi/UB from 1954 to 1958, while Tiomno and Elisa (as well as Plinio Süssekind Rocha and José Leite Lopes) were teaching there. In 1959, Swieca went to USP as a graduate student, working with *Werner Guettinger*, a theoretical particle physicist. Guettinger, recognizing Swieca's talent for both abstract mathematics and for its physical applications, arranged for him to spend some of his time as a graduate student at the *Max-Planck-Institut für Physik und Astrophysik* in Munich, Germany, at that time still directed by Werner Heisenberg. Swieca returned to USP to submit and successfully defend his thesis in 1963, after which he joined the group of *Rudolf*

Haag (the 'Haag School') at the University of Illinois/USA as a postdoctoral assistant, for three years. He then returned to USP, and shortly thereafter, in 1968, he was awarded the prestigious '*Prémio Moinho Santista*' . He was the second physicist to receive it, following Jayme Tiomno, who in 1957 was the first physicist to be awarded the prize. At USP, Swieca joined Tiomno's research group for particle physics.

Brazil was by that time in the grip of the dictatorship, and after Tiomno's and Schenberg's forced dismissals, Swieca left USP in 1970 and transferred to the calmer environment of PUC/Rio. He stayed there for nearly 8 years. In the second half of the 1970s, he suffered from coronary disease and had to undergo a bypass operation, a difficult and risky procedure in those days. His doctors suggested that a quieter workplace might be better for his health, and he transferred in 1978 to the Federal University in *São Carlos* (UFSCar, in the State of São Paulo). Its surroundings were indeed more bucolic, but there was infighting in the department there, which was probably a source of stress for Swieca, and he also suffered from depression after his surgery, leading to his suicide at age 44, in late 1980. He was considered to be one of the most talented of the younger theoreticians in Brazil and was revered by many colleagues. His name is now associated with a series of summer schools in theoretical physics sponsored by the Brazilian Physical Society (SBF). Swieca's life and his contributions to modern physics have been recounted in published memoirs.[7, 8]

Nicim Zagury was born on March 9th, 1934 in Brazil and studied physics at the FNFi/UB from 1955 to 1958. His professors included Jayme Tiomno and Elisa Frota-Pessôa at the FNFi and the CBPF, where he worked formally from 1959–1967. He was a co-author of Tiomno's important paper predicting the K' meson in 1961 ([JT35]; see Chap. 10, 'A New Particle').

In 1962, he applied for leave from the CBPF and went to study at the University of California, San Diego, on a CNPq fellowship, completing his MS in particle physics there in 1963 and his Ph.D. in 1965, both in the group of *William S. Frazer*, a particle physicist who had joined the Berkeley faculty in 1959 and had a joint appointment at the new San Diego campus from 1960–1981. Zagury spent a postdoctoral sojourn at Harvard in 1966/67, then returned to the CBPF. However, he was seen as member of Tiomno's group, and was boycotted by the CBPF administration. In addition to that, the salaries were extremely low at the CBPF by this time. He joined the PUC/Rio Physics Department, but tried to continue collaborating with the *Centro*. There, however, he was subjected to an administrative investigation by its President, Admiral Otacilio Cunha (for working without a formal position). Finally, he left the CBPF for good and remained on the faculty of

PUC/Rio until 1994, when he moved to UFRJ. He was a Guggenheim Fellow at the University of California/USA in 1974/75, and Visiting Professor at the *École Normale Superieure* in France in 1992. Zagury specialized in particle physics, magnetism and statistical physics, and later has worked in quantum optics and quantum information theory. Today, he is Emeritus Professor at UFRJ.[9]

Antonio Luciano Leite Videira was born in Lisbon, Portugal on June 12th, 1935. He moved with his parents to Brazil at age 7, in 1942, and lived there until 1986. He studied physics at the FNFi/UB from 1954 to 1958 as a student of Jayme Tiomno, Plinio Süssekind Rocha, Elisa Frota-Pessôa and José Leite Lopes, among others. Videira began research at the CBPF with Jayme Tiomno in 1956, remaining until 1962, then spent several years in the USA (at Princeton University, 1962–64, and at Carnegie-Mellon University, 1965–67) before returning to submit his doctoral thesis at USP/São Paulo in 1967. The thesis, available from USP,[10] is entitled '*Baryonic Magnetic Moments in an SU(6) Model*', and his thesis advisor was Jayme Tiomno. After two years at USP, and after Tiomno's and Schenberg's forced retirements, he moved to PUC/Rio in 1969, following in the steps of his colleague Zagury, and remained on the faculty there until 1987. He co-authored two publications with Jayme Tiomno, including the K'-meson paper ([JT35]; see Chap. 10) and a second article, [JT58] from 1978, published while both were at PUC. In 1970, he began a collaboration with the Department of Mathematical Physics in Lisbon, and later divided his time between Brazil and Portugal, moving his main workplace back to Portugal in 1986, where he occupied the Chair of Mathematical Physics at the *Universidade de Évora* from 1988 until his retirement in 2005. He was also professor at the Military Academy in Lisbon, and dedicated himself to teaching and research activities in the History of Science in his later years. We have encountered him in earlier chapters as the author of several scientific-historical articles and memoirs (e.g. [11–13]). Videira passed away on October 15th, 2018 in Portugal.[14]

Erasmo Madureira Ferreira was born on October 8th, 1930 in Rio de Janeiro. He studied physics at the FNFi/UB from 1948–1952; he would have encountered Plinio Süssekind Rocha, Elisa Frota-Pessôa and José Leite Lopes as his professors there, and perhaps also Jayme Tiomno towards the end of his studies. He worked at the CBPF from 1953 until August, 1967, although he was on leave from there in 1957–1961, doing graduate work at the Imperial College London (ICL, University of London) under Abdus Salam, with a fellowship from CAPES. He was thus one of Abdus Salam's first graduate students after the latter had set up his group at ICL in 1957. Ferreira obtained his MS in 1958 and his doctorate in 1961, with a thesis on '*Kaon-Deuteron Scattering*'. He returned to Rio de Janeiro, working at the UB/UFRJ, where he completed his *Livre-docência* (qualification as a university teacher) in 1967, then joined the faculty in the Physics Department at PUC/Rio that same year. He remained on the faculty there until 1994, when he moved back to UFRJ as a tenured professor. He spent a soujourn as Visiting Scholar (Guggenheim Fellow) at Stanford/USA in 1975/77, as Visiting Professor at the *Université Claude Bernard,* Lyon/France as CNPq Fellow in 1985/86, and as Visiting Scientist at CERN, Geneva/CH as a CNPq Fellow in 1992/93. His field of specialization has been High-Energy Physics, particles and fields. He is now Emeritus Professor at UFRJ.[15]

These four physicists, all students of Jayme Tiomno and Elisa Frota-Pessôa of the 'first generation', formed the nucleus of the theory group at PUC/Rio in the early 1970s and distinguished it as a progressive institution in that era, in spite of its relatively small size. They also made concerted efforts to help Tiomno and Elisa find employment as researchers and teachers in their own country after their dismissals in 1969, and in particular after they returned from their forced but brief exile in Princeton, in July 1972.

A Harmonious Environment

While Tiomno would have preferred a larger department where he could again start a major school of physics, with a maximal effect on progress in establishing and supporting research in Brazil, the smaller size and the inner harmony of the Physics Department at PUC were advantageous in their own way, especially during the troubled years of the 1970s, when Tiomno tended to be depressed as a result of the derailment of his career and his plans by the dictatorship. In fact, towards the end of his tenure at PUC, in the late 1970s, he was able to produce a number of articles in his new major research area, gravitation, field theory and relativistic astrophysics (cf. [JT58]–[JT65]). Admittedly, his scientific output was at a low point in the period shortly after he returned from Princeton and began work at PUC—in the four years between 1974–1977, he published only one article, and it can be considered to be a delayed product of his Princeton stay.

Jayme Tiomno occupied himself with some success at PUC during that period in teaching and in setting up and guiding his research group, even though he could not officially serve as a thesis advisor, and his own research output was minimal in those years. This must have been particularly disturbing for him, since he had only recently overcome the barren period in his research during the years 1965–1971; that had been due to the poor conditions at the *Centro* and to his engagement at the UnB/Brasília, as well as his efforts in the *concurso* for the position at USP in 1966/67. In those 7 years, he had produced only four scientific works, one of them being his thesis for the *concurso*, two others articles together with Bollini and Giambiagi, a result of their collaboration during his first longer visit to the ICTP/Trieste, and the last one an article published by Tiomno alone, but based on the earlier collaboration at the ICTP ([JT44]) (Fig. 14.2).

As Tiomno himself emphasized in his letter to John Wheeler in June, 1972, it had been his stay at Princeton and the interactions with younger physicists

Fig. 14.2 Jayme Tiomno lecturing, 1979 (during the International Einstein Year, the 100th anniversary of Einstein's birth). *Source Acervo* IF/USP. Reproduced with permission [Online at: http://acervo.if.usp.br/bio15]. Also published in *Ciência e Cultura*

there that had 'broken the spell' and allowed him to write a dozen articles within less than two years. But that period of productivity came to an abrupt end when he went back into the hostile environment of 1970s Brazil, where he was to some extent *persona non grata*, at least to the ruling military establishment.

Of this period in his career, Bassalo and Freire[16] say the following:

> … the damage caused by such vicissitudes had not yet come to an end, and even threatened Tiomno's health … Indeed, the hostility that some of the military exhibited toward the professors affected by the AI-5 caused Tiomno to suffer from a nervous depression, for which he had to submit to psychotherapy (-analysis) in 1974/75….

The protected environment at PUC and his interactions there with congenial colleagues, many of them his own students from an earlier time, finally again broke the barrier, and from 1978 on, continuing until well after his (formal) retirement in 1990, he was able to produce a steady stream of research results and scientific commentary.

Tiomno himself says of this period,

> "Afterwards [*after the stay in Princeton*], I returned to Brazil, and at that time I received an invitation [*to work*] at PUC. I have been there since 1973 [*this interview took place in 1977*]. My work at PUC is a phase that is, let us say, relatively modest, since PUC is a small institution, which doesn't offer the same possibilities for growth as the other institutions where I have participated. Well, it has more possibilities for growth than the *Faculdade de Filosofia* did back in that period when it was really fossilized. But, for example, in comparison to *Fundão* at present [*the new campus of UFRJ, the successor to the UB*], PUC doesn't have the same potential for growth. Nor in comparison to Brasília or São Paulo. In São Paulo, even working alone, only with the Chair [*of Advanced Physics*], I had a larger group [*from which the Department of Mathematical Physics later developed*] than here [*at PUC*]. In comparison, I have only a small group here.
>
> "In the USA, I changed from [*working on*] elementary particles – I had stopped doing research for several years [*before going there*] – to working on General Relativity. Here, I work with a small group, not having the conditions to do what most arouses my enthusiasm, which is to work for growth, *en masse*, for physics in Brazil. Having had three sufficiently good experiences – the CBPF, UnB, and USP – here, I content myself with giving graduate courses and directing my small group, without any immediate perspective of experiencing any major growth.
>
> "On the other hand, I have the satisfaction of being in a place where I have many [*former*] students as colleagues. The Department of Physics is quite

homogeneous; thus, things are usually very calm. In a much larger place, there are greater internal frictions and disputes...".[5]

Little by little, the situation for Tiomno and Elisa was becoming normalized. "Jayme is working very hard at the moment, and he says that he is feeling very good. Now I can begin to take care of my own life. There is news, I will tell it to you...".[17] Elisa was also able to obtain a position and a laboratory at the Catholic University in Rio de Janeiro. In 1975, based on an agreement between her and the two institutions, Elisa was allowed to move the Laboratory of the Nuclear Emulsion group from USP to PUC/Rio; there, she carried out studies of elementary-particle decays and reactions. That agreement was made possible by financial resources that Ernst Hamburger was able to mobilize, working with agencies that supported scientific research, both for buying equipment and for paying a fellowship for Elisa. After 1975, she also began to work part-time at the Department of Experimental Physics at USP, developing a research program in nuclear physics. She describes her situation there in a much later interview[18]:

In 1972, we returned to Rio; at that time, we received great support from PUC/Rio. Father Collins was the director of the Department of Physics there at that time, and he invited us. In spite of our having nothing at all to do with Catholicism, they helped us a lot – they thought in a similar way as we did. And for them, it also was advantageous, since all [*of our*] articles appeared as work from PUC. There, they gave me a room and computer use, and everything that I wanted. I set up a laboratory for myself in that room, with personnel paid by USP and materials from there. A laboratory which belonged to USP and to PUC. I was able to receive funding from USP, since Ernst Hamburger [*at USP*] managed to find financial reserves from which he could pay me.

Tiomno and Elisa spent some of their leisure time in a small house that they had purchased in a town on the coast to the northeast of Rio de Janeiro, 3 hours drive from the city, called *Arraial do Cabo*, near *Cabo Frio*. It was a thinly-populated place, without electricity and with water pumped by hand from a well on the property. At night, they lit gas lanterns, and the refrigerator was an icebox in which they put large blocks of ice bought locally. For Tiomno, this was a sort of return to the tranquil times in the small cities of his youth in *Minas Gerais*. He passed the time there caring for his garden or busying himself with his "fifi".

Science at PUC in the Late 1970s

Tiomno's first scientific paper published as a result of his tenure at PUC appeared only in 1978, and was entitled '*A Cartesian Operator Algebra for Expansion of Tensor Quantities and Equations in a Spherically Symmetric Background*'. It was published in the *Revista Brasileira de Física* (now the *Brazilian Journal of Physics*) with his co-author Antonio Luciano L. Videira ([JT58]). It is evidently to Videira's credit that he was able to break the spell hanging over Tiomno and encourage him to again participate actively in research and the publication of his results. The article deals with details of mathematical physics and is not wide-ranging or visionary in terms of physical insight, but it was a good introduction back into active research. The same year, Tiomno published a somewhat more 'physical' paper with a Masters student from his group at PUC, Ricardo Amorim; it bore the title '*Charged Point Particles with Magnetic Moments in General Relativity*' and it was also published in the *Revista Brasileira de Física* ([JT59]). It relates to some of Tiomno's earlier work on point particles in GR, and contains essentially the results of Amorim's Masters thesis.

All of Tiomno's remaining publications during his tenure at PUC, in 1979/80 ([JT60–64]), were co-authored with C. G. Bollini and J. J. Giambiagi, with the exception of the article on '*Hidden Singularities in Non-Abelian Gauge Fields*' ([JT63]), published in 1980 with Bollini as the only co-author. We may recall that those two talented Argentine theoreticians, who had recently escaped from the even more oppressive dictatorship in their home country, were both working at the CBPF after 1977. This was ironic in that they were able to work where Jayme Tiomno would have preferred to be, but at the same time it is a quirk of history that his earlier efforts to found and build up the *Centro* had enabled it to shelter two deserving, persecuted colleagues with whom Tiomno could carry on a highly successful collaboration. Thus, once again his efforts, although they had been frustrated by the dictatorship, had nonetheless come to fruition, in a way from which even he himself could benefit. They published five articles, all related to gauge fields (except for the fifth one, [JT64], with Bollini and Giambiagi (1980); it deals with Wilson loops). For completeness, we also include here Tiomno's sixth and final joint paper with Bollini and Giambiagi, entitled '*Wilson Loops in Kerr Gravitation*' and published as a Letter to *Nuovo Cimento* in 1981 ([JT65]). It was written after Tiomno had left PUC in 1980, but it belongs to the series begun while he was still working there.

The four articles of [JT60]–[JT63] deal with 'gauge fields'; they belong in the category of fundamental quantum field theory/mathematical physics.

The idea of a 'gauge theory' dates back to Maxwell's theory of electromagnetism in the mid-19th century. In that theory, electromagnetic phenomena can be alternatively described in terms of *vector fields* (the electric and magnetic fields) or of *potentials*—the electric (scalar) potential and the magnetic (vector) potential. While the fields have definite values at every point in space and time, the potentials are *relative*, and their zero-points can be arbitrarily chosen. Thus, the potential of the Earth's surface is usually chosen as the zero of the electric potential, and electric circuits are 'grounded', i.e. electrically connected at some point to the Earth's surface. However, any other potential value could just as well serve as the zero of the potential; an experimenter inside a metal box at a potential of 1 million volts above 'ground potential' feels perfectly normal and can measure small potential differences within that environment, unaffected by the large potential drop to 'ground'. All of the laws of electrostatics are thus invariant with respect to changes in the zero of the electric (scalar) potential (just as the laws of mechanics are invariant with respect to translations of the spatial coordinates, i.e. moving the zero (origin) of the coordinate axes linearly in space, and also to translations of the zero of time). In Chap. 4, we pointed out that such invariances lead to *conservation laws*—invariance with respect to *spatial* translation to the conservation of *linear momentum*, and invariance w.r.t. *time* translation to the conservation of *energy*. In a similar way, invariance w.r.t. 'translation' of the electric potential, called (global) *gauge invariance*, leads to the conservation of *electric charge*.

This concept can be generalized and applied to a whole class of theories that are invariant with respect to 'gauge transformations', which are examples of a certain type of symmetry transformation. (There are also more complex gauge transformations, in which the potential is shifted by a position- and time-dependent function. If the functions are properly chosen, the fields will still be invariant under such a transformation. This is called *local* gauge invariance). Such symmetry transformations can be represented mathematically by *symmetry groups*, and they give rise to vector fields, called the 'gauge fields'. Thus, the electric field is the gauge field of the gauge group corresponding to invariance under zero-point shifts (global or local) of the electric potential.

In a quantum field theory (QFT), these gauge fields also give rise to particles, the *quanta* of the gauge fields, called the *gauge bosons*. An example is quantum electrodynamics (QED), describing electromagnetism in terms of quantum physics; it is a gauge theory, whose potential is the scalar (electric) potential and the vector (magnetic) potential, combined into a 4-vector which satisfies Lorentz invariance, i.e. obeys Special Relativity. Its (massless) gauge boson is the *photon*, the interaction quantum of the Electromagnetic interaction. QED was formulated between 1930 and 1950 by Heisenberg, Dirac, Feynman, Schwinger, Tomonaga and Dyson, among others. [Note that QED is an *Abelian* theory,[19] and its gauge boson is required to be massless by gauge symmetry, corresponding to the gauge group U(1).]

Analogously, in the period 1955–1975, quantum field theories describing the Weak interaction (and combining it with the Electromagnetic interaction, in the *electroweak unification*, due to Weinberg, Glashow and Salam), and the Strong interaction (*quantum chromodynamics* (QCD), due to Gell-Mann, Gross, Politzer, Wilczek and others) were formulated, and together they form the foundation of the Standard Model. It contains, in addition to the *photon*, three (massive) *intermediate vector bosons* (the $W^\pm$ and the Z^0) which mediate the Weak interactions, and eight vector bosons, the *gluons*, which mediate the Strong interaction, distinguished by their quantum property 'color charge', and presumably also massless.

[The *Higgs field* gives mass to particles with a *weak charge*, i.e. to the intermediate vector bosons. It carries no electric or color charge, and thus leaves the photon and the gluons massless. The Electromagnetic and the Weak interactions belong to the same (gauge) theory, the electroweak theory, but a symmetry breaking occurs, and the weak part of the theory is non-Abelian, as is quantum chromodynamics. The Weak interaction is also *chiral*, as we have seen in Chap. 9.]

The articles published by Tiomno with Bollini and Giambiagi (BGT) in 1979–81 are mostly fundamental works in QFT, i.e. mathematical physics, without an immediate application to a specific theory. In the first of them, '*On the Relation between Fields and Potentials in Non-Abelian Gauge Fields*' ([JT60]), the authors discuss the problem of *ambiguous gauge fields*, where there is not a single, unique potential (up to a gauge transformation, of course) corresponding to the field. These had been pointed out by T. S. Wu and C. N. Yang a few years earlier. BGT find a mathematical condition for any gauge group which is satisfied by all groups having such ambiguous fields, and they give examples of families of ambiguous fields, also termed 'gauge copies'.

In their next paper, entitled '*Singular potentials and analytic regularization in classical Yang-Mills equations*' (1979; [JT61]), BGT discuss a 'regularization procedure' for finding well-defined solutions of the Yang-Mills equations. Yang-Mills theories were introduced by C. N. Yang and Robert Mills in the mid-1950s; they are a class of gauge theories which were used in formulating the electroweak unification and quantum chromodynamics. 'Regularization' is a mathematical procedure for modifying the integrals which occur in

QFT so that they do not diverge, i.e. do not become infinite at high energy or momentum values. Such divergences produce infinities in the calculable quantities and render the theories nonphysical. One such method of avoiding the divergences is 'dimensional regularization', in which the integrals are first formulated in higher $(4 + \varepsilon)$ dimensions, then the divergences are localized in the $1/\varepsilon$ term, and divergent parameters are used to cancel the resulting divergences in the limit $\varepsilon \to 0$, i.e. upon returning to the physical case of 4 (3 spacelike, 1 timelike) dimensions. Bollini and Giambiagi had established this method while they were still in Argentina in 1971/72, around the same time that it was also published by the Dutch physicists Gerard't Hooft and Martinus Veltman, who used it to regularize the electroweak unified theory and were awarded the Nobel prize in 1999 for their part in its development (see the article by Bietenholz and Prado).[20] Bollini and Giambiagi were thus experts on regularization and collaborated with Tiomno in this work on its application to a certain class of theories.

In their article on '*Gauge Field copies*' ([JT62], 1979), BGT deal with theories having more than one unique potential corresponding to a given field; these are related to the ambiguous fields described above. When two distinct potentials lead to the same gauge field, they are called 'copies'. In this article, the authors propose a method for finding all the possible copies of a given gauge, with no constraints on the gauge. They give some examples, including one in which not only the field strength but also the current strength is the same for two 'copies'. This was an important topic in the late 1970s. Their next paper in the series, published in *Il Nuovo Cimento* in January, 1980 by Tiomno and Bollini, was called '*Hidden singularities in non-Abelian gauge fields*' ([JT63]). They deal here with a particular subgroup of non-Abelian[19] gauge theories which have 'singular points', i.e. points in their parameter space where quantities derived from the theory become infinite. For such theories, the regularization procedure must be chosen carefully to ensure a well-defined potential and field. The authors give an example and show that the field in a subspace containing the singularity will depend on the regularization procedure, even if it does not do so globally.

The last two articles in the series published by the BGT collaboration in the early 1980s dealt with a different topic, somewhat less abstract than the previous four, which had treated gauge fields. This was the *Wilson loop*, introduced by Kenneth G. Wilson in 1974 (Wilson was also the originator of Renormalization Group theory, which proved useful for solving various problems involving phase transitions, condensed-matter and particle physics, for which he was awarded the 1984 Nobel prize in Physics). The Wilson loop (WL) in a gauge theory is a gauge-invariant variable which is formed along a closed curve (loop) and represents an elementary excitation of the quantum field, localized on the loop. The WL was introduced in the hope of improving the interpretation of QCD in the strongly-interacting regime. It provided motivations for modern *string theories* and *loop quantum gravity*, both very active fields of theoretical physics at present. In these works, BGT deal with 'instanton' solutions of Yang-Mills theories, i.e. solutions that are localized in space and time (analogous to *solitons* in a classical field theory on a Euclidian space), which they had also considered in [JT61]. In the first of these articles, entitled '*Wilson loop and related strings for the instanton and their variational derivatives*' and published in *Il Nuovo Cimento* in October, 1980 ([JT64]), they compute the Wilson loop variables for instanton solutions, using them to obtain derivatives for open and closed strings.

In the last article written by BGT as a collaboration, submitted as a Letter to *Il Nuovo Cimento* in February, 1981 ([JT65]), they again deal with Wilson loops, but now in a different context, that of General Relativity, in particular its Kerr solution,[21] which applies to a rotating black hole.

The BGT collaboration base this paper methodologically on the previous one [JT64], citing their calculation of the Wilson loop variable there, but now apply it to the configuration of the gravitational field described by Kerr's solution. In that case, the Wilson loop has a clearcut geometrical meaning: it is the displacement of a vector along the closed loop (e.g. around a rotating black hole). The paper can be summarized as follows: "The value of the gauge-invariant Wilson loop is computed for the gravitational field corresponding to the Kerr metric, which contains the Schwarzschild metric as a special case. Expressions for the Wilson loop corresponding to a closed curve are obtained, which are related to the parallel displacement of a vector along a closed path, and it is noted that a value of the Wilson loop equal to 4 is a necessary, but not sufficient, condition for the vacuum or flat state. It is shown that the metric curvature is related to the derivatives of the Wilson loop. The behavior of the Wilson loop inside the Schwarzschild sphere is also examined".[22]

Just why the BGT collaboration was terminated at that point is not entirely clear. In the end, Jayme Tiomno returned to the CBPF in the early 1980s, and perhaps he was simply too busy to continue working with Bollini and Giambiagi, setting up his new groups in that environment with 'more possibilities'. Both Bollini and Giambiagi remained at the CBPF for several more years, Giambiagi as head of the High Energy Physics department until 1985 (his successor in that function was Jayme Tiomno). That same year, Bollini returned to Argentina, to the University of La Plata, where he had begun his scientific career and had done perhaps his most important work. Giambiagi stayed on at the CBPF until his death in 1996.[23]

Tiomno resigned his position at PUC/Rio in August 1980, some months after returning to the *Centro*. He was very grateful to PUC, and in the letter asking for a temporary release from his position there, he wrote:

> One of the conditions accepted by the administration of the CBPF is that throughout the duration of my contract with that Center, I may continue my collaboration, conforming to your criteria, with this University, to which I owe such a great debt of gratitude for the help which it gave me at a difficult time in my life.[24]

Notes

[1] Nobre and Videira (2018).

[2] On December 7th, 1972, a request for information on Tiomno was made to the Intelligence Division of DOPS (Search request No. 2143/72/SI/BB/GB). Archive on Political Policy, sector INF, Number 118, p. 419; [APERJ].

[3] Quoted by Bassalo and Freire (2003) from the deposition made to them by Jayme Tiomno and Elisa Frota-Pessôa in August, 2003 (their Ref. [8]).

[4] Elisa Frota-Pessôa to Caden and Haity Moussatché, January 19th, 1974; [HM].

[5] Tiomno (1977).

[6] *Sarah de Castro Barbosa* was able to complete her masters thesis at PUC/Rio in 1973, working under Erasmo Madureira Ferreira there (cf. the section on 'the PUC Quartet'). Her thesis title was '*Kaon-Nucleon Interactions at Low Energies*'. She was known at the time by her married name, Sarah de Castro Barbosa Andrade; she was the wife of the film director Joachim Pedro de Andrade from 1962–75. Her prominence as his wife may have contributed to her dismissal by the AI-5, since she was younger than most of those dismissed and not scientifically well known at that time.

[7] Schroer (2010): a scientific memoir of Swieca's life's work (in English).

[8] Marino (2015): a more recent memoir of Swieca's life (in Portuguese).

[9] Source: *CNPq, Curriculo LATTES*.

[10] The thesis may be found at https://bdpi.usp.br/single.php?_id=000712251.

[11] Videira (1997, 2017).

[12] Videira (1980).

[13] Swieca and Videira (2013).

[14] See the obituary notices from the University of Évora, Portugal, October 18th, 2018; online at https://www.spf.pt/news/morreu-no-passado-dia-15-antnio-luciano-leite-videira-professor-catedrtico-jubilado-da-universidade-

de-vora; and from the CBPF on January 21st, 2019, online at https://portal.cbpf.br/pt-br/ultimas-noticias/pesquisador-titular-do-cbpf-relembra-a-vida-e-obra-de-ex-professor-2.

[15] Source: *CNPq, Curriculo LATTES.*

[16] Bassalo and Freire (2003).

[17] Elisa Frota-Pessôa to Caden and Haity Moussatché, October 1974; [HM].

[18] Frota-Pessôa (1990).

[19] The term '*Abelian*' refers to the question of whether the *order* is significant in which operators or elements of a group are applied. If A and B are two elements of a group, they can form a product A · B. If it is equal to the product B · A, the elements are said to *commute*, and their *commutator* (A · B − B · A) is zero. The group is then termed 'Abelian'. This is trivial if A and B are simply real numbers; real-number multiplication is commutative. However, many groups can be represented by matrices, and these do not necessarily commute. If not, the group is then called *non-Abelian*. Some physically-relevant theories (e.g. Yang-Mills theories) are non-Abelian, and the Standard Model is based upon such theories (electroweak interactions, QCD).

[20] Bietenholz and Prado (2014); see also https://www.df.uba.ar/es/actividades-y-servicios/hemeroteca/6613-juan-jose-giambiagi (in Spanish), and backnote [24] in Chap. 17.

[21] Kerr (1963).

[22] See [JT65]; quoted here from the abstract of that paper.

[23] His brother Mario, a chemist, continued working at the CBPF until his own death in 2002. See the memoir by Mario's wife Myriam Giambiagi (2001). Mario and Myriam were also friends of Jayme Tiomno and Elisa Frota-Pessôa, and of Haity and Caden Moussatché.

[24] Jayme Tiomno to Carlos Mauricio Chaves (at that time head of the Department of Physics at PUC/Rio), dated August 7th, 1980; [JT].

15

The Amnesty—Return to the *Centro*

The End of AI-5—Leaving PUC

During the decade of the 1970s, at the same time as various repressive measures, including arbitrary imprisonment, torture and assassinations, a 'slow, gradual and secure' process of relaxation of dictatorial controls was also begun by the regime in Brazil. In December of 1978, President-General Ernesto Geisel of the Brazilian Republic permitted the Brazilian Congress to rescind the infamous AI-5, the Act which had restricted constitutional liberties. This did not signal the end of the dictatorship, which continued until March of 1985, when the '*Nova República*' was founded; but it did represent an increased freedom of organization and demonstration rights. In the following August (1979), an Amnesty Law was passed. This was a process planned and controlled by the regime, which permitted the return of exiles, the restoration of civil rights to politicians who had been removed from their offices, freeing of political prisoners and the reinstatement to their former positions of employees who had been dismissed or forcibly retired (although their right to vote in presidential elections was not restored at that time). The amnesty also applied to agents of the dictatorship, such as torturers or those who had carried out political character assassinations. It was thus a strategy for installing a negotiated amnesty while the dictatorship was still in force, in contrast to what occurred later in some other Latin American countries, e.g. in neighboring Argentina.

The Amnesty Law, to be sure, required that those who had been dismissed or retired would have to request their reinstatement within a time limit of 120 days, and their petitions would then be evaluated by a commission.

© Springer Nature Switzerland AG 2020
W. D. Brewer and A. T. Tolmasquim, *Jayme Tiomno*, Springer Biographies,
https://doi.org/10.1007/978-3-030-41011-7_15

Many of the professors who had been forcibly retired refused to request their own rehabilitation. The head of the *Instituto de Física* at USP at that time was *Hersch Moyses Nussenzveig*, a colleague and friend of Tiomno, Elisa and Leite Lopes. He sent a letter to Tiomno inviting him to request his reinstatement and saying that he would be very welcome. But Tiomno refused to accept that invitation: "What was carried out without my participation must be corrected without my participation—only then will I re-enter the scene".[1] Mario Schenberg took the same position, namely that he would not accept having to solicit his own return to the University, and would return only if requested by that University, where he had been prohibited from even entering the campus. But in spite of not having been formally reinstated, Schenberg simply began giving lectures, participating in seminars and mentoring students, and no-one dared to protest. In a similar manner, Elisa refused to request her reinstatement at UFRJ in Rio de Janeiro.

The CBPF had in the meantime been trying to recover from the serious crisis into which it had been plunged, both academically and financially. In 1974, the private foundation had expired and the CBPF was converted into a public research institute associated with the CNPq. Soon after the publication of the Amnesty, all of the members of the *Centro* signed an open letter sent to the CNPq president requesting the reinstatement of Tiomno, Elisa and Leite Lopes.[2]

Their return was not a simple matter. There was considerable bitterness at their having been dismissed from the *Centro* which they themselves had founded. For example, in the first half of the 1970s, when Mario Novello returned from his postdoc position in the UK, the group of Gravitation and Cosmology had been established at the CBPF. That group invited Tiomno to participate in their seminars, but he refused, saying that he couldn't set foot in the *Centro*. The solution was that on a weekly basis, a group of around 10 researchers left the CBPF and went to PUC to hold the seminar there, so that Tiomno could also participate in it.[3] The CNPq approved their return after the Amnesty, and the director of the CBPF, Roberto Lobo, attempted to implement a strategy of rehabilitation of its former members which would overcome their resentments. On November 23rd, 1979, the CBPF held a 'Public Session of Moral Reparation and Re-Initiation of Collaboration', to which former members who had been removed by the AI-5 or by administrative decision of the CBPF, as well as those who had resigned in protest or simply left in disgust were invited. It was not an easy occasion; indeed, it was full of emotion: "...within the walls of this Center, where I have spent many of the best moments, but also the worst, of my scientific life", said Jayme Tiomno in his talk at that event.[4] After so many years of being banned from even entering the *Centro*, the result of arbitrary decisions taken by the administration of the old CBPF, he was back again!

Some sort of cathartic process was however necessary. It was not possible to simply forget the past and carry on as if nothing had happened. Tiomno, in his talk,[4] listed in some detail the injustices that had been perpetrated at the 'old *Centro*' under President Admiral Otacilio Cunha and the Scientific Director at that time. He mentioned the many people who were forced to leave, their fellowships and sometimes even their work contracts arbitrarily cancelled simply because of a letter from some military official in the regime. This included many of the students that he and Elisa had taken to Brasília, for whom they had obtained fellowships on returning to Rio, but who were left exposed when Tiomno and Elisa were in Trieste in 1966/67, and after they went to USP in 1968—their ex-students were subject to reprisals. Similar cases continued to occur well into the 1970s. It was necessary to affirm and reaffirm what had happened in order to have a realistic possibility of working together again in the new *Centro*:

> I will come to a stop here… hoping to have represented sufficiently the finality of this description of the detrimental atmosphere that characterized that former CBPF, which has now passed away, and hoping also that this will assist in leading the new, modern CBPF along paths that will enhance, and make positive contributions to, the development of physics in Brazil.[4]

About this period, Elisa remarked later on[5]:

> When the amnesty was announced, we would have been required to petition the UFRJ (the former FNFi) to take us back as faculty members. We decided not to petition them, because we found that absurd. The CBPF had a completely different attitude, they invited us back with all honors. It wasn't the point that we were seeking honors; but it is rather depressing, after an unjust dismissal, to have to petition an institution to allow one's return. Besides that, there were a number of conditions. Among them was the stipulation that we couldn't ask for full-time positions—which they gave us in the end. We couldn't be professors on an equivalent basis to the others; we would always be professors with restrictions. It was an absurdity, in particular because the people who had fought the hardest for the Faculty were those who had been dismissed. At the time of our dismissals, we thought, 'We're going to go back!'. But after the amnesty, we were thinking, 'No! I'm not going back!'. I wrote them a letter, explaining the reasons why I was not going back.

Indeed, for many professors and researchers, their return was not easy, since apart from resentments and old animosities, it was not simple to re-enter an institution after an absence of more than 10 years. For example, according to the rules and norms of the CNPq, Tiomno's *curriculum vitae* would have to be evaluated by a researcher at the CBPF to decide whether he was qualified to be included as its member. This task devolved upon his ex-student Mario Novello, who, with some embarrassment, wrote his recommendation. To the younger members of the CBPF, Tiomno and Elisa were almost mythical characters who had participated in the creation of the *Centro* and lived through its times of glory and decline, and they were physicists of international renown who deserved great respect. At the same time, the situation caused an uncomfortable sense of guilt in some of them, not for having done something wrong themselves, but simply because they had been hired by and were working at the institution that had dismissed them.

At that time, Leite Lopes was professor at the *Université Louis Pasteur* in Strasbourg, in its *Centre de Recherches Nucléaires*. After conversing with Tiomno and Elisa, he accepted his return to the CBPF, and also decided to accept his readmission to the University (UFRJ), on condition that he could continue to spend part of his time in Strasbourg. But he also felt the difficulty of returning, like the other professors: "the truth is that we are marked, we are former exiles, meaning that we were previously condemned, and at both the CBPF and at the University, I will only be tolerated", he wrote to his fellow exile and friend Haity Moussatché.[6]

In August of 1980, Tiomno requested permission from PUC to return to the CBPF, expressing his gratitude to PUC for their support "at a difficult time in my life".[7]

In his last joint article with Bollini and Giambiagi, submitted for publication in February, 1981, he indeed gave his professional address as the CBPF, but added a footnote stating that he was 'on leave from PUC/Rio', as a form of recognition of the help which that institution had given him.

As another part of this process of recognition and rehabilitation, his colleagues at PUC, Nicim Zagury, Jorge André Swieca and Antonio Luciano Videira, organized an homage to Tiomno for his 60th birthday; it took place during the *II Encontro Nacional de Partículas e Campos* ('2nd National Meeting on Particles and Fields'), in September 1980 under the auspices of the *Sociedade Brasileira de Física* in the little town of *Cambuquira*, in Minas Gerais. It fell to Videira to make a speech about the life and works of Jayme Tiomno, without his imagining that Tiomno's scientific productivity was still very far from approaching its end.

Jayme Tiomno was initially designated as head of the Scientific Department, corresponding to the position of Scientific Director of the 'old *Centro*'. However, when the private foundation had been founded in 1949, the

Centro's President had not been a physicist—instead, he was a figure with national prestige, who had important functions of representation, while the Scientific Director was responsible for guiding the scientific development of the institution. Now in fact the CBPF no longer had a President, and its Director Roberto Lobo *was* a physicist and wanted to be responsible for the scientific conduct of the *Centro*, causing some friction with Tiomno. Apart from that, since the CBPF had become part of the CNPq, there was no longer an organizational structure in which each member's functions were formalized. In the end, after two years, Tiomno asked to be dismissed from his function as head of the Scientific Department, and a new organizational plan was adopted, eliminating that position.[8]

Tiomno was also very active in the discussions that took place at the Centro about the criteria for evaluation and promotion of its members. He took a very firm position regarding the importance of publication of research, perhaps influenced by his American experience; and furthermore, he wanted to see publications in the most important journals. He soon became known as a very strict and rigorous judge who expected results. The physicist Luiz Alberto Oliveira, one of Mario Novello's students, still recalled much later, even after his own retirement, how he was struck with fear when he learned that Tiomno would be on his PhD committee [*comment made to one of the authors*]. On the one hand, Tiomno's rigor became a kind of guarantee of excellence for the *Centro*, while on the other, it caused some disaffections among the younger CBPF members.

Jayme Tiomno, having organized research groups in elementary-particle physics at the CBPF, UnB and USP between 1950 and 1969, decided, once back at the CBPF, to join *Mario Novello* and *Ivano Damião Soares* (previously a student of Mario Novello's) in expanding the Gravitation and Cosmology group, in line with the shift in his research emphasis since his 1971/72 stay in Princeton. They set up the 'Department of Relativity and Particles' (DRP) within the CBPF, and Tiomno headed it for a few years. That Department was almost completely theoretical. Some of its members however, including *Alberto Franco de Sá Santoro* and *João dos Anjos*, among others, had expressed their interest in becoming involved in experimental work and joining large centers for high-energy physics.

Early in 1982, Tiomno attended a meeting in Cocoyoc, Mexico, organized by Leon Lederman, director of Fermilab in the USA, to discuss how to stimulate the formation of experimental groups in particle physics in Latin America. There, he heard about the case of an experimental group in Mexico that was initially organized and directed by theoretical physicists, who then became experimentalists. He was convinced that the same could be done at the CBPF with the help of Fermilab. To test this idea, he organized the Experimental High Energy Group, and in 1984 it sent three theoretical physicists from the CBPF and one from USP to Fermilab, "to suffer the transmutation", in his words. This was successful, and as a result, an agreement between the CBPF and Fermilab was signed in 1985. In 1988, this group formed a new department at the CBPF, called LAFEX. Tiomno himself however remained in the Department of Relativity and Particles.

Back at the CBPF—Research in the 1980s

During the decade after he returned to the *Centro*, Jayme Tiomno published more scientific works than during any other comparable time period in his career. In fact, in terms of publications, his later years were his most productive, contradicting Dirac's opinion that a (theoretical) physicist over 30 'might just as well be dead'. Not only was their number greater—over twice as many publications as during the difficult decade of the 1960s, which saw the decline of the CBPF and difficulties at the FNFi, the exhausting and unsuccessful project in Brasília, his new beginning at USP and then his forced retirement—but in addition, the variety of topics that he dealt with was astonishing. He took up some old threads, such as Special Relativity theory (SRT), in which he had been interested many years before, and hadron physics—the Strong interactions, which had been his main research subject in the 1950s and 1960s; but he also continued his research in field theory, gravitation, and cosmology. He even revisited his old favorite theme, the Weak interactions, and also wrote several papers touching on the history and sociology of physics. He let his political opinions be known, and harvested a certain amount of scorn as a result.

The Physics of Special Relativity

The first of Tiomno's new research directions began when he discussed the topic of Special Relativity in 1980, in his lecture to the Brazilian Physical Society (SBF) in *Cambuquira*/MG during the session in honor of his 60th birthday. He had evidently been thinking about it for some time, although he had published no work on it. His active entry into the field was stimulated shortly thereafter by a paper that appeared in 1982 in the journal *Foundations of Physics*, written by D.G. Torr and P. Kolen, in which they propose a method using two atomic clocks to measure the velocity of light in a novel manner, with the background intention of showing the validity of the Lorentz ether theory (LET) over that of Einstein's Special Relativity theory (SRT). Tiomno collaborated with *Waldyr A. Rodrigues*, his former student at USP in 1968, who had become a professor at Unicamp, to write a critique of the proposed Kolen-Torr experiment. He was thus returning to a field (SRT) in which he had been interested earlier in his career but had not pursued.

Waldyr Alves Rodrigues Jr. was born in 1946 in a small town in the interior of the State of São Paulo. He studied physics at USP under Mario Schenberg, completing his Bachelor's in 1968, just as Tiomno took up his professorship. Rodrigues then went to the University of Turin in Italy, and obtained his doctorate in theoretical nuclear physics there in 1971. On returning to Brazil, he took a faculty position at Unicamp, where he worked with César Lattes on the detection of cosmic-ray events; however, he gradually shifted his field of interest to mathematical physics, and finally moved to the Institute of Mathematics, Statistics and Scientific Computation in 1981. He remained there until his retirement in 1998, then worked in France for some time. Rodrigues passed away in April, 2017.

Between 1982 and 1985, Tiomno and Rodrigues together published three articles on SRT ([JT66, 73, and 76]), culminating in their reply to Kolen and Torr in *Foundations of Physics*. It was submitted to that journal in September, 1982, intended as a (prompt) response to the Kolen-Torr article in the same journal. The editor sent their manuscript for refereeing to one of the authors whose work they were criticizing, and he rejected it. Tiomno wrote to John Wheeler in 1983, asking his opinion, and Wheeler in turn wrote to the journal's editor,[9] defending Tiomno's scientific integrity and supporting publication of the paper. It finally appeared in mid-1985.[10]

Tiomno continued working on the LET *vs.* SRT question, publishing seven more articles over the next four years, five of them in collaboration with *A.K.A. Maciel*, his former student at PUC.

Arthur Kos Antunes Maciel began his undergraduate studies at PUC/Rio in 1970, and no doubt experienced Jayme Tiomno as a professor there from 1973 on. He remained at PUC to complete his Masters, under the guidance of Humberto Brandi, with a thesis in solid-state theory (color centers), on the '*Calculation of the electronic structure and hyperfine parameters of a U2 center in the alkaline halides*'. After finishing his MS in 1977, he went to Oxford/UK on a CNPq fellowship, and obtained his doctorate there in 1981, working with Jack Paton, with a thesis on '*Hadronic Couplings*'. Following a postdoctoral stay at the CBPF in 1983–1985, he became an Associate Researcher there, leaving in 1995 to go as Visiting Scientist to Fermilab in the USA, supported by the collaboration in Experimental High-Energy Physics. There, he directed the D-Zero experiment for two years. In 1999, he obtained a professorship at Northern Illinois University/USA, where he remained until 2005. He then returned to the CBPF, becoming a Tenured Researcher in 2007. Maciel retired formally from the CBPF in November, 2017.

At the SBF meeting in Cambuquira in 1980, Tiomno had summarized his ideas on questions related to Special Relativity theory.[11] He considered the possibility that, in contrast to Einstein's SRT, where there is no *privileged* frame of reference that would be 'at rest' with respect to the universe, it might be possible to define an *absolute rest frame*, analogous to the 'ether' of classical physics, which was originally held to be the medium for the propagation of electromagnetic radiation as described by the Maxwell theory. As a plausible definition of such a frame, he suggested the cosmic microwave background (CMB) radiation, the remnant of the radiation originating during the initial expansion of the early universe (the 'Big Bang'), which has been expanding and cooling over the intervening 13.5 thousand million years. The frame in which the CMB is isotropic might be considered as the 'absolute rest frame' of the universe. The CMB was discovered by A. Penzias and R.W. Wilson in 1964 (Nobel prize 1978) and has been the subject of intense study ever since.

A 'boost' from one inertial frame to another, moving at a different velocity, would still conform to Lorentz invariance, so this alternative approach is called 'Lorentz ether theory' (LET). Tiomno and his co-authors suggested various experiments to distinguish between SRT and LET, and gave a critique of existing experimental results. They treated a rigid body rotating within a moving frame, and how its shape would be influenced by transforming to other frames. Experiments of the Michelson-Morley type were considered, among others. Maciel and Tiomno published a summary as a *Physical Review Letter* in 1985, entitled '*Experiments to Detect Possible Weak Violations of Special Relativity*' ([JT80]); and Tiomno listed some additional considerations in a paper published as a conference report in the *Revista Brasileira de Física* ([JT77]); this latter publication was a review of previous works on the subject, given at the Symposium in honor of the 70th birthday of Mario Schenberg in August 1984.

Tiomno also wrote an invited paper for the IVth *Marcel Grossmann Meeting on Relativistic Astrophysics* (MG-IV), held in Rome in June, 1985, and published in its Proceedings the following year. It was entitled '*Experimental Evidence against a Lorentz Aether Theory*' ([JT86]). Another two articles by Maciel and Tiomno, on the theoretical analysis and on experimental aspects of absolute spacetime theories (i.e. LET's) appeared in *Foundations of Physics* in 1989 ([JT94], [JT95]). Tiomno's last article on the subject, also published together with Maciel, appeared in *Foundations of Physics Letters* in 1989

([JT99]). It was a reply to a critique of their earlier papers of that year, pointing out that effects due only to coordinate transformations are arbitrary and have no physical significance. Their promised review of the subject, called '*Test Theories of SR: A General Critique*', was apparently never published, but it is available as a preprint from the CBPF.[12]

Strong Interactions Revisited

During the period from 1984–1990, Tiomno also worked in the area of hadron physics (Strong interactions, SI). In July of 1983, he had attended the 3rd*ICTP Workshop in Particle Physics* in Trieste, an indication of his renewed interest in SI. In that field, he published together with several younger physicists; this was part of a larger collaboration that centered around the Italian theoretician *Enrico Predazzi*. The story of this long collaboration, whose roots go back to the 1950s and involve Gleb Wataghin, then recently returned to Turin after over 15 years in Brazil, is told in detail in a tribute to Predazzi for his 60th birthday, written by two of his Brazilian collaborators, *Ignácio Bediaga* and *Francisco Caruso*.[13]

Enrico Predazzi was born in Italy in 1935. He was educated in physics at the University of Turin, beginning his studies there soon after Gleb Wataghin had returned from his stay at USP, in Brazil. Predazzi was influenced by Wataghin, like so many young physicists before him, even though they didn't work together directly. Predazzi obtained his *laurea*, roughly equivalent to the MS, in 1958 and became a research assistant in Turin, moving up through the ranks and eventually, as associate professor, going to Chicago for two years in 1964; there he worked with *Yoichiro Nambu*, among other researchers. Nambu had at that time recently suggested *color charge* as a property of the Strong interaction, and he received the Nobel prize in 2008 for his contributions to symmetry-breaking theories.

Predazzi's long collaboration with Brazilian physicists was due primarily to two people: One was *Henrique Fleming*, a Brazilian born in 1938, who went to Turin from USP as a doctoral student in 1967, and worked with Predazzi there. They continued to collaborate after Fleming's return to Brazil, and Fleming (today professor *emeritus* at USP) invited Predazzi to spend some time as a visitor in São Paulo following his sabbatical at the University of Indiana/USA in 1969.[13], [14]. Predazzi was additionally invited (by *Erasmo Ferreira*, who was a student and later a colleague of Tiomno's at PUC) to give a course at a summer school in theoretical physics to be held in Rio in early 1970. There, he became aware of the difficult situation in Brazil due to the decline of the CBPF and the punitive measures of the dictatorship. He spoke to Jayme Tiomno about the wisdom of going to USP after the recent forced dismissals, and Tiomno encouraged him to go there and support his young theory group, left leaderless by his dismissal. Predazzi was able to recruit a number of other visitors to USP, based on his own experience, and this proved to be the salvation of the theory group there, as Jayme Tiomno also later emphasized.[14]

The second key figure in Predazzi's 'Brazilian connection' was *Alberto Franco de Sá Santoro*, born in 1941 in *Manaus*/AM. He was one of the students who began his studies at the UnB in the mid-1960s, and he completed his first degree at UFRJ in 1969.[15] He obtained his doctorate from the *Université de Paris VII* in 1977, and met Predazzi at a summer school while he was in France. He subsequently worked at the State University in Rio de Janeiro (UERJ), and at the CBPF. He was one of the members of the group interested in experimental particle physics, and went to Fermilab, sponsored by the Experimental High-Energy physics project at the CBPF. He asked Predazzi to take over some of his theory students in the early 1980s when he was shifting to experimental physics, and two of them—the doctoral student *Ignácio Bediaga* (who had studied at PUC from 1974–1980, while Tiomno was a professor there; today, he is a professor at the CBPF), and the masters candidate *Francisco Caruso* (born in Rio in 1959, studied at UERJ from 1976–1980, today professor *emeritus* at UERJ)—went to Turin in 1984. They obtained their degrees there and continued collaborating with Predazzi, in Italy and in Brazil, in the following years.

It was thus natural that Jayme Tiomno, after his return to the CBPF in 1980, and especially as Project Coordinator for High-Energy Physics, would also collaborate scientifically with the group around Predazzi, Santoro, Bediaga and Caruso in the subsequent years. Their joint papers furthermore included *Antonio Carlos Baptista Antunes*, *Moacyr H.G. Souza*, and the French physicist *Jean-Louis Basdevant* as co-authors. The result was a flurry of publications in the years 1984–1987 ([JT71, 74, 78, 79, 84, and 87]). The last of those articles, entitled '*A new Hadronization Scheme: The case of explicit Charm Decay*', was published in *Nuclear Physics* by Tiomno together with Basdevant, Bediaga and Predazzi. After that publication, Tiomno stopped working in the field of hadronic physics, in which he had begun research at USP 40 years earlier, in 1946/47. Their work in the 1980s dealt with the 'new physics' ushered in by the *November revolution*[16] of 1974, when the discovery of the first '*charmonium*' particle, the J/ meson, gave evidence for the *charm* quark, completing the second generation of quarks.

Gravitation, Field Theory, and Cosmology

At the same time, Tiomno continued his work on field theory, gravitation, and cosmology, which he had begun with Mario Schenberg at USP in 1946, and which had received a strong reviving impetus during his second Princeton stay in 1971/72. In the 1980s, he wrote eight articles on field theory and gravitation, beginning in 1983 with a paper published together with *Mario Novello* and *Ivano Damião Soares* in the *Physical Review* ([JT67]). Other collaborators included *Antonio Fernandes da Fonseca Teixeira*, *José Duarte de Oliveira*, *Marcelo José Rebouças*, *Bartolomeu Donatila Bonorino Figueiredo*, *Fernando Deeke Sasse*, *Jose Ademir Sales de Lima*, and *Mauricio Ortiz Calvão*, all of them younger members of the Gravitation and Cosmology Group at the CBPF. This collaboration continued through 1992.

Mario Novello was one of Elisa and Tiomno's students who went along to the UnB in 1965 (and he was also a member of their 'fan group' "GOELTI 64"). Born in Rio de Janeiro in 1942, he completed his MS at the CBPF in 1968, working with Leite Lopes, and went to the *Université de Genève* in Switzerland to obtain his doctorate, with a thesis on the *'Algèbre de l'espace-temps'* ('Space–time Algebras'), under Josef Maria Jauch in 1972. Back at the CBPF, he founded the Gravitation and Cosmology group in 1976, while Tiomno and Leite Lopes were still banned from working there, and later on he was 'César Lattes ICRANet Chair' (Brazilian representative for the ICRANet, Pescara/Italy; see Chap. 13, section on Remo Ruffini) after 2008 (Fig. 15.1).

Ivano Damião Soares obtained his first degree in electronic engineering in 1969, then completed his MS in physics at the CBPF in 1972, and he obtained his doctorate, also at the CBPF with Mario Novello, in 1976 on the topic *'A Study of the Neutrino-Gravitational Interaction'* (cf. Fig. 16.5 in Chap. 16). Novello and Soares had already begun research in gravitation theory and cosmology while Tiomno was still at PUC in the 1970s, and on his return to the CBPF, he began to collaborate with them in those fields. His collaboration with Ivano Soares in fact continued until well after his 'official' retirement.

Tiomno's publications dealt in particular with *'Gödel's universe'*, a solution to the Einstein field equations found by Kurt Gödel in 1949 (while Tiomno was his neighbor in Princeton!). Tiomno's first article on this topic, published in 1983 with Novello and Soares, was entitled *'Geodesic Motion and Confinement in Gödel's Universe'* ([JT67]).

[*Geodesics* are the shortest distance between two points in a given space-time. In flat, homogeneous Euclidean space, a straight line suffices; in a

Fig. 15.1 Jayme Tiomno, Mario Novello, and Elisa Frota-Pessôa, 1991. *Source* Private archive, [EFP]

curved space (e.g. on the surface of the Earth, roughly spherical), geodesics are curved lines ('great circles'). In more complex spaces (or spacetimes), they can be complicated curves, along which light rays propagate and particles can pass. In *Riemannian spacetime*, which Einstein used for his initial formulation of General Relativity (GR), the geodesics are curves, since the spacetime itself is curved, and its *curvature* generates gravitation. In *Einstein-Cartan spacetime*, investigated by the French mathematician *Élie Joseph Cartan* in the early 1920s and later rediscovered by Einstein (as 'teleparallel' space) and used in his search for a unified field theory, the spacetime has a *torsion* ('twisting'), which is responsible for gravitation. *Gödel's universe*[17] describes a closed, rotating, stationary and homogeneous cosmos containing closed-loop, time-like worldlines (which could represent paths that would permit time travel!— This is one of the paradoxes introduced by even such a staid, non-quantum theory as GR)].

Later works in that same year (1983: [JT69], [JT70]) dealt with the geometric properties of the Gödel metric (in *Einstein-Cartan spaces*, the 'teleparallel' space used by Einstein in the late 1920s), published together with Teixeira and Duarte; and in *Riemannian spacetime* (the curved spacetime of standard GR), with Rebouças. Another article on (*non*-homogeneous) Gödel-type models appeared in *Il Nuovo Cimento* in 1985 ([JT81]), also written together with Rebouças.

Two further publications, with Soares in 1984, and with Figueiredo and Soares in 1989, treated the coupling of microscopic matter to the 'vorticity' in an Einstein-Cartan spacetime (i.e. to its *torsion*, which produces the 'matter vorticity') ([JT72], [JT93]). The last two papers in this series appeared in 1990 and 1992, co-authored with Soares and Calvão, and with Figueiredo and Soares, respectively. They dealt with '*Geodesics in Gödel-Type Spacetimes*' ([JT100]) and with the '*Gravitational Coupling of Klein-Gordon and Dirac Particles to matter vorticity and spacetime torsion*' ([JT101]). The latter paper, published in the specialized journal *Classical Quantum Gravity*, was Tiomno's final publication on gravitational theory.

In the field of cosmology, Tiomno published 7 papers during the later 1980s, beginning in 1986 with an article on the '*Homogeneous Cosmos of Weyssenhoff fluid in Einstein-Cartan Space*', published with Duarte and Teixeira in the *Physical Review* ([JT85]); and continuing with two papers on '*Antipodal Universes*' ([JT89], [JT90]), with I.D. Soares and F.D. Sasse, in 1988. (The second of these, entitled '*Neutrinos in Antipodal Universes: Parity Transformation and Asymmetries*', harks back to Tiomno's early work on the Weak interactions).

Tiomno's last work on cosmology, entitled '*Thermodynamical Analysis of Cosmological Models*', was published in 1989, together with *José Ademir Sales de Lima*, who is currently Professor of Astronomy at USP, São Paulo (see Fig. 17.1 in Chap. 17). He was Jayme Tiomno's doctoral student at the CBPF in 1985–1990, when he wrote four articles together with Tiomno, on heat flow and thermodynamics in various cosmological models ([JT88, 91, 96, 97]). They were all published in 1988 and 1989. Ademir Lima's thesis topic was '*Geometro-Thermodynamics of Cosmological Models*', and these four publications represent works-in-progress from his doctoral thesis.

In addition to these various purely scientific works, Tiomno produced 6 publications during the 1980s which can be regarded as scientific-historical and scientific-sociological, based on his own experiences during his long career. These included [JT68, 75, 83, 92, 98, and 102]. The first and last of these dealt with his much earlier and successful involvement with the Weak interactions, which was revived in the mid-1980s, as we shall see in the following section.

The Weak-Interactions Interlude

In 1983, plans were being made by David Cline in Wisconsin to hold a meeting at a conference center near Racine/WI, USA on '*50 Years of Weak Interactions*' (dated from Fermi's initial publication of his 'four-fermion interaction' in 1933/34[18]). John Wheeler heard of those plans and remembered Jayme Tiomno's contributions in the early period of the Weak interactions (for which the Universal Fermi Interaction was the principal model at the time). He wrote to Tiomno in February of 1984, reaching him in Italy where he was visiting the University of Padua, and entreated him to take part in the conference and to represent his own claims to authorship of some of the initial ideas that contributed to the theoretical development of the field. (To emphasize the urgency of the matter, he quoted Napoleon, speaking of his Russian campaign: "…*chaque heure perdue était irréparable, fut retardée de deux mois…*"—"every hour lost was irreparable, a setback of two months …"). Tiomno took this to heart and registered for the conference, and he submitted a paper entitled '*The Early Period of the Universal Fermi Interactions*' ([JT68], 1984). It was published in the Proceedings[19] of the Conference by the University of Wisconsin and is available from the CBPF *Notas de Física*. The Conference had originally been scheduled for November, 1983, and its postponement was a boon for Tiomno, who otherwise would have missed it.

All in all, there were four major conferences on the history of the Weak interactions in the US between 1980 and 1993. The first was a Fermilab Symposium on the *History of Particle Physics* held in 1980.[20a] The Racine Conference took place from May 29th to June 1st, 1984 at the Wingspread Conference Center, near Racine/WI. A year later, another Symposium was held at Fermilab, entitled '*From Pions to Quarks. Physics in the 1950s*'.[20b] And in February, 1993, there was a fourth meeting, held in Santa Monica/CA on the topic '*Discovery of Weak Neutral Currents: The weak interaction before and after, 1993*'.[21] Evidently a need was felt within the physics community to discuss and summarize the history of weak-interactions theory. Tiomno attended only the second of those four meetings, but his article from it was later reprinted in revised form in the Proceedings of the Santa Monica Conference[21] ([JT102], 1994).

The Racine Conference was attended by many of the participants in the early work on Weak interactions, among them Robert E. Marshak (1916–1992), a colleague of Tiomno and Wheeler's, who gave a talk there on the Universal V–A theory, originated with his student E.C.G. (George) Sudarshan in late 1956 and early 1957. Marshak's lecture is also reprinted in[21]. In its original version, he had neglected to mention Tiomno's early work on the subject.

The diagram in Wheeler and Tiomno's second 1949 article ([JT6]; cf. Chap. 6) had initially been known as the 'Tiomno-Wheeler triangle'; later, Giampietro Puppi was also associated with the diagram (with no reference to Wheeler), and it came to be called the 'Puppi-Tiomno triangle'. And finally, it was referred to simply as the 'Puppi triangle'. That name for the triangle was used for example by Robert Marshak in his article '*Particle physics in rapid transition during the period 1947–1952*', published in the Proceedings of the 1980 Fermilab Symposium (which appeared in 1983).[20a] This issue of authorship greatly bothered Wheeler, and led him to attempt to set the record straight.

Tiomno, in his own paper for the Racine Conference, emphasized that earlier concept in his work with Wheeler in 1948/49, namely the 'Tiomno triangle' (cf. Chaps. 6 and 7), in which he had illustrated in a clearcut manner just what was meant by a *universal* Fermi interaction. He referred to it ironically as "the case of the mutating triangle"; it had 'mutated' in the years following its publication to being called the 'Puppi triangle', although Puppi, in his paper of December 1948, had not included a diagram at all.

After the Racine conference, as a consequence of Tiomno's presentation, Marshak wrote to him[22a] (with a copy to Wheeler), enclosing a corrected version of the paper he had given there. "Your citation of my reference to the 'Puppi's triangle' in my article in the Brown-Hoddeson book[20a] is understandably scornful…", he wrote. He excused his error of attribution as being due to his lack of contact with the literature in recent years because of heavy administrative duties, so that he had simply taken the information from a current bibliography. In a second letter, on September 28th, 1984,[22b] he took up the topic of the universal V–A interaction,

> I hope the enclosed version does justice to your contributions to the universal V–A program. It is clear to me that you were aware of the γ_5 transformation before the rest of us (e.g. your thesis), and that if you had not been so isolated in Brazil, you would have made the same comprehensive analysis of the experimental situation as George [*Sudarshan*] and I did, and probably arrived at the same V–A conclusion.

Marshak's remark about Tiomno's 'isolation in Brazil' refers to the latter's unawareness of the discussions surrounding the experimental situation in Weak processes (see Chap. 9, backnote [37]). In the revised (printed) version of his lecture,[21] Marshak made the following statements about Tiomno's contributions:

> The papers that were closest in spirit to the chirality invariance underlying the V–A theory were written by Tiomno and by Stech and Jensen. Tiomno invented the idea of 'mass reversal invariance (the idea that the Dirac equation is invariant under the transformation $\psi \rightarrow \gamma_5 \psi$, $m \rightarrow -m$), and postulated the invariance of the weak current $\psi_1 O_\mu \psi_2$ (where O_μ is the S, V, P, A or T operator) under the *simultaneous* transformation (to conserve parity): $\psi_1 \rightarrow \pm \gamma_5 \psi_1$, $\psi_2 \rightarrow \pm \gamma_5 \psi_2$. Tiomno found that if the signs in the γ_5 transformations are the same, O_μ has to be a combination of V and A, whereas if the signs are different, the combination has to be S, P, T. This clear cut separation into two classes of Fermi interactions was interesting but still quite different from the idea of applying separate chirality invariance to each Dirac field, which led to parity violation and the V–A interaction.
>
> ... We were aware of these papers at the time that we wrote ours but we chose not to refer to them because of their limitation to the parity conservation case. In hindsight, we consider these papers as valuable contributions to the chirality invariance approach and wish to correct the record on this score.
>
> ... Unbeknown to us, Tiomno's paper on '*Nonconservation of Parity and the Universal Fermi Interaction*' was sent to *Nuovo Cimento* in early July 1957 and published in October. He went beyond Salam in trying to reconcile the accepted (S,T,P) combination for the interaction with the (V,A) muon interaction by postulating opposite helicities for the neutrino and thus ended up with a somewhat inelegant and incorrect UFI.
>
> ... With this complicated set of facts, how does one settle the priority question in which historians of science are interested? In this instance, perhaps the simplest solution is to quote Feynman, who said a decade ago: 'We have a conventional theory of weak interactions invented by Marshak and Sudarshan, published by Feynman and Gell-Mann and completed by Cabibbo—I call it the conventional theory of weak interactions—the one which is described as the V–A theory'.

Leite Lopes expressed a pessimistic opinion of the consequences of being a researcher outside the mainstream scientific countries; in fact, he was in general very critical of the scientific relations between 'mainstream' and 'peripheral' countries. In his view, "This is only an example of the well-known tendency in the community of physicists of industrial countries for the rarity of quotation of articles the authors of which are out of one of the groups of these countries". (see Chap. 6, backnote [28]).

The neglect of Tiomno's contributions was likewise acknowledged by Chen Ning Yang, who changed his own manuscript for the 1985 Fermilab Symposium accordingly.[23] The same confusion about the authorship of the triangle and Tiomno's other early works had been unintentionally perpetrated by Yang in his presentation at that Symposium, where he referred to the 'Puppi triangle'. Correcting his mistake, Yang wrote to Tiomno on May 10th, 1985,[23] saying that he had met two of Tiomno's students/colleagues at the Fermilab Symposium, where he had given a talk on the work on the UFI in 1948/49. He had noted in his talk that several groups independently came up with the idea of a universal strength for the weak interactions, and that the first to mention it was Puppi. Tiomno's colleagues gave him a copy of Tiomno's paper [JT68] from the Racine (Wingspread) Conference,[19] which convinced Yang to change the written version of his Fermilab talk and to recognize Tiomno's early contributions.

In a similar fashion, Tiomno's 'mass reversal invariance' mutated (with Abdus Salam's help) to be termed 'chiral invariance', the name later used by Sudarshan and Marshak (cf. Chap. 9).

Nevertheless, the 'Tiomno (/Wheeler/Puppi) triangle' had merely a symbolic significance: Originally, it was an heuristic device calling attention to the *universality* of the Fermi-type interaction, but that function had done its job by around 1955. Later on, to Wheeler, Tiomno and some others, it became a symbol of the lack of recognition of Tiomno's early work on the UFI, and that is why he mentioned it again in his paper for the Racine Conference. But as we can see from Marshak's remarks quoted above, Tiomno's real and important contributions were the introduction of the γ_5 operator and of mass-reversal (or chiral) invariance, and this was finally acknowledged by both Marshak and Yang.

George Sudarshan published his own view of the events of the 1950s as a contribution to the 1985 Fermilab Symposium.[20b] He was evidently rather embittered by the lack of recognition of his early discovery of the Universal V–A interaction, which was based on a thorough review of the experimental situation and a good understanding of the theory. The chain of events leading to the lack of recognition of the priority of Sudarshan and Marshak's work is detailed by Marshak in his talk from the Racine Conference (reprinted in[21]), and again in[24]. Sudarshan's final remark in his lecture at the Fermilab symposium was pessimistic: "It has been a sad but enlightening experience to recognize that the universality of science does not imply unbiased acclaim for scientific truth and a true history of science, and that if one has neither powerful alliances nor influential sponsors, he should learn to do science for its own sake and not be depressed by lack of appreciation". This statement could indeed just as well be applied to Jayme Tiomno. Sudarshan was nominated nine times for a Nobel prize, and apparently narrowly missed receiving it on at least two occasions.[25] He died in April, 2018.

The Racine Conference in any case stimulated Tiomno to make a contribution to the history of the theory of the Weak interactions, and brought him some deserved recognition for his early work on the subject, thanks to his old mentor, John Wheeler.

The *'Pará* incident', 1986

In 1986, Jayme Tiomno was invited to give the Semester Inaugural Lecture at the *Universidade Federal do Pará* (UFPA) in *Belém*/PA, in the Amazon region, founded in 1957. His talk was given in mid-February. The Department of Physics at UFPA had been supported by the CBPF, following the *Centro's* tradition of support for universities in the less-developed Brazilian states that had begun with Luis Freire and *Pernambuco,* and continued with *Ceará* and *Rio Grande do Sul.* This program involved in particular establishing scholarships and fellowships to bring students to the CBPF or to USP, and eventually to send them abroad for post-graduate work, after which they could return home and help to establish the new university there.

The initiative for Tiomno's invitation is credited to the physicist *José Maria Filardo Bassalo,* who in 1965 joined the group attempting to establish the Physics Institute at UnB, as assistant to Roberto Salmeron and Jayme Tiomno, and completed his Bachelor's in physics there. He became a key figure in the relations between the CBPF and the UFPA.[26]

Tiomno was very concerned about the paths that the universities as well as many research institutions in Brazil had taken after the university reform in 1968. The dismantling of the professorial system (*catedrática*), although desirable in itself, brought a great deal of chaos to the universities, since it put an end to scientific and academic authority, placing a researcher with years of experience and great international recognition on the same level as young people just starting their university careers. The dictatorship had at the same time destroyed the merit system within the universities and had wiped out their principles of academic governance. Thus, Tiomno decided to give an autobiographical lecture, in which he could make use of his own career and experiences as examples upon which to discuss the political, social and educational developments in Brazil during the previous 50 years, and to draw conclusions about what should be done in the future to guarantee the quality and continuity of Brazilian universities. His lecture took place soon after the end of 21 years of dictatorship and authoritarianism, which could not simply be forgotten: "... surprisingly, Brazil is a country without a memory, in which the same discussions and proposals, even the same attempts at finding solutions, are repeated cyclically, with a total lack of awareness of the previous [*cycles*], thirty or even fifty years later".[27]

Indeed, this is a characteristic not only of Brazil, and not only of many other countries in the world, but apparently of humanity in general; and if Jayme Tiomno could see the world today, he would no doubt agree that precisely this characteristic is making itself felt all over, to the detriment of our societies and future and perhaps of humanity itself.

In his lecture, Tiomno points out the extraordinary influence of 'good professors and poor professors', in Brazil and elsewhere, and the value of study abroad (in Europe, particularly), as well as the development of their home universities made possible by the students returning from such studies abroad; and of an idealism based not on simple nationalism, but on the recognition of the need to contribute to the national development. The departmental system and the difficulty of imposing it to replace the professorial system without due preparation (i.e. without establishing a merit system) produced many of the problems faced by the universities by 1986. In his conclusion, Tiomno lists seven points which he considers to be fundamental to the necessary reform of the universities:

1. "The university cannot receive blank checks without offering real services to the local and national community;
2. Democracy does not give precedence to mediocrity, nor is it the enemy of excellence;
3. If the universities do not impose quality of teaching and research upon themselves, they will be replaced by military colleges, reduced to the function of diploma factories;
4. The universities must diversify themselves and experiment with various solutions—'let a hundred flowers blossom, and a hundred schools of thought...';
5. An increase in the number of positions at the universities has as its precondition the formation of a nucleus of competent professors and researchers, based on a merit system;
6. That increase also presupposes new opportunities for the students who wish to follow mid-level careers;
7. Every type of scholarship and fellowship must be accompanied by a requirement of participating in and aiding teaching and instruction".[27]

Tiomno's lecture was summarized by the local newspapers, and the conservative Senator *Jarbas Passarinho* reacted by publishing a criticism in the newspaper '*O Liberal*' a few days later (on February 23rd, 1986), which he titled '*Remembering Lacerda*' (referring to *Carlos Lacerda*, journalist, author and right-wing politician).

The good Senator Passarinho was a supporter of the dictatorship—in fact, he had been one of its active members; he had been Minister of Education, among other functions. much decorated as a public official. When the AI-5 was approved by the Congress, he was the author of the embarrassing phrase, "At this moment, Mr. President, all the scruples of conscience can go to Hell". He was evidently angered by Tiomno's lecture and its characterization of the dictatorship and its effects on the universities in Brazil. Passarinho himself gives the impression of being an elitist masquerading as a populist, while trying to accuse Tiomno of just that, and of giving a trivial lecture, of placing himself in the foreground; and he hints at Tiomno's hidden, presumably communist or socialist agenda.

All together, Passarinho's article is a character assassination of the politically-motivated sort. He proceeds to praise the presumed benefits of the dictatorship in improving the previously desolate academic landscape in Brazil, closing with a renewed attack on Jayme Tiomno's veracity and alleged motives.

In Tiomno's reply, "*Considerações sobre o artigo 'Lembrando Lacerda' de Jarbas Passarinho*", ("Considerations on the article '*Remembering Lacerda*' by Jarbas Passarinho"), published on March 1st in the same newspaper, Tiomno points out the insidious nature of Passarinho's insinuations and denies having a political agenda himself. He refers to Passarinho as 'the journalist', which indeed characterizes his role as author of the critical article, but must have been demeaning to the former Minister, Governor and current Senator. Tiomno divides Passarinho's article into three parts and systematically refutes his claims in each part. He emphasizes that simply increasing funding to the universities (for 'buildings and grounds', as cited by Passarinho) will not solve their problems, and that the forced 'democratization' (the departmental system) was not sufficiently prepared and had opened the door to mediocrity, since no merit system was put in place. He points out that Passarinho, while accusing him of telling untruths, gives no evidence or concrete points to support that assertion. Finally, he defends himself against Passarinho's accusation that he had spoken badly of the UFPA without knowing the details of its situation (with which, as Tiomno remarks, Passarinho had tried to drive a wedge between him and the UFPA).

The Executive Council for Teaching and Research (CONSEP) of UFPA issued an official statement[27] addressing Tiomno and offering the support of the Council against the accusations made by Passarinho. The Council had furthermore resolved to have the entire text of Tiomno's lecture published in a newspaper 'of large circulation'. They reaffirm their support for having invited Tiomno to give the lecture, and for its content; and they condemn Passarinho for reaching 'precipitate and emotional conclusions' without knowing the full content of the lecture. Their letter was sent to both Tiomno and to Passarinho, and it represents a statement of solidarity with Tiomno.

Somewhat later (in 1987), J.M.F. Bassalo published an article[28] explaining the reasons for inviting Jayme Tiomno as commencement speaker, intending it to provide information to a public who were possibly confused by the dispute that appeared in the newspapers, and unsure of just who this professor from Rio de Janeiro really was.

It would appear that Jayme Tiomno emerged victorious from this episode, as viewed from the long term, and the public attack on his integrity and motivations can be regarded as one of the final convulsions of the retreating authoritarianism which had accompanied the years of the dictatorship. The 'trade-off' between the negative impact of the repression by the military regime and the (alleged) benefits of their increased funding for education and research has been considered by Brazilian historians of science.[29] Tiomno himself was very clear on this point. His warning in this lecture, delivered over 30 years ago, that "Brazil has no memory", seems all the more important in view of current events, in Brazil and elsewhere in the world.

Towards the End of the Decade

In 1982, after his retirement from UFRJ, Leopoldo Nachbin also returned to the CBPF. In 1985, when the first post-dictatorship government had been established, the Ministry of Science and Technology was set up. Its new Minister, Renato Archer, began an effort to bring back those scientists who had gone abroad, or, for those who were firmly established in their new institutions, to at least encourage cooperations with Brazilian research groups. The researchers at the CBPF also began a movement to bring Leite Lopes back from France, where he had remained up to that point, and to make him Director of the *Centro*. Finally, more than 40 years after their friendship as students at the *Faculdade Nacional de Filosofia*, those four, Jayme Tiomno, Elisa Frota-Pessôa, José Leite Lopes and Leopoldo Nachbin—who had dreamed of creating an institution devoted to physics research and education—were once again reunited in their old spiritual home: the CBPF (Fig. 15.2).

In the later 1980s, Jayme Tiomno could once again dedicate himself to his passion, advancing physics education and research in Brazil. He accepted several administrative duties at the CBPF, in spite of his many research activities described above. From 1985 through 1988, he was *Project Coordinator* for High-Energy Physics at the CBPF, leading to the founding of the 'LAFEX' Department (international cooperations for experimental HEP, beginning at Fermilab in the USA).

Tiomno was also *Project Coordinator* for Algebraic Computing at the CBPF from 1985 to 1992, and helped achieve the modernization of the necessary computing facilities. Somewhat later, from 1988 to 1996, he served as President of the Executive Council of FAPERJ (the state science funding agency for the State of Rio de Janeiro).

Fig. 15.2 The 'Three Musketeers', 50 years later (*from left*): Jayme Tiomno, Elisa Frota-Pessôa, José Leite Lopes, and 'd'Artangnan', Leopoldo Nachbin, in 1989. *Source* Private archive, [EFP]

Tiomno, in 1985, also published a revised version of his 1968 Inaugural Lecture as a newly-installed Professor at USP[30]; and in 1988, an account of the founding of the ICTP in Trieste in 1964, in which he played a significant role.[31] The following year, he contributed a chapter entitled '*Collaboration in Physics in Latin America* (Round Table)' to the book *Proceedings of the 3rd Symposium on Pan-American Collaboration in Experimental Physics*, edited by R. Rubinstein and A. Santoro.[32]

After all of these activities, Jayme Tiomno experienced his seventieth birthday in April, 1990, and it was commemorated in a *Festschrift* called '*Frontier Physics* ', edited by his former students and/or collaborators *Samuel W. Mac-Dowell*, *Herch Moysés Nussenzveig*, and *Roberto Aureliano Salmeron*, and published the following year.[33] It will be our starting point for the next decade, the last of the old century, and the subject of the following chapter.

Notes

[1] Jayme Tiomno to José Goldemberg, vice chairman of the Physics Institute at USP, December 12th, 1979; [JT].

[2] Report of the *Departamento Geral de Investigações Especiais* ['General Department of Special Investigations'] of the Political Police, August 30th, 1979; [APERJ—Series DGIE, number 2719-D, p. 124].

[3] From the interviews with Mario Novello for this book by ATT, August/September 2019.

[4] Jayme Tiomno, Talk given at the '*Public Moral Reparation Session and Re-Initiation of Collaboration*', CBPF, November 23rd, 1979. Available online from the CBPF.

[5] Frota-Pessôa (1990).

[6] Leite Lopes to Caden and Haity Moussatché, January 27th, 1982; [HM].

[7] See the last section of Chap. 14, and Backnote [24] there.

[8] Leite Lopes to Caden and Haity Moussatché, January 27th, 1982; [HM].

[9] John Wheeler to Alwyn van der Merwe (Editor of *Foundations of Physics*), Sept. 13th, 1983; [JWA]. See also Bassalo & Freire (2003).

[10] [JT76].

[11] Jayme Tiomno, Lecture given at the session in honor of his 60th birthday; in Proceedings of the *II. Encontro Nacional de Partículas e Teoria de Campos*, Cambuquira, MG (Sept. 1980), *Sociedade Brasileira de Física* (unpublished).

[12] A.K.A. Maciel and J. Tiomno, '*Test Theories of Special Relativity, A General Critique*' (1988); preprint available from the CBPF, *Notas de Física*, CBPF-NF-066/88 (unpublished).

[13] Bediaga & Caruso (1996).

[14] cf. Chap. 12, 'Dark Times' (Backnote [18]); and Chap. 17, Predazzi's memoir of Tiomno in its final section.

[15] See Caruso et al., (2011).

[16] The 'J/ψ particle' was simultaneously discovered by Samuel Ting's group at Brookhaven/NY, who called it the *J particle*, and by Burton Richter's group at Stanford/CA (SLAC, SPEAR storage ring), who named it the ψ. This meson is an example of '*charmonium*', since it consists of a particle-antiparticle pair of *charm quarks*. The discovery of the J/ψ meson heralded the November 1974 'revolution', giving rise to 'new physics' (i.e. the physics of hadrons containing 'charm' quarks, of the second quark generation). Richter and Ting shared the 1976 Nobel prize "for their pioneering work in the discovery of a heavy elementary particle of a new kind".

[17] See Gödel (1949). Gödel originally wrote this work as a contribution to the *Festschrift* for Einstein's 70th birthday in March, 1949. Einstein is said to have disappointed him by remarking that it had no application to the real world. For a general discussion of Gödel's universe, cf.Rindler (2009).

[18] Fermi (1934a,b).

[19] Cline & Riedasch (1984): Proceedings of the Racine Conference, 1984.

[20a] Brown & Hoddeson (1983): Proceedings of the Fermilab Symposium held in May, 1980.

[20b] Brown, Dresden & Hoddeson (1989): Proceedings of the Fermilab Symposium held in May, 1985.

[21] See Mann & Cline (1994), the Proceedings of the Santa Monica Conference held in 1993. Tiomno's article is on pages 99–109, Marshak's begins on p. 110. This Proceedings volume includes reprints of many previous articles by most of the contributors to the early development of UFI/Weak-interactions theory. Tiomno's article in it is also available online: https://doi.org/10.1063/1.45451. The quotation from Feynman in Marshak's article is from a conference in Philadelphia in 1974; Feynman was trying to 'make amends' to Sudarshan for the (unintended) lack of recognition of his work. Marshak repeated this quotation in a personal communication to Jagdish Mehra, in 1990. See Mehra (1994), Chap. 21, endnote 40.

[22a] This quotation and Marshak's response are requoted from Bassalo & Freire (2003); they are in a letter from Robert E. Marshak to Jayme Tiomno, dated July 27th, 1984; [JT].

[22b] Robert E. Marshak to Jayme Tiomno, Sept. 28th, 1984; [JT].

[23] C.N. Yang to Tiomno, May 10th, 1985; [JT]. See also Chap. 6, Backnote [37].

[24] Marshak (1997).

[25] cf. www.tifr.res.in/TSN/article/Sudarshan.pdf.

[26] Bassalo was born in *Belém do Pará* in 1935. He studied civil engineering at the former Engineering School of *Pará* (later absorbed into UFPA) from 1954–1958, and he was active in supporting the development of physics

in *Pará* through the CBPF, as an instructor at the Physics and Mathematics Group at UFPA, the nucleus of the later Department of Physics, from 1961–1965. In 1965, he joined the group establishing the Physics Institute at UnB, as assistant to Roberto Salmeron and Jayme Tiomno (an experience which he describes in Bassalo (1998) and Bassalo (2012), quoted also in Chap. 11). He completed his Bachelor's in physics in Brasília in 1965, and then obtained his MS and his doctorate at USP in 1973 and 1975, both in the group of *Mauro Sergio Dorsa Cattani*. The subject of his doctoral thesis was '*Cálculo Quântico do Alargamento e Deslocamento da Linha 4^3 s–2^3p do Hélio em um Plasma*' ('Quantum Calculation of the Broadening and Shift of the 4^3 s–2^3p Line of Helium in a Plasma'), i.e. a topic in the theory of atomic/plasma spectroscopy. Participating successfully in *concursos* in 1978 and 1989, he became Adjunct Professor and then Tenured Professor at the Physics Department of UFPA, continuing there until his retirement in 2005. He has written a number of books on the history of science.

[27] See Document CBPF–CS–005, 1986 for a reprint of Tiomno's lecture ([JT83]), as well as supporting documents.

[28] Bassalo (1987).

[29] See e.g. Freire, Videira & Ribeiro Filho (2009).

[30] [JT75].

[31] [JT92].

[32] [JT98].

[33] MacDowell et al. (1991).

16

Researcher Emeritus—The 1990s

Frontier Physics

On April 16th, 1990, Jayme Tiomno celebrated his 70th birthday. Among the accolades offered to him by his colleagues were a session in his honor at the annual meeting of the Brazilian Physical Society (SBF) in *Caxambu*/MG, September 26–30, 1990, and an International Symposium dedicated to his birthday, held at the CBPF on November 8–9, 1990. Out of that latter meeting, a *Festschrift* entitled '*Frontier Physics*' grew; it contained the talks given at the symposium, and was edited by *Samuel Wallace MacDowell*, *Herch Moysés Nussenzveig*, and *Roberto Aureliano Salmeron*[1], as we have mentioned (Fig. 16.1).

Roberto A. Salmeron was Tiomno's colleague during the early years of the CBPF and from the UnB/Brasília project, and one of his best friends.

Samuel W. MacDowell is a member of a well-known family from Brazil's north, evidently immigrants from Scotland at a much earlier time. He had illustrious ancestors in *Pará* and *Pernambuco* during the nineteenth century. Born in *São Lourenço da Mata* (near Recife) in March, 1929, MacDowell was one of those talented students discovered in Recife by Luis Freire in the early 1950s and sent to Rio, to the FNFi/UB and the CBPF. There, he was one of Tiomno's and Leite Lopes' first students (and co-author with Tiomno of an article on polarization by nuclear scattering: [JT24], 1956). That article was from his Masters thesis with Tiomno. He then went to the University of Birmingham/UK, where he obtained his doctorate in 1958, working with quantum pioneer Rudolf Peierls, with a thesis on quantum physics. He returned to Brazil and became a professor at the CBPF in 1963, just when

© Springer Nature Switzerland AG 2020
W. D. Brewer and A. T. Tolmasquim, *Jayme Tiomno*, Springer Biographies,
https://doi.org/10.1007/978-3-030-41011-7_16

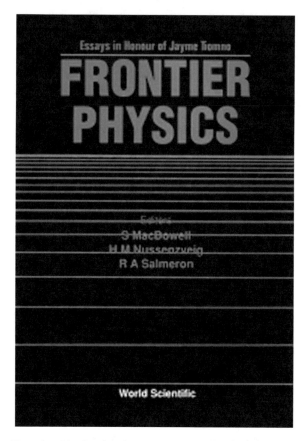

Fig. 16.1 The 'Frontier Physics' book cover. *Source* World Scientific; used with permission. See https://www.worldscientific.com/worldscibooks/10.1142/1371

many of the earlier faculty were leaving. From late 1963 to mid-1965, he was on leave as a Fellow at the IAS in Princeton, then joined the faculty in Mathematical Physics at Yale University/USA, where he spent the rest of his career, retiring in 2001. During most of his scientific life, he specialized in the theory of elementary particles (hadron physics), but also worked in gravitation theory, fundamentals of quantum mechanics, and electroweak theory.

Herch Moysés Nussenzveig was born in São Paulo in January, 1933, and studied physics as an undergraduate at USP, graduating in 1954. He continued there to obtain his doctorate in 1957, and then went to Birmingham as a postdoctoral assistant in Peierls' group, where he was a contemporary of MacDowell's. He spent additional postdoctoral stays at Zurich, with Res Jost, and at Utrecht, with Léon van Hove. During that time, he was on leave from a position as adjunct professor at USP, but went to the CBPF as professor from 1962–1968, spending longer sojourns at the IAS in Princeton (1964–1965), and at the University of Rochester (1969–1975). In 1975, he returned to USP as professor, and there he became head of the *Instituto de Física*, overseeing the founding of the Department of Mathematical Physics, based on the group set up there in 1968/69 by Jayme Tiomno. In 1984, after a final attack on USP by the declining dictatorship, he moved to PUC/Rio, remaining on its faculty until 1994, when he transferred to UFRJ, finally retiring there at age 70, in 2003. Nussenzveig changed his area of interest from particle physics to quantum optics late in his career, and had great success in that field, collaborating with, among others, the French group of Serge Haroche, who shared the 2012 Physics Nobel prize for his work on the "measurement and manipulation of individual quantum systems"[2] (Fig. 16.2).

The book '*Frontier Physics*' is a fitting memorial to Tiomno's scientific career, which was entering a less intensive phase when the volume appeared. The three editors wrote in the Foreword:

Fig. 16.2 A group of colleagues, August 2003. (*Standing, from left*): Roberto Salmeron, Jayme Tiomno, Herch Moysés Nussenzveig; (*seated*): Sonia Salmeron, Elisa Frota-Pessôa, Micheline Nussenzveig, and Maria Laura Leite Lopes (José Leite Lopes' former wife). *Source* Private archive, [EFP]

Jayme Tiomno's life, enthusiasm and dedication are sources of inspiration to us and to future generations of physicists and it gives us great pleasure to participate in this celebration. We wish him many happy returns[3].

Its table of contents, listing the contributors to the symposium and the *Festschrift*, reads like a list of Tiomno's many collaborators during his by then fifty-year-long career in research and teaching. His early collaborators from Princeton, Chen Ning Yang and John Wheeler, open the book with general chapters. Yang's is a memoir of Tiomno's early scientific life, and is particularly relevant as a reflection on his early work and his collaborative style of research:

> I first learned of the name Tiomno in early 1949 when as an instructor at the University of Chicago I was informed by A. Ore that Tiomno and Wheeler had done something similar to what T. D. Lee, M. Rosenbluth and I had just finished working on. Tiomno and Wheeler's work and ours, and several other papers around that time, served to classify elementary particle interactions into four classes, a clarifying development in those years.
>
> Later on in the fall of '49, when I moved from Chicago to the Institute for Advanced Study in Princeton, I met Tiomno personally. He was at that time at Princeton University. We discussed many topics in physics and wrote a paper together in early 1950. The paper was entitled 'Reflection Properties of Spin-½ Fields and a Universal Fermi-Type Interaction'. That was probably the first time that the term universal Fermi-type interaction was used...
>
> The content of this paper turned out to be not particularly relevant. However, some considerations in it later were of crucial importance for my work (1956) with T. D. Lee on parity non-conservation...
>
> Six years later these computations of 1950 led directly, in my work with T.D. Lee, to considerations of all 10 terms simultaneously (see Ref. 7 of the paper by Lee and Yang[4]), a development that allowed us to calculate various parity violation effects in ß-decay, leading to the breakthrough experiment on ^{60}Co by C.S. Wu[5].
>
> I am very happy to be here today to help celebrate Jayme's seventieth birthday. He has contributed many things to physics and many things to the development of physics in Brazil. It is a career he could be proud of...[6] [*References added*].

Wheeler's chapter is a scientific/philosophical reflection on the emergence of the universe. These introductory chapters are followed by contributors from the early days of the Weak interactions, including Lederman, Sudarshan and Marshak, and a chapter by Tiomno's earliest collaborator and longtime friend, José Leite Lopes. Then come a series of articles by his former students and collaborators from the 1970s and 1980s, as well as colleagues from the CBPF and other Brazilian institutions, and from other Latin American countries (Argentina, Mexico) who had worked on similar topics to those studied by Tiomno. These are followed by contributions from some of his collaborators during his second stay at Princeton. The book closes with an article by Salmeron on a current topic in experimental high-energy physics: Relativistic heavy-ion collisions.

This symposium and the resulting book no doubt gave great satisfaction to Jayme Tiomno, whose efforts in research and scientific education had so often gone unrecognized or been frustrated (but had nevertheless again and again borne fruit, against all expectations!).

However, along with the celebrations, they also marked a transition in his career, owing to his compulsory retirement (for the second time!)—this time due to the Brazilian law mandating retirement from public service at age 70, and not to an arbitrary government repression, as in 1969. Nevertheless, to accommodate its Founders, all approaching or past 70, the CBPF created a new category, called '*pesquisador emérito*' (researcher *emeritus*), which would permit the holders of that title to continue giving lectures and classes and directing students' work, even without a formal contract or an active position. In a ceremony on August 27th, 1992, the title was conferred upon Jayme Tiomno, Elisa Frota-Pessôa, José Leite Lopes, Leopoldo Nachbin, Hervásio Guimarães de Carvalho and Francisco Mendes de Oliveira, beginning a tradition which has been continued with later generations of researchers at the CBPF.

Tiomno (like the others) became an 'elder statesman' of science, still working but no longer pursuing research and teaching with such energy and purpose as in previous years; he however remained active in scientific organizations. From 1991, he was president of the Special Commission of the CNPq for guiding the interactions of Brazilian groups with major international laboratories for High Energy Physics, and in 1992/94, for three years, he was a member of the Commission for the Selection of Academicians of the *Academia Brasileira de Ciências* (ABC). During this period (until 1997), he was also a member of the Executive Committee of the 'International Society of General Relativity and Gravitation', headquartered at the University of Victoria, Canada.

Expressions of recognition continued to come in from all over the world. In 1993, the Department of Mathematical Physics at the *Instituto de Física* of USP, honoring the group that he had founded in 1968, named a lecture room *Sala de seminários Jayme Tiomno*. In 1994, he received a high honor from his native country, the '*Grã Cruz da Ordem Nacional do Mérito Científico*', the medal 'Grand Cross' of the National Order of Scientific Merit, presented to him by the President of the Republic (at the time *Itamar Franco*, who served as 33rd President of Brazil after the previous President, Fernando Collor de Mello, resigned in 1992). In 1995, Tiomno received the Physics Prize of the *Third World Academy of Sciences* (TWAS), and traveled to Trieste to accept the prize and deliver his Award Lecture ([JT104]). Shortly thereafter, he was elected to Permanent Membership of that academy.

[The TWAS was founded in Trieste in 1983 on the initiative of Abdus Salam. It serves as a complement to the ICTP, providing an academy devoted specifically to encouraging and recognizing scientific research in the developing countries, and to providing an internationally prominent voice for science in the 'third world'. It is located on the ICTP campus in Trieste, and is supported by the Italian Ministry of Foreign Affairs, the ICTP itself, and a number of academies, ministries, and organizations from other countries, as well as some companies with an interest in developing science. It began operations in 1985, and its name was changed in 2013 to 'The World Academy of Sciences for the advancement of science in developing countries'].

The Marcel Grossmann Meetings

In 1975, just three years after Jayme Tiomno had left Princeton for the second time and had taken up his professorship at PUC/Rio, his colleagues and collaborators *Abdus Salam* and *Remo Ruffini* founded a new conference series in Trieste, dedicated to bringing together the growing community of researchers in gravitation and relativistic astrophysics, which had to a considerable extent emerged from Wheeler's school in Princeton. They named the conferences '*Marcel Grossmann Meetings*' (MG), after the Swiss mathematician, fellow-student and friend of Albert Einstein, who provided Einstein in 1912/13 with the mathematics that he needed to finish his General Theory of Relativity. Without Grossmann, Einstein would certainly not have been able to complete the theory by 1915, and it might well have been published by David Hilbert, instead of Einstein[7].

The conferences in the MG series are held every three years at varying locations; in addition to Trieste/Italy, where the early meetings were held, there have been several—including the two most recent, MG-XIV (2015) and MG-XV (2018)—in Rome, and others in various cities on most of the continents. Given his association with their founders and his own work in field theory, gravitation and cosmology, it was natural that Jayme Tiomno would have an interest in those meetings.

He however did not attend the first two, MG-I (1975) and MG-II (1979), both in Trieste; they took place during the *leaden years* in Brazil when he may not have felt free to travel, and he had also not done any new, relevant research since leaving Princeton. By the time of MG-III, held in Shanghai/China in 1982, Remo Ruffini invited Mario Novello, from Tiomno's DRP Department at the CBPF, to participate in the meeting; Tiomno himself did not attend the MG-III Meeting. Shortly thereafter, Ruffini and Abdus Salam, together with several other prominent scientists, founded the ICRA organization (*International Center for Relativistic Astrophysics*), based in Pescara, Italy. By 1985, Tiomno's time had come, and he gave an invited paper ([JT86]) on the LET–SRT question at MG-IV, held that year for the first time in Rome. Remo Ruffini and Abdus Salam were no doubt pleased to give him the opportunity to attend an MG conference, and Novello's enthusiasm also played an important role in convincing him to attend.

After the meeting in Rome, Tiomno resumed his contacts with his former collaborators, and in 1994, the year of MG-VII, held in Stanford/CA,

Fig. 16.3 José Leite Lopes, Jayme Tiomno, and Remo Ruffini at the MG-X conference in Rio de Janeiro (2003), after the presentation of the Marcel Grossmann Institutional Award for that year, given to the CBPF. *Photo* courtesy of the CBPF

Fig. 16.4 Jayme Tiomno and José Leite Lopes, as founders of the CBPF at the MG-X Awards Ceremony. *Photo* courtesy of the CBPF

USA, he was a member of the 'Committee of the Americas', part of the international coordinating committee for the conference; but he again did not attend the meeting itself—by then, as researcher *emeritus*, he was reducing his overseas travel and participation in conferences, and encouraging some of his former students and collaborators to actively take over those functions, including Mario Novello and Marcelo Rebouças, among others. Prior to the MG-X Meeting planned for 2003, they offered to host the conference at the

CBPF, and Mario Novello served as chairman of its local organizing committee. At that meeting, the Institutional Award, presented at each conference to an outstanding institution in research and teaching, was given to the CBPF *'for its role as a teaching and research institution and as a place originating fundamental physics ideas in the exploration of the universe'*, and presented to its founders, César Lattes, José Leite Lopes, and Jayme Tiomno, who were all present (Figs. 16.3 and 16.4).

That same year, ICRANet, the international networking arm of ICRA, was officially chartered; cf. Chap. 13. Ruffini needed at least three states to create it, and at the time, he had just two: the Vatican and Armenia. Brazil's membership in ICRANet[8] was also prepared at the 2003 Meeting, and Mario Novello later became its representative for Brazil (a post now occupied by *Ulisses Barres de Almeida* of the CBPF). Tiomno was not in favor of Brazil's joining the ICRANet, perhaps because of the friction with Ruffini that he had experienced much earlier in Princeton, or due to other reservations. However, Brazil formally agreed to join in 2005, and its membership was finalized in 2011.

So although Jayme Tiomno himself was not strongly engaged in these meetings, his influence continues to be felt in them through the *Centro* that he helped found, and through his former students and collaborators, who are actively involved in them, both scientifically and organizationally.

The Physics of Relativistic Rotating Systems

By the time he was named researcher *emeritus* in 1992, Jayme Tiomno's projects and collaborations on Special Relativity, hadronic physics, and on field theory, gravitation, and cosmology were all winding down, but he was not ready to give up his physics research altogether. One of his collaborators on gravitation and field theory had been *Ivano Damião Soares*, and with him, Tiomno started a new collaboration in the mid-1990s, on a topic in which he had probably become interested during his research on the Lorentz ether theory (LET) in the 1980s. In that work, he had used a rotating body within a (translationally) moving inertial frame as a model to consider how to distinguish the LET from conventional Einsteinian Special Relativity theory (SRT). This quite possibly aroused his interest in relativistic rotating systems (an old subject, dating back over 85 years, but one which had not been brought to a satisfactory conclusion in all of that time). In particular, interest in an early aspect, the *Sagnac effect*, and in one of its newer refinements (the *Mashhoon effect*), had persisted over the years.

[The *Sagnac Effect* was described by Georges Sagnac in 1913[9]. It involves the interference of light in a *ring interferometer*, in which two light beams split from a single light source are passed—one clockwise and the other counter-clockwise—around the same loop, then recombined and allowed to interfere. It is extremely stable, since the two sub-beams travel along precisely the same path. When the whole apparatus is rotated at an angular velocity ω, … a phase shift between the recombined sub-beams is observed. This case had already been treated relativistically by Max von Laue in 1911[10]. Sagnac took the observed shift to be evidence for an absolute frame of reference, the 'ether', in contradiction to Einstein's SRT, but this was later seen to be incorrect. Since the advent of fiber-optic light guides and of solid-state lasers, Sagnac interferometers can be made very compact and robust, and can be used as 'light gyroscopes' which are suitable for example in space applications. The effect has also been demonstrated with quantum particles (e.g. neutrons[11] and electrons[12], and even atoms of superfluid helium), which can likewise show interference effects due to their quantum–mechanical wave character. A much later development is the *Mashhoon Effect*[13], described in 1988 by Bahram Mashhoon. It is a small additional phase shift due to the coupling of the spin of a particle used in the Sagnac interferometer (e.g. the spin-½ of neutrons or electrons) to the orbital motion of the rotating apparatus, analogous to the spin–orbit coupling observed for atomic electrons, where their spins couple to the orbital angular momentum of their quantum states within the atom].

In the mid-1990s, Jayme Tiomno and Ivano Soares set out to give a correct and complete derivation of the Sagnac and Mashhoon effects in the relativistic limit. Their first article on this topic was published in the *Physical Review* D in 1996[14], entitled '*The Physics of the Sagnac-Mashhoon Effects*' ([JT103]). It was in many ways a theoretical *tour de force*, and made use not only of insights from Tiomno's earlier work on the LET—SRT controversy, but also of his 1970 paper on the equivalence of Lorentz transformations and Foldy-Wouthuysen (F-W) transformations ([JT44])[15]. The latter paper was published shortly before his departure for Princeton in 1971, and was a result of his collaboration with Bollini and Giambiagi at the ICTP in 1966/67, as he mentions in it. In fact, he states that the paper was written in 1969 at the ICTP, which he gives as his professional address at the time. Tiomno's final paper on gravitational theory, published together with B. Figueiredo and I.D. Soares in 1992 ([JT101]), is also cited and made use of in their 1996 article on the physics of rotating systems (Fig. 16.5).

[The F-W transformation was introduced by Foldy and Wouthuysen in 1950 as a way of simplifying the Hamiltonian (total-energy operator) in

Fig. 16.5 Jayme Tiomno and his young collaborator Ivano D. Soares, at the CBPF in 1995. This was the period when they were working on their article about the Sagnac and Mashhoon effects. *Source* Private archive, [EFP]

applications of the Dirac equation. Tiomno was familiar with it owing to his thesis work using that equation. It fell into disuse in the later twentieth century, but has been revived in the past 15 years due to its effectiveness in treating problems in vector optics. In its original form, it was a unitary (length-preserving) transformation to a new basis within the Hilbert space underlying the solutions of the Dirac equation. Tiomno made use of *generalized* Lorentz and F-W transformations in his 1970 article ([JT44]), and showed that they are mathematically equivalent, apart from a numerical factor].

In their 1996 paper, Soares and Tiomno took up the problem of rotation within a moving inertial frame obeying Special Relativity. Such a rotation produces effects not seen in purely inertial frames with only translational motion, and that gave rise to Sagnac's original claim that he had detected the ether with his ring interferometer. Indeed, all experiments carried out on the Earth are in a rotating frame, due to the Earth's axial rotation, and that was employed in the neutron experiments[11].

The phase shift observed in the Sagnac effect is proportional to $\omega \cdot A$, where ω is the (angular) rotational velocity and A the enclosed area within the ring interferometer, both expressed as vectors. The 'dot product' means that only the parallel components of the two vectors contribute to the effect (i.e. if the rotational axis were perpendicular to the area vector, that is lying within the plane of the ring, there would be no effect at all). The Mashhoon effect is proportional to $- \omega \cdot S$, where S is the (vector) spin of the particle.

Soares and Tiomno begin their paper by stating that, "With the exception of the paper by Dresden and Yang[16] [*where a semi-classical derivation of the Sagnac effect is given*], all derivations of the Sagnac-Mashhoon effect are incorrect or incomplete". They go on to say that, "The origin of this problem lies in the use of nonphysical frames to describe the motion of the experimental apparatus. We also show that two sets of effects are to be distinguished in experiments with the rotating apparatus. The first is a typical wave effect, related to the fact that the frame of the rotating experiment is noninertial, while the second set is due to the dragging of the particles of the experiment by the rotating apparatus. The origin of the effects in both cases is due to *active* Lorentz transformations realized on the system, taking it from rest to a state of uniform rotation about an axis"[14]. They use a similar procedure as Dresden and Yang[16]—but fully relativistic, not semi-classical—and consider a cylindrical apparatus, rotating around its cylinder axis. They identify the origin of the Mashhoon effect and show that it can determine whether free spin-½ particles in a rotating frame behave as gyroscopes. They also derive the Sagnac effect, due to the 'drag' on the particles in the rotating frame, and finally show that the frames used in previous articles in the literature would yield <u>no</u> Sagnac effect, although they are not mathematically incorrect. They conclude with the statement, "A non-null Sagnac effect indicates that the experiment is rotating with respect to the inertia rest frame, defining a rotating frame connected to the inertia rest frame by an active (instantaneous) Lorentz boost" (i.e. a transition to a frame moving with a different velocity—including rotation—within SRT).

Tiomno returned to this subject several times in the following years; in his talk in Trieste in 1996, when he accepted the Physics Prize from the TWAS (Third-World Academy of Sciences), he discussed an aspect of his work with Soares, which he called '*Spin-rotation Coupling*' ([JT104]). And his last scientific publications included a contribution to the *Festschrift* for his friend Roberto Salmeron[17] for the latter's 80th birthday in 2002, with the more general title '*The Physics of Rotating Experiments*' ([JT105]). His *very* last publication appeared in the Proceedings of the 28th Meeting of the SBF on Particles and Fields, held in *Águas de Lindóia*/SP in September, 2007; it was entitled '*On the exact relativistic dynamics of the Sagnac effect*', and was co-authored with M.O. Calvão, F.D. Sasse, and I.D. Soares ([JT108]). This was Tiomno's final 'topic of interest', and it occupied his attention for a decade beginning around 1995. Even at his by-then advanced age—he was 80 in the year 2000—he was able to pull together the strings of previous works and apply them to a new problem, one which had occupied physicists for nearly a century.

Private Life

The 1990s and their retirement offered Jayme Tiomno and Elisa Frota-Pessôa an opportunity to slow down a bit and enjoy their other interests and their family, by now large and still growing. They were in the USA in 1993 and visited the Wheelers, who were back in Princeton (Fig. 16.6). They also traveled in Europe in the mid-1990s, when Tiomno accepted the Physics Prize from the TWAS.

Barra and **Itaipú**—Following the Amnesty and their reinstatement as 'normal citizens' in Brazil, Jayme Tiomno and Elisa moved out of the city, in January, 1980, to a new apartment in the beautiful region of *Barra da Tijuca*, an originally remote and previously undeveloped area to the southwest of Rio de Janeiro, with a long and impressive beach, still within the city limits of the city of Rio. On old maps, that region was termed '*O Sertão Carioca*' (a reference to the wilderness in the interior of the State of *Bahia*, implying that it was the 'wilderness of Rio')[18]. Today, it is sometimes called 'the Brazilian Miami'[19].

Fig. 16.6 John A. Wheeler and Elisa Frota-Pessôa in Princeton, 1993. Elisa and Jayme Tiomno were traveling in the USA. *Source* Private, [EFP]

Tiomno and Elisa moved there fairly early during its development process, leaving the apartment in the *Lagoa* district of Rio which they had acquired in 1960 (cf. Chap. 9). Their new apartment was located in a modern but not very tall building, only 50 m from the beach—a beautiful and very favorable site. Elisa, in particular, was an avid collector of art and supported many local artists by buying their paintings, and the walls of their apartment were soon filled with pictures, in a kind of 'Petersburg Hanging' (Fig. 16.7).

In November of 1981, they also moved out of their small summer house in *Arraial do Cabo*, in the region of *Cabo Frio*, on the coast around 100 km northeast of Rio, which they bought after returning from Princeton in 1972 (Chap. 14). Their new summer lodgings were located in a development much nearer to the city, across the bay to the southeast of the smaller city of *Niterói*, in the village of *Itaipú* (a name derived from a native language, meaning 'ringing stone'). There, they had an informal and convenient house with an ample garden and a small pool, and room for visitors.

Fig. 16.7 Elisa and Jayme in their *Barra* apartment, 1990. We can see some of the many paintings which filled the walls in the background. There were more and more as time passed. *Source* Private archive, [EFP]

Elisa and Jayme had always enjoyed entertaining guests, and now they had the time and places to do so (Fig. 16.8). They often invited their former students and collaborators to visit their home and discuss physics, politics, music and art, and their lives and careers. They both enjoyed following the progress of their former students and were happy with their successes, which reflected the success of their own efforts on behalf of education.

Fig. 16.8 A relaxed group on the sofa in *Barra*, 1992. From left to right: Elisa Frota-Pessôa, Jayme Tiomno, Leopoldo Nachbin, José Leite Lopes, and Roberto Salmeron. This group comprises the 'Three Musketeers' of their student days, as well as their 'd'Artangnan', in addition to Salmeron, who was visiting from France. *Source* Private archive, [EFP]

Jayme and Elisa not only enjoyed inviting guests to their own home, they also planned excursions to rustic but comfortable resorts, mainly in the nearby regions of the State of Rio de Janeiro, for example the *Alcobaça*, in *Teresópolis*, north of Rio; or the *Arvoredo*, at an old coffee plantation from the nineteenth century, to the northwest in *Barra do Piraí*, where they invited family and friends to accompany them. They often did this during the school vacation period in July, when their grand- (and great-grand-) children would be able to accompany them (Fig. 16.9).

Fig. 16.9 Elisa, her oldest granddaughter Carla (*Carla Mattos Roberts*), and Jayme, in their apartment in *Barra*, 1995. Carla is a biochemist and is currently on the faculty at Northeastern University in Boston/MA, USA. *Source* Private archive, [EFP]

Notes

[1] See MacDowell et al. (1991).

[2] H. Moysés Nussenzveig, pdf: '*Os Três Mandemantos*' (in Portuguese), available online at: http://www.sbfisica.org.br/v1/home/images/acontece-na-sbf/2018/agosto/acontece-moyses-2018-08-14.pdf. See also Freire, Videira & Ribeiro Filho (2009).

[3] S. MacDowell, H.M. Nussenzweig, and R.A. Salmeron, from the Preface to the book '*Frontier Physics*', the *Festschrift* for Tiomno's 70th birthday; cf. MacDowell et al. (1991).

[4] Referring to Lee & Yang (1956).

[5] Wu et al. (1957).

[6] C.N. Yang, Chapter '*To Jayme Tiomno*' in the book '*Frontier Physics*', the Festschrift for Tiomno's 70th birthday; cf. MacDowell et al. (1991). Quoted with permission from Prof. Yang and from World Scientific.

[7] See Graf-Grossmann (2018), the biography of Marcel Grossmann by his granddaughter, Claudia Graf-Grossmann.

[8] ICRANet is the international network organization of ICRA [the *International Center for Relativistic Astrophysics* (RA)], with headquarters in Pescara, Italy; see Chap. 13, section on Remo Ruffini. It is planned that ICRANet will establish centers in various countries where research in RA is carried out, and its member organizations also include universities in several countries; see https://en.wikipedia.org/wiki/ICRANet. After the official founding of ICRANet in 2003, Brazil announced its intention to join, and its membership was formally confirmed in August, 2011. Elisa and Tiomno's former student *Mario Novello* was nominated to the 'César Lattes ICRANet Chair', as representative of Brazil. Plans were made, with the support of the Brazilian Ministry of Science, to renovate the old Casino in Urca, long since closed as a gaming hall, and to install the Brazilian ICRANet Center there; see Giacconi (2012) (Report of the ICRANet Scientific Committee meeting in December, 2012). This would be an interesting use of the building, which has already entered the history of science through the naming after it of a physical effect (the cooling of compact stellar objects by neutrino emission) by Mario Schenberg and George Gamow in 1940 (see Chap. 5, Backnote [4]). The present status of this project is however unclear.

[9] See Sagnac (1913).

[10] Laue (1911).

[11] Werner, Staudenmann & Colella (1979); see also Greenberger & Overhauser (1979).

[12] Hasselbach & Nicklaus (1993).

[13] Mashhoon (1988).

[14] [JT103]: J. Tiomno and I.D. Soares, in the *Physical Review* D, Vol. 54, p. 2808 (1996).

[15] J. Tiomno, '*Equivalence of Lorentz Transformations and Foldy-Wouthuysen Transformation for Free Spinor Fields*', in: *Physica*, Vol. 53, pp. 581–601 (1971); cf. [JT44] .

[16] Dresden & Yang (1979).

[17] I.D. Soares and J. Tiomno, '*The Physics of Rotating Experiments*', in Aldrovani et al. (2003).

[18] See the website https://www.multirio.rj.gov.br/index.php/leia/reportagens-artigos/reportagens/3608-barra-da-tijuca-o-sertao-que-virou-a-miami-brasileira.

[19] The *Barra* region was fairly inaccessible from the city in the early 20th century, requiring either a long detour through *Jaacrepaguá*, to the northwest of *Barra*, or else following a narrow, winding road along the rocky coastline, the '*Estrada do Joá*'. Although some effort was made in the 1930s to settle the area, it was only in the late 1960's and early 1970s that the project gained momentum. That was accelerated in 1969, when the then-governor of the State of Rio de Janeiro, *Francisco Negrão de Lima*, invited the urban planner

Lucio Costa, of Brasília fame, to prepare a plan for the development of Barra. (This also explains the name of the main street along the beach, where Jayme and Elisa's apartment was located: '*Avenida Lucio Costa*'. Previously, it was called *Avenida Sernambetiba*).

His plan made use of an expressway, BR 101, which within the urban region of Barra is called *Avenida das Americas*, as the main axis of a sort of garden city, fronted by the Atlantic and the beach. The region near the ocean would become the '*Jardim Oceânico*' (the Oceanic Garden), a second region to the northwest would be named '*Sernambetiba*', and a third nucleus in between would be a small urban center, the '*Centro Metropolitano*'. Lucio Costa even envisioned a connection to the city via a subway line. It was finally constructed just prior to the Olympic Games held in Rio de Janeiro in 2016, and has greatly improved the traffic situation to and from *Barra*. A motorway was also built, tunneling under the giant boulders which form a natural barrier, to connect Barra with the *Lagoa* region, nearer to the city center, and development then proceeded rapidly, but not necessarily according to Lucio Costa's plan. (He later distanced himself from the project). In the meantime, Barra has become rather densely populated, with many modern buildings, and there have been (unsuccessful) attempts to make it an independent city.

17

Conclusion—The New Century. A Life for Science

Looking Back

The turn of the new century marked Jayme Tiomno's 80th birthday, on April 16th, 2000. Once again, as in 1990, he was publicly honored, both by professional institutions such as the Brazilian Physical Society (SBF), and also more privately: The small town in *Minas Gerais* where he began his secondary education, *Muzambinho*, invited him [through its municipal government, the *Câmara Municipal da Cidade*, and its mayor (*Prefeito*)] to visit there and accept the title of Honorary Citizen. He was accompanied by Elisa and his three sisters, as well as his brother-in-law Israel Rosenthal (the husband of his sister Riva), and was treated to a formal ceremony naming him '*Cidadão Honorário de Muzambinho*', a reception, and a visit by the city councilors and mayor, who presented him with a gift basket from the region.[1] The family experienced a warm welcome in their old home town, after more than 60 years. This was a gesture that gave pleasure not only to Tiomno himself, but to his siblings and to Elisa as well.[2]

The Brazilian Physical Society, of which Tiomno was a co-founder, held a session in honor of his 80th birthday at its annual meeting in *Caxambu*, Minas Gerais, in April of 2001 (a year late, but no matter!). At the same meeting, his former student Erasmo Ferreira (one of the 'PUC Quartet') was also honored for his 70th birthday (Fig. 17.1).

In the following years, Tiomno received a number of awards and honors. He was honored at the 35th anniversary of the founding of the *Instituto de Física* at USP (IF/USP) in 2004, and in 2005, he received the title '*Pesquisador emérito*' (Researcher *emeritus*) from the CNPq (now in Brasília).

© Springer Nature Switzerland AG 2020
W. D. Brewer and A. T. Tolmasquim, *Jayme Tiomno*, Springer Biographies,
https://doi.org/10.1007/978-3-030-41011-7_17

Fig. 17.1 Panel at the annual meeting of the *Sociedade Brasileira de Física* (SBF) in *Caxambu*/MG, April 2001; Session in honor of Jayme Tiomno's 80th birthday. From left: Sergio Joffily, Marcelo Otacílio Caminha Gomes, Jayme Tiomno, Elisa Frota-Pessôa, Ivano Damião Soares, and José Ademir Sales de Lima. *Source* SBF. Used under a Creative Commons Attribution License. From: Brazilian Journal of Physics, Vol.31, No.2. São Paulo, June 2001. See https://dx.doi.org/10.1590/S0103-97332001000200001

In September of that year, the National Institute for Space Research (INPE) organized the First INPE Advanced Course Roadmap for Cosmology, in honor of Jayme Tiomno. That same year, he was awarded two additional medals: the '*Centenário de João Christovão Cardoso*', from the Institute of Chemistry, UFRJ; and the medal '*Carlos Chagas Filho do Mérito Científico*', granted by the *Carlos Chagas Filho* State Agency of Rio de Janeiro for Scientific Funding (FAPERJ). The medal was presented at an impressive ceremony in the *Teatro Municipal* in Rio (Fig. 17.2), commemorating the 25th anniversary of the founding of FAPERJ.[3] The following year, Tiomno received the title of '*Aluno Eminente*' ('eminent pupil') from his old high school, the *Colégio Pedro II* in Rio de Janeiro.

Later on, in 2010, the National Observatory organized the *Jayme Tiomno School of Cosmology*. Furthermore, an exceptional homage to Tiomno and Elisa came from their former students at the *Faculdade Nacional de Filosofia*. In March, 2002, Carlos Alberto da Silva Lima, Mario Novello, Sergio Joffily, José Carlos Valladão de Mattos, Marcelo Gomes, Maria Helena Poppe de Figueiredo, Sonia Frota-Pessôa and Miguel Armony commemorated the 40th anniversary of their matriculation at the FNFi in 1962; and, in honor of Elisa, they created the *Grupo dos Oito da Elisa* (GOEL-62; the 'Group of eight of ELisa'), recalling the four semesters when she was their professor for the introductory physics course, 'General and Experimental Physics I–IV' (Fig. 17.3).

Fig. 17.2 Jayme Tiomno (*right*) at the ceremony for the 25th anniversary of FAPERJ, with the mathematician Jacob Palis (*left*; President of the ABC at the time), after Tiomno received the prize medal '*Carlos Chagas Filho de Mérito Científico* '. December 14th, 2005, *Teatro Municipal*, Rio de Janeiro. *Source* Private archive, [JT]

Two years later, in 2004, they enlarged the scope of their group, adding Jayme Tiomno, who had begun giving the advanced course on electromagnetism to their class in 1964. Thus the group was now renamed as the '*Grupo dos Oito da Elisa* ***e do Tiomno*'** (GOELTI-64: the 'Group of Eight of ELisa and TIomno'). All of those eight students had accompanied Tiomno and Elisa on the adventure of establishing the Physics Institute at the University of Brasília in 1965. The physicist Carlos Alberto da Silva Lima, one of the principal initiators of the group, made the following comment in his address to them during one of their meetings:

Fig. 17.3 A gathering in Barra in 2003, where Tiomno's last collaborator Ivano Damião Soares (*at upper left*), as well as some of the members of the GOEL-62 group (*behind, from left after Soares*): Sergio Joffily, Carlos Alberto da Silva Lima, and Mario Novello, followed by da Silva Lima's physicist wife Miriná da Sousa Lima, were guests of Jayme Tiomno and Elisa Frota-Pessôa (*foreground*). Source: Private, [EFP]

> The enthusiasm and the will to support the transformation of our society moved some of us to follow our two mentor-teachers to the desert of Brasília in 1965, already completely dominated by the military forces that had beset our country the year before, and which would persist for the following 20 years".[4]

The group continued to meet every year, each time with more former students and colleagues attending, to pay homage to their mentors, Tiomno and Elisa. But that wasn't limited to the organized meetings of the group; they also frequently encountered their ex-fellow students at the couple's apartment in Barra or at their house in Itaipú. Elisa and Jayme, even at their advanced age, continued to exert a certain magnetic attraction after so many years (Fig. 17.4).

Tiomno himself wrote a last scientific–historical memoir in 2005, recalling the early days of particle physics in Brazil during the period 1946–1961, and

Fig. 17.4 A meeting of the GOELTI-64 group in Rio, 2004. *From left*: Maria Helena Castro Santos (in 1964, Poppe de Figueiredo), Sergio Joffily, Marcelo Otacilio Caminha Gomes, Sonia Frota-Pessôa, Elisa Frota-Pessôa, José Carlos Valladão de Mattos, Miguel Armony, and Carlos Alberto da Silva Lima. Missing from the group picture is Mario Novello. *Source* Private, [EFP]

his own role in those beginnings ([JT107]). It sheds some light on his early work in Weak interactions and his prediction of the K' meson.

John Wheeler's 90th Birthday Celebration

An event of some significance for Jayme Tiomno was the 90th birthday celebration of his first mentor at Princeton, and lifelong supporter, *John Archibald Wheeler*. It was held in July, 2001, in Princeton. Tiomno was of course invited, but the written invitation was lost for a time, since it was sent to the address of the Brazilian Academy of Sciences (ABC—of which Tiomno was indeed a member; but evidently no-one at the Academy office had his personal address at hand). When he finally received the invitation, time was short, and he was unable to apply for travel funds from the usual funding agencies—once again, the ABC was able to help, quickly and unbureaucratically.[5] So Jayme Tiomno, himself over 80, was able to attend the celebration for Wheeler's 90th birthday in 2001.

In addition, a Symposium with the title '*Science and Ultimate Reality*', in honor of Wheeler's 90 years, was also held in Princeton from March 15th–18th, 2002. There were a number of contributors from a variety of fields, ranging from fundamentals of quantum mechanics to cosmology and complexity theory. Jayme Tiomno did not give a lecture at the birthday Symposium; he was, after all, nearly 82 years old by that time, and had not been very active in research in the previous decade—he had, however, worked on a 'last topic', the complex physics of relativistic rotating systems (cf. Chap. 16), and he could have spoken on that subject; but he perhaps decided to leave the lectures to younger researchers. The fact that he was able to attend and to meet with many former colleagues and friends, in particular with John Wheeler himself, was its own reward (Fig. 17.5).

The Symposium was documented in a *Festschrift* published in 2004, entitled '*Science and Ultimate Reality—Quantum Theory, Cosmology, and Complexity*'.[6] The scientific topics covered were quantum physics—theory and experiment; cosmology, including gravitational theory; and the theory of complex systems (emergence, life, self organization…).[7]

The symmetry of Wheeler's career, and its three 'phases', is remarkable; the first two lasted almost precisely 20 years each, while the third ('*Everything is Information*') coincided rather closely with his 10 years as Senior Professor in

Fig. 17.5 Tiomno, Wheeler and Ruffini in March, 2002, at the birthday symposium for Wheeler in Princeton. *Source* Private archive; [JT]

Austin, Texas.[8] And Wheeler began his second phase '*Everything is Fields*' with a two-semester lecture course on Relativity, and his third with a course on Quantum Information. His philosophy of passing the torch on to his students and collaborators—"If you would learn, teach!"—was very similar to that of Jayme Tiomno and Elisa Frota-Pessôa.

The *Festschrift* for Roberto Salmeron

Another of Tiomno's long-term associates and friends celebrated his 80th birthday in June, 2002: that was *Roberto Aureliano Salmeron*. He was honored by the publication of a *Festschrift*, which appeared in 2003 and was entitled '*Roberto Salmeron Festschrift: A Master and a Friend*'.[9] This time, Jayme Tiomno wrote an article for it, together with Ivano Soares; it was called 'the *Physics of Rotating Experiments*'. *Emanuele Quercigh,* researcher at CERN and the INFN *Padova,* wrote a review of the book for the *CERN Courier,* the magazine of the laboratory where Salmeron worked for over 10 years:

> This book is a token of the admiration and friendship felt for Roberto Salmeron on the occasion of his 80th birthday. Some of the authors regard him as both their teacher and mentor, and others as a respected colleague and friend. The book is thus a tribute to the scientist, the professor and the friend.
>
> There is scarcely a domain in Brazilian science and culture where Roberto has not played an important role. He is remembered for example by his support for the *Instituto de Fisica Teorica* in São Paulo in the 1950s. He also proposed the establishment of the first ICFA Instrumentation School in Brazil, as well as the creation of a synchrotron light laboratory in Campinas. Then came his support for the development of instrumentation to be used in experiments both at CERN and at Fermilab.
>
> The 32 contributed papers cover a wide spectrum of topics, from general relativity and cosmology to the interaction between science and society. They touch in turn upon the physics of neutrino masses and mixings, the study of cosmic rays and particle physics, and the search for the quark-gluon plasma, the origin of masses, the development of nanotechnology and several quantum-physics issues.
>
> They also provide vivid glimpses of Roberto's personality, such as his enthusiasm for teaching physics, the desire to develop science for the benefit of all society, and his awareness of the social responsibilities of scientists. A dramatic article by Michel Paty evokes their common struggle, in 1963–1965, in

building the Institute of Physics of the University of *Brasília* and defending its freedom against the intervention of the dictatorial regime. This was a fight that ended in the resignation of most of the faculty members and Roberto's exile—an example of his courage and determination in defending the dignity of science. Long live Roberto!.[10]

The Nobel Prize Question

John Wheeler repeatedly called attention to Jayme Tiomno's early contributions to the Universal Fermi Interaction, even proposing him for a Nobel prize in 1987,[11] along with E.C.G. Sudarshan and R.E. Marshak, who first suggested the universal V-A interaction,[12] and C.S. Wu, who initiated and guided the first experimental proof of parity non-conservation in the Weak interactions.[13] This would have been a unique 'four- (sub-) continent award', to the North American Marshak, the Brazilian Tiomno, the Indian Sudarshan and the Chinese Wu (Fig. 17.6).

His proposal was however doomed to failure, firstly because the Weak interactions were by then 'old hat'; they had received Nobel prizes in 1957 (to T.D. Lee and C.N. Yang, for their suggestion of parity non-conservation); and in 1979 (to Sheldon Glashow, Abdus Salam and Steven Weinberg, for their contributions to the electroweak unification)—and secondly, because it would have violated the '*Rule of Three*', which limits the number of Nobel prize recipients in each field (except for the Peace Prizes) to a maximum of three in a given year (a rule which many now consider to be outdated).[14] In any case, there was an exciting new development by 1987, the discovery announced the previous year of *high-temperature superconductivity*, and the Physics Nobel prize was awarded in 1987 to J.G. Bednorz and K.A. Müller, who had made that discovery.

Elisa Frota-Pessôa, among others, also speculated[15] on the possibility that Jayme Tiomno might have received the Nobel prize if in 1957 he had considered her 1950 article[16] on the branching ratio of pion decays leading directly to electrons, and thus might have published the correct universal V–A Weak-interactions theory before any of the other contenders. But this is also highly unlikely—due to the unclear priority discussion at the time,[17] none of the various authors of the V–A theory received the prize, and they were apparently not even nominated for it to any great extent. Had Jayme Tiomno published the correct theory in his 1957 paper,[18] he would have been just another contender, along with Sudarshan and Marshak, who had demonstrably been the first to suggest the idea (if not the first to publish it), Feynman and Gell-Mann, who were generally given credit as its authors

CENTER FOR THEORETICAL PHYSICS

THE UNIVERSITY OF TEXAS AT AUSTIN

Austin, Texas 78712·(512)471-3751   February 6, 1987

Professor Stig Lundqvist
Chalmers University of Technology
University of Göteborg
Institute of Theoretical Physics
S-412 96 Göteborg
Sweden

Dear Professor Lundqvist:

Through my colleague, San Fu Tuan at the University of Hawaii, I have
had the good fortune to learn of his nomination of Madame C. S. Wu with
R. E. Marshak and E. C. G. Sudarshan for the 1987 Nobel Prize in physics for
their part in

The Universality of the Weak Interaction.

I have just received from Professor Tuan the documentation for this nomination and want to communicate to you and through you to our other colleagues
on the Committee of Selection my __full__ __support__ for this nomination, with one
essential addition, the inclusion of Professor Jayme Tiomno of Centro Brasileiro
de Pesquisas Físicas (CNPq, RJ 22290, Brazil). I enclose Tiomno's own long-
after-the-event summary of what he did and what he published, from which his
own originality and priority in a vital part of the discovery become inescapably
clear:

The Magnitude of the Coupling is the same in Beta Decay, in
Mu-Meson Decay and in the Charge-Exchange Interaction of a
Mu-Meson with a Nucleon.

I do not see any reasonable way to give more recognition to the details of
angular distribution than the magnitude of the interaction. All four names together form a splendid, natural unit.

The unusualness of a four-person award, by overstepping the normal limita-
tion so rightly adhered to so long by the Committee of Award, would doubly
emphasize to every thinking person the wonderful binding power of science,
linking in this way North America, South America, India and China.

I send every good wish.

Sincerely yours,

John Archibald Wheeler

enclosure: Jayme Tiomno, "The Early Period of the Universal
            Fermi Interaction," pp. 82-92, in 50 Years of
            Weak Interactions, edited by David Cline and
            Gail Riedasch, University of Wisconsin-Madison,
            1984

zd

WP. Series II
Box Th-To - Folder Tiom

Fig. 17.6 The letter sent by J.A. Wheeler in 1987 to *Stig Lundqvist*, past chairman of the Nobel Committee for Physics, recommending that the physics prize for 1987 be awarded to C.S. Wu, E.C.G. Sudarshan, R.E. Marshak and Jayme Tiomno for their work on the Weak interactions. *Source* Wheeler archive, Princeton (APhS); see also Freire & Bassalo (2008).

(neither of them had yet received the Nobel prize at that time, and when they did, it was not for the V–A theory), and J.J. Sakurai, who published the correct theory but apparently only after seeing the ideas of other authors.

A similar speculation was published by Daniel Piza[19] after a 2006 interview with Jayme Tiomno and Elisa Frota-Pessôa. Piza—and Elisa—conflated three topics related to Tiomno's work on the Weak interactions which in fact are not directly connected: (i) Tiomno's thesis ([JT13], 1950), which contained important new ideas but was not really Nobel prize material; (ii) the question of parity conservation, held by most physicists before 1957 to be a basic tenet of nature (Yang—and his mentor Fermi—were exceptions; see backnote[32] in Chapter 9); and (iii) the choice as to which current-operator combination (S + P–T or V–A) would give the correct description of the fundamental Weak interactions. Tiomno never seriously considered parity non-conservation before 1957, and was not a candidate for the Physics Nobel prize that year, bestowed upon Lee and Yang for precisely that discovery. Elisa's 1950 publication[16] was furthermore not relevant to that discussion. It *was* relevant to the question of the correct operator combination, but that, as we have seen above, was not in itself a prize-worthy topic.

There have been many discussions about 'missed' Nobel prizes, and those discussions in Brazil are often underlaid by the inference that scientists from 'third-world' countries are systematically overlooked. It has for example also been suggested that César Lattes (as well as Giuseppe Occhialini) should have been co-recipients of the 1950 Physics prize which was given to Cecil Powell alone.[20] This has been emphasized in particular by Leite Lopes, who took a rather critical view of the worldwide organization of science and the role assigned in it to scientists from the third world (or the 'scientific periphery').[21] On the other hand, most authors maintain that the prize was awarded to Powell alone for his development of the photographic-emulsion particle detection method, and not for its first important application to the pion discovery (although the latter certainly aroused international interest in the method). Given that the 1950 prize was awarded for the development of the emulsion method, however, one can make a strong case that Marietta Blau should also have shared in that prize, since it was her pioneering work that stimulated Powell to begin his research program on photographic ('nuclear') emulsions in the first place.[22]

Roberto Salmeron was critical of the idea that it was an injustice that Lattes and Occhialini did not share in the 1950 prize with Powell. Salmeron wrote to Tiomno and Elisa, including the draft of a paper that he had written which mentions that topic:

I have also sent my article '*Desafios da física no século XXI*' [*Challenges for physics in the twenty-first century*]; in Sect. 1.2, I gave an historical resumé of some topics which in Brazil are distorted and presented in a completely unrealistic manner… I am referring to the Nobel prize awarded to Powell for developing the technique of nuclear emulsions: in Brazil, a legend was created according to which Powell received the prize for the discovery of the pi [*meson*], and this represents an injustice toward Lattes and Occhialini.[23]

Another case in which the 'third world effect' seems to have played a role concerns the 1999 Nobel Physics prize, awarded to the Dutch theoreticians Gerard 't Hooft and Martinus Veltman, who demonstrated the renormalizability of the electroweak unified theory in a series of papers in the early 1970s. The Argentine theoreticians Juan José Giambiagi and Carlos Guido Bollini[24] in fact had developed the method of *dimensional regularization* before 't Hooft and Veltman, although in a less complete form; however, they received practically no recognition, only a brief mention in one of the Nobel lectures.[24] It has furthermore been argued that there is a bias against women scientists in dealing out the Nobel prizes.[25]

These debates are not simple and are laden with emotions. What seems to be a consensus is that it is important to recognize past failures in order to reform (and modernize) the process of selecting Nobel laureates. Indeed, both Jayme Tiomno and César Lattes remained very discreet on this issue, and neither of them ever publicly expressed disappointment at not having been prizewinners.[26]

John Wheeler, for his part, continued to give moral support to Jayme Tiomno throughout the rest of his own life. Even as late as 1998, Wheeler said of Tiomno, "I always think of Tiomno as one of the most unappreciated of physicists. His work on muon decay and capture in 1947–1949 was pathbreaking and would still merit recognition by some suitable award."[8]

Some Reasons for Being a Scientist

In an interview given in 2005 for a chapter[27] in a book published by the CBPF, entitled '*Algumas razões para ser um cientista*' ('Some reasons for being a scientist'),[28] which contains statements by 25 prominent scientists, including Jayme Tiomno, Elisa Frota-Pessôa, and Roberto Salmeron, from within Brazil and from abroad, a third of them Nobel prizewinners, Tiomno made some remarks about his motivations for becoming a physicist and his experiences as an academic teacher and researcher in 20th-century Brazil. Tellingly, his chapter is called '*Trabalho duro*' ('Hard work'). The summary of Tiomno's interview gives some insights into his life's philosophy:

> According to Prof. Jayme Tiomno, hard work and being prepared for anything are what guaranteed him so much professional success. In his own words, 'It is necessary to be aggressive, not in the personal sense, but in the sense of presenting oneself as ready for anything. [*One needs*] to look for what is outside what the others are doing.' … [*Although he*] worked during his career mainly as a theorist, it was practice in the laboratory and the contact with professors and researchers that motivated him [*initially*] to follow an academic career, combining research, giving courses and mentoring students. He relates that he had contact with modern physics only after he went to São Paulo, after he had finished his undergraduate education.

> … He continued collaborating with universities all over the world, and helped to create diverse scientific institutions in Brazil and abroad. He was one of the founders of the Institute of Physics at the University of Brasília, a pioneering project which could not withstand the arbitrary actions common during the years of the military regime in Brazil. This is one of the few disappointments which Prof. Tiomno experienced with physics. His discomforts, so far as they existed, were more due to human problems than to professional ones. There is hardly anyone else who speaks with such calmness and tranquility about his career. The secret, in his work and in the political accomplishments of his generation, became evident when he said, '*tínhamos a convicção de que íamos chegar lá* ' ['*We had the conviction that we were going to get there!*']. [Interview and article by *Carolina Cronemberger*].[28]

Last Years

Jayme Tiomno's health had not been very stable from the mid-1990s on—like his father Mauricio, he suffered from lung problems, perhaps in part inherited, but no doubt exacerbated by his smoking in his early years, and he also suffered from a kidney disorder. He had however been treated with some success, and Elisa made sure that he got proper exercise and took his medications regularly. He was thus able to lead a fairly active life in his late eighties, and continued to travel, at least within Brazil (Fig. 17.7).

In the years after 2005, Jayme's health seemed to stabilize, and he and Elisa continued their quiet life, no longer making long trips but still receiving guests, family and friends, in *Barra* and *Itaipú*. In April, 2010, he celebrated his 90th birthday quietly, apart from notes in the journals of some scholarly organizations[29].

Fig. 17.7 Family group at the *Hotel Fazenda Arvoredo, Barra do Piraí*/RJ, 2007. Front row (*seated, from left*): Lidia's husband, Renata, Suely, Roberto, Claudia. (*Behind, standing*): Flavio, Negrita; Sonia, her partner Noel, Lidia; Elisa, Jayme (*seated*); Laura, Matias, Gabriel, his girlfriend. (*Behind, in the door*): Laura's children Pedro and Bia, Claudia's husband Marcio[30,31]. *Source* Private, [EFP]

The photo above shows Jayme and Elisa with their extended family at a resort near Rio, in mid-2007, when he was already 87 years old.

In the early morning of January 12th, 2011, precisely 60 years to the day since their informal 'marriage' ceremony, Elisa awoke early and went to get water and Jayme's medicines for him to take. Both of them then went back to sleep. Only when Elisa again awakened some time later did she realize that Jayme had gone to sleep forever.[30] It was not quite a week before her 90th birthday.

Many messages of sympathy arrived from all over the world. An Internet search reveals several pages of death notices and reminiscences. There were, however, some voices suggesting that Jayme Tiomno's passing was not adequately noted in the Brazilian popular press.[26] In fact, media attention at that time was completely focused on a tragedy in the mountains of the State of Rio de Janeiro, where over 900 people had died as a result of a torrential rainstorm. Tiomno also continued to receive various honors, even posthumously. In October of 2011, the Institute of Physics at the University of Brasília organized the First *Simpósio Jayme Tiomno*. This meeting highlighted contributions from scientists who work in the areas of physics treated by Tiomno during his active scientific life. Various of his colleagues and former students appeared there, among them Elisa Frota-Pessôa and his youngest sister Silvia. More recently, in 2018, the Institute of Physics at USP established an annual scientific encounter called the '*Escola Jayme Tiomno de Física Teórica*' ('Jayme Tiomno School of Theoretical Physics'), with its second scheduled meeting in 2019.

His old friend and collaborator *Enrico Predazzi*, from the University of Turin, was among the many people who wrote obituaries, memoirs and eulogies for Jayme Tiomno, and his article in a Brazilian journal[32] is particularly appropriate and touching:

It was indeed a very sad news to hear that Jayme had died in the morning of January the 12th, 2011. Although I had been previously briefly acquainted with him, my first direct interaction with him and my most vivid recollection of him goes back to the dark days of the military dictatorship in Brazil when Jayme, along with many other distinguished Brazilian physicists (I believe this was the fate of more than 200 university professors), had just been ousted from his chair at the USP. At the time I was in the US on a sabbatical leave and a Brazilian friend had contacted me to invite me to spend some time in Brazil to help the many (then) young promising local physicists to overcome the momentous difficulties. It was a difficult time for me and an extended visit to Brazil made sense also in view of the traditional relationships between my *alma mater* (the University of *Torino*) and Brazilian physics. Before accepting the invitation, however, I felt my duty to check with senior Brazilian physicists how my going to Brazil would have been perceived. The chance was that someone would see it as an endorsement of the dictatorship, which was certainly not the case…

Among them had been Jayme, who enthusiastically had encouraged me to accept the invitation and who, when [*the*] time came, made indeed quite clear what my intentions had been and his support in this venture…

It was still later on (the eighties) when we got to collaborate and work regularly together. At the time we shared to some extent the responsibility of tutoring a by-now prominent Brazilian physicist who was taking his Ph.D. under my guidance in Torino. We ended up writing several papers together (6 to be more precise) on a subject which was then pretty hot. It was a time of rather close relationship and exciting collaboration. Jayme came several times to Torino and I was often going to Rio. In fact, I remember when, during one such visit, at dinner, he and Elisa begged our pardon for their turning on the TV set: the *Telenovela* of the moment was going on the air and the plot that had been left open the day before was so exciting that they were very eager to find out what would have happened. A nice and very warm human weakness that it is so nice to discover in important men...

... As all this shows, a complex, very knowledgeable personality and a great scientist. Most of all, a good friend; another friend that has left us.

His memory, however, will last and remain with us.[32]

Notes

[1] cf. Tolmasquim (2014), p. 146.

[2] See also Chap. 2, Backnote [6].

[3] The ceremony is documented on a web page, online at https://www.faperj.br/?id=655.2.9.

[4] Speech given by Carlos Alberto da Silva Lima during one of the meetings of the GOELTI-64 group; [EFP]. Quoted by Valladão de Mattos (2015), p. 408; see also Frota-Pessôa, Roberto (2019); and interviews with Sonia Frota-Pessôa (ATT, March 2019) and Mario Novello (ATT, August/Sept. 2019).

[5] See Bassalo & Freire (2003).

[6] Barrow, Davies & Harper (2004) (The title page and table of contents of this book can be found as a pdf at https://www.gbv.de/dms/goettingen/37147213X.pdf. See also the symposium summary at https://www.aps.org/publications/apsnews/200205/princeton.cfm).

[7] Among the speakers were *Lee Smolin* and *Fotini Markopoulou*, two younger physicists who were at the time a married couple and both working on quantum gravity. Neither of them had worked directly with Wheeler, but he was to a considerable extent their *spiritus rector*. Smolin had indeed been a postdoctoral fellow at the IAS in Princeton, but that was during Wheeler's tenure as Senior Professor at the University of Texas. Markopoulou was educated in the UK and received her doctorate with Christopher Isham at the Imperial College, London in 1998. At the time of the Symposium 31 years old, she was co-recipient of the 'Young Researchers Award' presented at the Symposium, for her paper on '*Planck-scale models of the universe*'. Their later experiences at the Perimeter Institute (PI) in Waterloo, Canada, of which they were both co-founders in 1999, are reminiscent of the history of the CBPF, and its evolution from having a highly democratic structure to an authoritarian one—albeit less dramatic in the PI case.

[8] cf. Wheeler & Ford (1998), p. 176. See also Misner et al. (2009) on Wheeler's scientific life.

[9] Aldrovandi et al. (2003).

[10] Emanuele Quercigh, in the *CERN Courier,* September 5th, 2004, Section '*Bookshelf*'. Online at https://cerncourier.com/bookshelf-47/.

[11] Wheeler to Tiomno, 1987; [JT]. See also Bassalo & Freire (2003), and Freire & Bassalo (2008).

[12] Sudarshan & Marshak (1958).

[13] Wu et al. (1957); See also [Hargittal 2012] on the Nobel prize question.

[14] See e.g. the blog by Joel Achenbach in the Washington Post of October 9th, 2013, where he writes of the Nobel physics prize for that year, conferred on Peter Higgs and Francois Englert for their part in elucidating the 'Higgs Mechanism' that imparts mass to some of the elementary particles in the Standard Model. The discovery at CERN of the 'Higgs boson', the field quantum of the Higgs field, announced in 2012, had made that prize imminent. Other early contributors to the work were however left out, eliminated by the 'Rule of Three' (and their lack of publicity). The article is available online at https://www.washingtonpost.com/news/achenblog/wp/2013/10/09/nobel-prizes-the-rule-of-three/?noredirect=on&utm_term=.4e08a849116e.

[15] Frota-Pessôa (1990).

[16] Frota-Pessôa & Margem (1950).

[17] See Chap. 9, section 'From the UFI to the Universal V–A Interaction', and Chap. 15, section on 'The Weak Interactions Interlude'.

[18] Jayme Tiomno, '*Non-conservation of parity and the Universal Fermi Interaction*', in *Il Nuovo Cimento* (1957); cf. [JT28].

[19] Piza (2006).

[20] In this connection, see also Ferreira Nascimento (2015).

[21] Leite Lopes treats this topic in his history of physics in Brazil, Leite Lopes (2004).

[22] cf. Sime (2012).

[23] This article by Salmeron was presented at the '*XXIX Congresso Paulo Leal Ferreira*', organized by the *Instituto de Física Teórica da UNESP*, September 18–20th, 2006, but apparently it was never published; cf. the letter from Roberto Salmeron to Jayme Tiomno and Elisa Frota-Pessôa, March 24th, 2008. Both the article and the letter can be found in the archive [JT].

[24] Bietenholz & Prado (2014). See also Chapters 12 (Backnote [5]) and 14 (Backnote [20]), as well as the blog of Francisco R. Villatoro (in Spanish): https://francis.Naukas.Com/2012/11/11/nota-Dominical-Los-Dos-Argentinos-Que-Descubrieron-Hace-40-Anos-Como-Calcular-En-4%CE%B5-Dimensiones/.

[25] A glance at the statistics of Nobel awards over the past 119 years, in particular for the Physics prize, would seem to support this contention. One

need only recall the controversial discussions *vis-à-vis* the fact that Lise Meitner was not recognized for the explanation of nuclear fission (nor for her many years of work with Strassmann and Hahn, which laid the groundwork for their 1938/39 discovery of fission), while Hahn alone received the Chemistry prize for that discovery in 1944. Likewise, one can argue that her nephew Otto Frisch, who shared importantly in the theoretical explanation of the fission experiments (and even in their experimental confirmation), as well as Fritz Strassmann, who was for many years an indispensable member of the Hahn-Meitner group and who carried out the decisive experiments leading to the fission discovery, should have also been co-recipients. Or that—quite possibly—that discovery was sufficiently important so that both a Chemistry prize (for Hahn and Strassmann) and a Physics prize (for Frisch and Meitner) should have been awarded—but again, the priority discussion was complex, and other theorists quickly developed their own explanations for the Hahn-Strassmann discovery, even before the publications of Frisch and Meitner appeared. Speculation is not too helpful here, and the release of the Nobel Committee files 50 years after the award in question has not really shed much light on the discussion (but compare S. Hanel: Hanel (2015)). See e.g. Sime (1996), and Stephanie Hanel, '*Lise Meitner—Fame without a Nobel Prize*', on the website of the Lindau Nobel Laureate Meetings, at https://www.lindau-nobel.org/lise-meitner-fame-without-a-nobel-prize/. See also Hargittal (2012) on the Nobel prize question for another Woman in Science.

[26] – cf. the opinion blog of Daniel Piza, where he suggests that Jayme Tiomno did indeed have regrets on this score: Piza (2011).

[27] cf. [JT106].

[28] Cronemberger et al. (2005).

[29] See for example the lecture organized by the Institute of Astronomy, Geophysics and Atmospheric Sciences at USP on October 13th, 2010, in honor of Tiomno's 90th birthday, online at https://www.usp.br/agen/?p=36625.

[30] Personal communications from Sonia Frota-Pessôa, 2018/19; cf.[EFP].

[31] Renata and Lydia are Roberto's daughters (Renata with his first wife, Vasni, and Lidia with his second wife, Suely [*Braga*]). Carla [*not in photo*], Claudia and Laura are Sonia's daughters. Flavio was the grandson of Jayme and Elisa's cook in the 1990s. He came to live with his grandmother at their apartment in *Barra* when he was five, and after his grandmother's death, when he was 12, he stayed on, informally adopted. Jayme and Elisa supported him through college. In this photo, he is accompanied by his wife Negrita; he now lives in São Paulo and works in IT. Matias and Gabriel are Claudia's older sons.

[32] See Predazzi (2011); quoted with permission of the author.

18

Epilogue

Elisa lived on for nearly eight years, surrounded by her family and in her familiar environment. She stopped going to *Itaipú* after some years, living in her apartment in *Barra*, full of memories. She gradually withdrew from the outside world, but remained friendly and welcoming to her family.

In late 2018, she contracted pneumonia, and after several weeks in hospital, where all the medications tried proved ineffective, she passed away on December 29th, 2018, three weeks before her 98th birthday. She was nearly the last of the founding scientists of the CBPF. Elisa Frota-Pessôa became an example for Brazil as a pioneering Woman in Science (Fig. 18.1).

Science, and in particular physics, has changed enormously in Brazil since Jayme Tiomno made his decision to enter the old *Universidade do Distrito Federal* in 1939, to major in the not-very-popular subject of physics. His goal, and that of his colleagues of the same generation, to educate younger physicists and to establish a favorable ambience for research in their country, was undoubtedly successful in spite of all the difficulties that they had to overcome. Today in Brazil there is an organized community of physicists, with numerous graduate courses and various post-graduate programs, reputable journals and a significant participation in multinational projects. In 2017, for example, institutions in Brazil offered 41 doctoral courses in physics, producing more than 350 Ph.D.'s each year.[1] According to the SJR – *Scimago Journal & Country Ranking*, in 2018 Brazil occupied the 15th place worldwide in physics publications.

In Tiomno's own words[2]:

© Springer Nature Switzerland AG 2020
W. D. Brewer and A. T. Tolmasquim, *Jayme Tiomno*, Springer Biographies,
https://doi.org/10.1007/978-3-030-41011-7_18

Fig. 18.1 Elisa on her 96th birthday, in January 2017, with her daughter Sonia. *Source* Private, [EFP]

This gives me a very great satisfaction, the fact that I can see that [*my former students*] are able to produce [*research results*] directly and through their own disciples, much more than what the few dozen or the hundred articles that I might have written had I remained abroad as a professor at this or that university would have contributed. That wouldn't have meant so much for Brazil, the output of such self-perpetuating work.

Jayme Tiomno's and Elisa Frota-Pessôa's long and active lives were truly *lived for science*, and *lived for Brazil*.

Notes

[1] *Coordenação de Aperfeiçoamento de Pessoal de Nível Superior* (CAPES): Quadriennial Report of Evaluation in the areas of Astronomy and Physics, covering the period 2013–2016: Quadriennial Evaluation 2017; see https://www.capes.gov.br/images/stories/download/avaliacao/relatorios-finais-quadrienal-2017/20122017-ASTRONOMIA-E-FISICA-quadrienal.pdf.

[2] Tiomno (1977).

Appendix A: A Brief History of Brazil

The Colonial Period, 1500–1822

Brazil, the largest nation in South America, both geographically and in terms of population and economy, was "discovered" by the Portuguese explorer *Pedro Álvares Cabral*, who landed near the site of the modern city of *Porto Seguro* on April 22nd, 1500. It has been estimated that between 1 and 5 million indigenous people were living there at that time. The country was claimed and developed as a Portuguese colony, beginning in the northeast and moving northwards and southwards along the coast, over the following century. The Portuguese initially respected the *Treaty of Tordesillas* (1494) between the Portuguese Empire and the Crown of Castile and León, which gave to the Spanish crown the lands west of the Tordesillas meridian and to Portugal those east of the meridian. (The meridian is about 46.5° W longitude, passing somewhat to the east of the modern city of *Belém* at the mouth of the Amazon and extending down through *São Paulo* in the south). Later claims by the Portuguese to the western part of Amazonia were tolerated by the Spanish, giving Brazil its modern shape: roughly triangular, with its greatest width around the latitude of Natal (5.7° S), tapering off to the narrow tip formed by the modern state of *Rio Grande do Sul*, and ending at 33.5° S, near Buenos Aires and Montevideo.

Adventurers from other European colonial powers (Britain, France, the Netherlands) attempted early on to exploit resources from the nominally Portuguese colony (in particular *pau-brasil*, the wood of the Brazilwood tree, used to make a red dye, which gave the country its name). The French also occupied regions near Rio de Janeiro and around São Paulo for some time in the

© Springer Nature Switzerland AG 2020
W. D. Brewer and A. T. Tolmasquim, *Jayme Tiomno*, Springer Biographies,
https://doi.org/10.1007/978-3-030-41011-7

mid-16th and early 17th centuries. Beginning around the middle of the 16th century, sugar from sugar cane became a major export item, produced on plantations near the coast using slave labor.

In the mid-17th century, A Dutch incursion occupied many of the sugar plantations in the northeast, even including the colonial capital, *Salvador da Bahia*, until the intruders were repelled in 1652. Many slaves were imported from Africa, and there were numerous slave rebellions during the 17th and 18th centuries, the most famous in inland Bahia (*Quilombo dos Palmares*), lasting until the end of the 17th century.

In the late 17th century, gold was discovered in south-central Brazil (*Minas Gerais*), and was mined for over 100 years, in the 19th century by a British company. As an additional agricultural product, coffee was introduced in the south in the early 18th century and became the new major export commodity; Brazil remains today the largest coffee producer in the world. The 'coffee cycle' was supplanted in the latter part of the 19th century by the 'rubber cycle', exploiting the latex produced by the *Pará* rubber tree, native to the Amazon region. The establishment of rubber plantations in Southeast Asia, mainly by the British, as well as the development of synthetic rubber, ended the Brazilian rubber boom in the 1920s, although it still played a certain strategic role during the 2nd World War.

Politically, pressure for independence from Portuguese rule and for the abolition of slavery increased in the late 18th century. The capital of the Portuguese Empire was moved to Rio de Janeiro during the Napoleonic Wars in 1808, when the court of the Portuguese King *João VI* was transferred there from Lisbon with the help of the British Navy (cf. Chap. 2). The King remained there for nearly 15 years, and left his son, *Dom Pedro*, as vice-regent of the Brazilian colony when he departed.

The Brazilian Empire and the Old Republic, 1822–1930

On September 7th, 1822, Pedro declared independence from Portugal (uttering "*O grito d'Ipiranga*", the Cry of Ipiranga: "Liberty or death!"), and achieved it after a brief war of independence, founding the Brazilian Empire, a constitutional monarchy, with himself as its first Emperor (later known as *Dom Pedro I*). His son, *Dom Pedro II*, ruled the Empire until his forced abdication in 1889, after he had ended slavery in 1888.

This marked the beginning of the Brazilian Republic, initially called the *Republic of the United States of Brazil*, later referred to as the *Old Republic* (1889–1930). The first Republican constitution established a presidential democracy with direct election of the president every four years, balanced by a parliament and the judicial system. In its early years, the Brazilian Republic was essentially an oligarchy, with the offices of the President and the State governors occupied mainly by wealthy coffee plantation owners and cattle ranchers, alternating between candidates from the states of *Minas Gerais* and *São Paulo* (the 'milk and coffee alliance'). Internationally, Brazil remained politically isolationist but was allied economically with its trading partners, mainly the large Western nations.

The Modern Era, 1930–Present

The Old Republic came to an end in 1930 with the military takeover in October of that year, which removed the serving president, Washington Luis, and his elected successor, Júlio Prestes, and led to the inauguration of *Getúlio Vargas* as president, on November 3rd, 1930. The uprising (led by the states of *Rio Grande do Sul*, Vargas' home state, as well as *Paraíba* and *Minas Gerais*), is referred to as the 'Revolution of 1930'. Vargas remained in office until October 29th, 1945. Early in his regime, the military rule, tending towards authoritarianism, was challenged by the *Constitutionalist Revolt* ("Paulista War") in 1932, an effort to restore a constitutional democracy (or else to secure the secession of the State of *São Paulo*), which was however rapidly defeated. It will play a role in our story. A third revolutionary change in late 1937 was introduced by Vargas himself (responding to an abortive Communist revolt in 1935), before the planned elections of 1938: He declared the '*Estado Novo*', increasing the powers of the central (Federal) government and its president, effectively installing a dictatorship.

Following the Second World War, in 1945, Vargas was deposed by still another military *coup*, and after a brief interlude with an interim president, the *Second Republic* was founded in early 1946. Its first president was *Eurico Gaspar Dutra*, who served for five years and was succeeded again by *Getúlio Vargas* (this time democratically elected!). Vargas however committed suicide while in office in 1954, when it became clear that his government had lost support, and he was followed by three short-term interim presidents until the election of the social democrat *Juscelino Kubitschek de Oliveira*, (JK), who took office in January, 1956. Kubitschek was a modern and liberal president who presided over the construction of the new capital city of *Brasília*.

Kubitschek however was not re-elected in 1960, but instead was succeeded by *Jânio Quadros*, who ran on a coalition ticket of the National Labor Party (PTN), the Christian Democratic Party, and the National Democratic Union (UDN), a relatively conservative lineup. Quadros resigned under mysterious circumstances, on August 25, 1961, after only 7 months in office, and was followed by his vice president, *João Goulart* ('*Jango*', from the Brazilian Labor Party (PTB), i.e. Vargas' party, national-conservative), although with reduced powers since the Congress had installed a Parliamentary Democracy after Quadros' resignation. This prevented a takeover by military authorities who regarded Goulart as too left-wing and not trustworthy.

A popular referendum in 1963 however re-established the presidential system, allowing Goulart to begin carrying out his planned reforms, and this led to a further military *coup d'état* on April 1st, 1964 (dated by the military regime from March 31st, to avoid the stigma of 'April Fools' Day'). This forced Goulart to resign and initiated the military dictatorship which persisted for over 20 years (with nominal officer-presidents representing the National Renewal Alliance or ARENA, the party of the military rulers).

After nearly 15 years of dictatorial control, the military gradually relinquished power. In the late 1970s, President-General *Ernesto Geisel* introduced the 'detente' or 'relaxation', repealing some of the more repressive measures taken by the dictatorship, and in the first half of the 1980s, in a move known as *the opening*, military control of the government was gradually reduced. Finally, in 1985, the first 'democratic' election of a president in over 20 years was held; however not as a direct election, but through an Electoral College in which supporters of the military regime still had a major share of power. This was the beginning of the *New Republic* (of the *Federative Republic of Brazil*, as the nation was renamed in 1967), which still remains in place today. During the term of its first president, *José Sarney*, a new democratic constitution was adopted, and finally genuinely democratic elections were held in 1989.

They were won by the young conservative *Fernando Collor de Mello*, the son of a northeastern oligarch family, who was however impeached for corruption in 1992. He was followed by *Itamar Franco*, of the PL (Liberal Party; previously, he had belonged to the PMDB, the major opposition party during the military regime). His term was marred by runaway inflation.

In 1994, a regular election was again held, and the social democrat and sociologist *Fernando Henrique Cardoso* was elected; he was a former colleague of Tiomno's in São Paulo who had been forcibly retired at the same time as Tiomno. He had introduced a new currency as Minister of Finance during the Franco administration and effectively ended the period of inflation. He was politically liberal but fiscally conservative. His two terms in office were followed in 2002 by two terms each of the Workers' Party (PT) presidents *Luiz Inácio Lula da Silva* and *Dilma Rousseff*. The latter was however impeached in the middle of her second term by an alliance of conservative forces in August 2016, and replaced by the conservative *Michel Temer*, who was succeeded in January 2019 by the winner of the recent elections, the far-right national-conservative populist *Jair Bolsonaro*.

The Brazilian economy has been thoroughly modernized since the mid-20th century. Brazil is currently energy-independent, producing enough oil and gas for its own needs and exporting the excess. A considerable fraction (nearly 90%) of its electrical energy is derived from hydroelectric plants, especially the large plant at *Itaipú*, on the *Rio Paraná* between Brazil and Paraguay.

Industry has grown rapidly and today includes a large production of steel, aluminum, automobiles, aircraft, appliances, electronics, and chemicals, accounting for nearly 30% of the GDP. The agricultural sector remains strong, providing over 60% of the GDP. The banking and financial sector is the largest in South America. In the past 5 years, however, there has been a turndown in economic growth and numerous finance and management scandals. Brazil is the 6th most populous country in the world, with more than 200 million inhabitants, and the 9th largest economy, but occupies a sad position on the inequality index (Gini), taking the 7th position from the 'bottom' in terms of income inequality.

Appendix B: Jayme Tiomno—Chronology and Publications List

Part 1. Chronology: Life and Career

1920—*April 16th:* Jayme Tiomno is born in *Rio de Janeiro* (RJ)/Brazil as the son of Mauricio Tiomno and Annita Aizen Tiomno.

1924—Tiomno moved with his family to the interior of Minas Gerais (MG).

1926—Attended the Municipal (elementary) School in *São Sebastião do Paraíso*/MG.

1932—Attended the *Ginásio Mineiro* (high school) in *Muzambinho*/MG.

1934—Moved back to Rio and continued high school at the *Colégio Pedro II*, Rio de Janeiro.

1936—Admitted to attend the two years' *curso complementar* [*university preparatory course*] for medicine.

1938—Admitted to study medicine at the *Faculdade Nacional de Medicina* of the *Universidade do Brasil* (UB), Rio de Janeiro.

1939—Matriculated as a physics major at the *Universidade do Distrito Federal* (UDF), RJ (transferred to *Universidade do Brasil* after the UDF was terminated).

—Student assistant to the Chair of Biophysics (Professor *Carlos Chagas Filho*) at the *Faculdade Nacional de Medicina*.

© Springer Nature Switzerland AG 2020
W. D. Brewer and A. T. Tolmasquim, *Jayme Tiomno*, Springer Biographies,
https://doi.org/10.1007/978-3-030-41011-7

1940—Required to choose between the two majors, and chose physics.

—Student assistant to the Chair of General and Experimental Physics (Professor *Joaquim da Costa Ribeiro*) at the *Faculdade Nacional de Filosofia* (FNFi), UB.

1942—*August 31st*: Brazil declares war on the Axis countries; the Italian and German professors in Brazil were deported, and Tiomno was called up for military service for the war effort.

—*December*: Obtained his Bachelor's degree in physics and was contracted as a teaching assistant to the Chair of General and Experimental Physics.

1943—*December*: Obtained his *licenciatura* (teaching certificate) in physics.

1944—Entered the *Center for Reserve Officer Training* (CPOR), where he could take time off to continue working as a teaching assistent at the UB.

1945—*May*: End of the war in Europe. Tiomno concluded the CPOR course and was discharged from military service.

1946—*March*: Transferred to the *Faculdade de Filosofia, Ciências e Letras* (FFCL) at USP, São Paulo, under a fellowship from USP, working in the group of Prof. *Mario Schenberg*.

1947—*February*: Returned to the FNFi in Rio and worked with *Guido Beck*, who was on leave in Rio from Córdoba, Argentina.

—*Summer*: Contracted as Assistant to the Chair of Theoretical and Mathematical Physics at the FFCL/USP, and returned to São Paulo.

—Submitted his first publication to an international journal, on proton-proton scattering, with J. Leite-Lopes: [JT4]. See also [JT7].

1948—*February*: Went to Princeton University/USA for post-graduate studies on a fellowship from the U.S. State Department, Office of Education; there he joined the group of J.A. Wheeler.

—Met with Albert Einstein, together with John Wheeler.

—Began planning the establishment of the *Centro Brasileiro de Pesquisas Físicas* (CBPF) in Rio de Janeiro, together with José Leite Lopes and Cesar Lattes.

1949—*January 15th*: Official registration of the CBPF as a foundation, in Rio de Janeiro.

—*February*: Tiomno received a scholarship from the Rockefeller Foundation in order to continue his studies in Princeton.

—*March*: Met with José Leite Lopes, Cesar Lattes, Hervásio de Carvalho and Walter Schützer at Princeton and planned the future organization of the CBPF.

—Published several articles with Wheeler (cf. [JT5, 6, 8, 9, 11]). Collaborated briefly with C.N. Yang and with David Bohm, and met Richard Feynman.

—*June*: Completed his MS in Physics.

—*Summer*: Attended the course on Field Theory at the Michigan Summer School, given by Richard P. Feynman.

—Elected Associate Member of Sigma Xi ("Society devoted to the promotion of research in science", USA)

1950 –*July*: Attended the Summer School at the University of Wisconsin.

—*September*: Completed his doctorate in physics (Ph.D.), with the thesis '*Theories of neutrino and the double beta decay*', under the guidance of Eugene Wigner.

—*October*: Returned to São Paulo and the Physics Department at USP, where he still held an assistantship.

1951—*January 12th*: Celebrated his 'unofficial' marriage to Elisa Frota-Pessôa, in Rio.

1952—*March*: Requested dismissal from his position at USP and moved definitively to Rio de Janeiro. Contracted as tenured professor of the *Centro Brasileiro de Pesquisas Físicas*, where he also organized the teaching laboratories and the collection of preprints '*Notas de Física*'.

—Contracted as adjunct professor of Theoretical Physics at the FNFi/UB in Rio.

—*March*: Elected to full membership of the *Academia Brasileira de Ciências*.

1954—*July*: Traveled with Leite Lopes to Europe to participate in the *Glasgow Conference on Nuclear and Meson Physics*, and took the opportunity to visit various institutions in Lisbon, Paris, London and Manchester.

—Became the professor responsible for the course in '*Theory of Electromagnetism*', and while Leite Lopes was traveling abroad, Tiomno temporarily assumed responsibility for the Chair of Theoretical Physics and Advanced Physics at the FNFi/UB, and as Head of the Department of Theoretical Physics at the CBPF.

1955—Served as Professor for the '*Final course for secondary-school teachers of physics*', *São José dos Campos*, State of São Paulo (established by the *Ministério da Educação e Cultura*, and the *Instituto Tecnológico da Aeronáutica—ITA*).

—Became General Secretary of the Brazilian Academy of Sciences.

1956—*September/October*. Traveled to the US to attend the *International Conference on Theoretical Physics*, in Seattle/WA, and took the opportunity to visit the University of California in Berkeley and to return briefly to Princeton.

1957—*April*. Traveled with his mother, Annita, to New York for her medical examinations, and to Rochester to attend the *Seventh Annual Rochester Conference on High-Energy Physics*.

—Traveled do Buenos Aires, Argentina, to attend the '*Simpósio sobre Partículas Elementares*' [Symposium, on elementary particles].

—Awarded the "*Prêmio Moinho Santista*" (national prize for the Exact Sciences).

1958—*September*. Visiting Professor for one year at Imperial College London, UK, where he worked with Abdus Salam. Elisa traveled with him.

—Attended the *Conference on Nuclear Physics*, in Oxford.

1959—*June*. Attended the Paris *Conference on Relativity and Gravitation*.

—*August*. Returned do Rio de Janeiro and reassumed his functions at the FNFi and the CBPF. Head of the Teaching Department at the CBPF.

1960—*August*. Attended the *Tenth Annual International Conference on High-Energy Physics* in Rochester, where he presented his preliminary results on the existence of the K' meson.

—*September*. Attended the General Assembly of the International Union of Pure and Applied Physics (IUPAP), in Ottawa, Canada, as representative of Brazil, and became a member of the IUPAP Commission on Physics Education.

—Docent at the *Escola Latino-Americana de Física* (cooperation between Brazil, Mexico, and Argentina), Rio de Janeiro.

—Assumed the duties of Scientific Director of the CBPF temporarily (in Leite Lopes' absence).

1961—*May*: Received public recognition for his prediction of the K' meson, and was elected "Person of the Year" by the Cultural Committee '*Ruy Barbosa*' of the *Instituto de Educação*, Rio de Janeiro.

—*July/August*: Attended the Summer School at the University of Wisconsin, Madison/WI, USA, and visited Brookhaven, conferring there with Yang and others.

—Head of the Theoretical Physics Department at the CBPF.

—Director of the Physics Research Section of the CNPq.

1962—*January/February*: Sojourn at the University of Wisconsin, where he collaborated with Abdus Salam.

—*February*: Attended the *American Conference of Academic Deans*, Cleveland/OH.

—*July*: Discussion leader for the '*Seminar on Theoretical Physics*' sponsored by the International Atomic Energy Agency, in Trieste, Italy.

1963—*April*: Member of the Expert Committee for planning the International Centre for Theoretical Physics (ICTP); attended its meeting in Vienna.

—*June/July*: Co-Chair of the First Inter-American Conference on Physics Education, titled '*International Conference on Physics in General Education*', sponsored by the IUPAP and held in Rio de Janeiro.

—Structured the curriculum in Meteorology at the FNFi/UB.

—Founding of the *International Centre for Theoretical Physics* (ICTP), in Trieste, Italy.

1964—Tiomno was responsible for teaching at the Chair of Theoretical Physics and Advanced Physics of the FNFi/UB, Rio de Janeiro.

—Associate of the *International Centre for Theoretical Physics*, Trieste, Italy.

1965—*March*: Establishment of the Physics Institute (*Instituto Central de Física*, IF) at the *Universidade de Brasília* (UnB), Brasília/DF; acting tenured professor of Theoretical Physics, Coordinator for the *Instituto de Física*, and responsible for the courses in *Theoretical Mechanics, Electromagnetic Theory, Theoretical Physics,* and the *Introduction to Quantum Mechanics*.

—*September*: Attended the '*International Conference on Elementary Particles*', Oxford/UK.

—*October*: Resigned from the University of Brasília, together with 210 other professors, corresponding to 80% of the faculty, due to the arbitrary actions of the Rector.

1966—*July*: Co-founder and vice-President of the *Sociedade Brasileira de Física* (SBF).

—*September*: Traveled to the ICTP, Trieste (with Elisa) for a one-year stay, where he collaborated with C.G. Bollini and J.J. Giambiagi.

1968—*January*: Contracted as Full Professor for Advanced Physics, FFCL/USP, São Paulo.

1969—*April*: Compulsory retirement from USP; Elisa was also retired from the UB.

—*July/September*: Traveled to Trieste, Paris, New York, Rochester, Princeton, Miami, Los Angeles, and Mexico, taking part in conferences, giving lectures, and investigating possibilities of working abroad for himself and Elisa.

—*October*: Dismissed from the CBPF, together with Elisa and Leite Lopes.

1970—Traveled to Buenos Aires and to Chile.

1971—*January*—Visiting Researcher at the *Institute for Advanced Studies*, Princeton, USA.

1972—*July*—Returned to Brazil and used his apartment as his office.

1973—*December*: Contracted as tenured professor at the *Pontifícia Universidade Católica* (PUC) in Rio de Janeiro, but not allowed to serve as mentor for student research or theses.

1977—*September 27th*: Official marriage of Jayme Tiomno to Elisa Frota-Pessôa.

1980—After the Amnesty, asked to be dismissed from PUC/Rio and was (again) contracted as tenured professor at the *Centro Brasileiro de Pesquisas Físicas*, Rio de Janeiro.

—Honored at the "Session dedicated to his 60th birthday", by the *Sociedade Brasileira de Física* (SBF) at its meeting in *Cambuquira*, MG.

1983—Traveled to Trieste in June/July to take part in the '*3rd ICTP Workshop in Particle Physics*' there.

1984—Attended the *Wingspread Conference* (Racine, Wisconsin/USA) on '*50 Years of the Weak Interactions*', and gave a paper on '*The early period of the Universal Fermi Interaction*'.

1985—Elected member of the *Academia Paulista de Ciências*, São Paulo.

—Project Coordinator for High-Energy Physics; accompanied the founding of the 'LAFEX' Department (international cooperations for experimental HEP) at the CBPF.

—Project coordinator for 'Algebraic Computing', also at the CBPF.

1988—Designated President of the Executive Council of FAPERJ, Rio de Janeiro (state science funding agency).

1990—Honored by an International Physics Symposium at the CBPF, in Rio de Janeiro.

—Honored at a 'Session dedicated to his 70th birthday', *Sociedade Brasileira de Física*, in *Caxambu*, Minas Gerais.

—Officially retired from the CBPF at age 70.

1991—Publication of the '*Essays in Honour of Jayme Tiomno*', *Frontier Physics* series, eds. S. MacDowell, H.M. Nussenzweig and R.A. Salmeron (World Scientific, Singapore, 1991).

—President of the Special Commission of the CNPq for guiding the interactions of Brazilian research groups with major international laboratories for High Energy Physics.

1992—Received the title "*Pesquisador Emérito*" (Researcher *emeritus*) of the CBPF, Rio de Janeiro.

—Designated member of the Commission for the Selection of Academicians of the *Academia Brasileira de Ciências*.

1993—Patron of the seminar room 'Professor Jayme Tiomno', chosen by the Departmental Council of Mathematical Physics, *Instituto de Física* at USP.

1994—Decorated with the "*Grã Cruz da Ordem Nacional do Mérito Científico*" [*Grand Cross of the National Order of Scientific Merit*] by the *Presidência da República*, Brasília/DF.

—Designated member of the 'Committee of the Americas', organizing committee of the '*VIIth Marcel Grossmann Meeting on Relativistic Astrophysics*' (MG-VII), Stanford/CA, USA.

1995—Received the Physics Prize of the *Third World Academy of Sciences* (TWAS) in Trieste.

1996—Elected Permanent Member of the *Third World Academy of Sciences*.

—Designated member of the executive committee of the "International Society of General Relativity and Gravitation", headquartered at the University of Victoria, Canada.

1999—Honored as a Founder of the CBPF on the occasion of the 50th anniversary of its founding.

2000—Awarded the title "*Cidadão Honorário de Muzambinho*" [*Honorary citizen of Muzambinho*] by the *Câmara Municipal da cidade*, Muzambinho/MG.

2002—Participation in the "Symposium on Science and Ultimate Reality", dedicated to the 90th birthday of John Archibald Wheeler, at the Merrill Lynch Center, Princeton/NJ, USA.

2004—Homage from the Physics Institute of USP on its 35th anniversary.

2005—Received the title *"Pesquisador Emérito"* (Researcher *emeritus*) from the CNPq, in Brasília.

—Event organized in honor of Jayme Tiomno: First INPE Advanced Course, '*Roadmap for Cosmology*', *Instituto Nacional de Pesquisas Espaciais* (INPE), São José dos Campos/SP.

—Awarded the medal '*Centenário de João Christovão Cardoso*' by the *Instituto de Química*, UFRJ, Rio de Janeiro.

—Awarded the medal '*Carlos Chagas Filho do Mérito Científico*' by FAPERJ, in Rio de Janeiro, at its 25th anniversary celebration in December, 2005.

2006—Awarded the título '*Aluno Eminente*' [distinguished pupil], by the *Colégio Pedro II*, Rio de Janeiro.

2011—Passed away quietly in his sleep at home in *Barra da Tijuca*, RJ, on January 12th.

Part 2. Publications and Contributions

[Notes: * indicates publications listed in the '*Academic Tree*'; § indicates articles included in Tiomno's own list of his 'papers of major importance' Tiomno (1994). Note that some entries have an alphabetic extension (e.g. [JT4a], [JT28b], etc.). These are items which are not referred to directly in the main text. Most of them are reports or manuscripts which were not published in the scientific literature, or else articles on Physics Education or the History of Science which appeared in *Ciência e Cultura*, the science magazine of the SBPC (*Sociedade Brasileira para o Progresso da Ciência*)].

1940s _____ (14 items).

[JT1] '*Sobre o teorema de unicidade da distribuição de cargas em condutores*' [*On the uniqueness theorem for the charge distribution in conductors*], in: F.N.F. 1, 2, 1942 (F.N.F. was the house journal of the FNFi).

[JT2] '*Sobre um problema da Teoria da Elasticidade*' [*On a problem in the theory of elasticity*], in: F.N.F. 3, 1942.

[JT3] '*Sobre as derivadas do campo de radiação do electron puntiforme com spin*' [*On the radiation-field derivatives of an oscillating point-like electron with spin*], in: *Anais da Academia Brasileira de Ciências* 19, 333, 1947 (with W. Schützer).

[JT4]* '*On the proton-proton scattering at 14.5 MeV*', in: the *Physical Review* 72, 731, 1947 (with J. Leite Lopes).

[JT4a] '*Sobre o teorema fundamental da álgebra hipercomplexa de Sobrero*' [*On the fundamental theorem of Sobrero's hypercomplex algebra*], 1947 (with L.

Nachbin); unpublished manuscript prepared in the first half of 1947 at the FNFi with Leopoldo Nachbin.

[JT4b] '*The deflection of light in a gravitational field*', 1947 (with M. Schenberg); unpublished manuscript prepared in Schenberg's group at USP, second half of 1947.

[JT4c] '*Test particles in General Relativity*' (with J.A. Wheeler); unpublished manuscript prepared in Wheeler's group, February–June, 1948.

[JT5]*§ '*Energy spectrum from mu meson decay*', in: *Reviews of Modern Physics* 21, 144, 1949 (with J.A. Wheeler).

[JT6]*§ '*Charge-exchange reaction of the mu meson with the nucleus*', in: *Reviews of Modern Physics* 21, 153, 1949 (with J.A. Wheeler)—republished in Japan (Phys. Soc. Japan, Series of selected papers on the Weak interactions, 1972; see Chap. 6, Note [29]).

[JT7] '*Distribuição angular na difusão proton-proton a 14.5 MeV*' [*Angular distribution from proton-proton scattering at 14.5 MeV*], in: *Anais da Academia Brasileira de Ciências*, 21, 56, 1949.

[JT8] '*On the coupling of pi and mu mesons*', in: the *Physical Review* 75, 1306, 1949 (with J.A. Wheeler) (Abstract M6, Proceedings of the APS New York Meeting, Jan. 26–29, 1949).

[JT9] '*On the spins of pi and mu mesons*', in: the *Physical Review* 75, 1306, 1949 (with J.A. Wheeler) (Abstract M7, Proceedings of the APS New York Meeting, Jan. 26–29, 1949).

[JT10]* '*On the Spin of Mesons*', in: the *Physical Review* 76, 856, 1949.

[JT11]* '*Guide to the Literature of Elementary Particle Physics, Including Cosmic Rays*', in: the *American Scientist* 37, 2, pp. 202–218, 1949 (with J.A. Wheeler).

1950s _____ (26 items).

[JT12]*§ '*Reflection properties of spin-½ fields and a Universal Fermi-type Interaction*', in: the *Physical Review* 79, 495, 1950 (with C.N. Yang).

[JT13]§ '*Theories of neutrino and the double beta decay*', Doctoral Thesis (Ph.D.) at Princeton University, USA, 1950 (unpublished; available from the CBPF). (In English).

[JT14]*§ '*On the connection of the scattering and derivative matrices with causality*', in: the *Physical Review* 83, 249, 1951 (with W. Schützer).

[JT14a] '*Singularidades da Matriz S no plano complexo*' [*Singularities of the S Matrix in the complex plane*], 1951 (with Leo Borges Vieira); unpublished manuscript.

[JT15]§ '*Gamma radiation emitted in the pi-mu decay*', in: *Anais da Academia Brasileira de Ciências* 24, 245, 1952 (with G.E.A. Fialho)—republished in Japan (Phys. Soc. Japan, Series of selected papers).

[JT16] *'Non-relativistic equations of charged particles with spin 3/2'*, in: *Notas de Física*, No. 9, 1952 (with A. Silveira; unpublished).

[JT17] *'On the connection of the scattering matrix with causality'* (with W. Schutzer); talk given at the *Symposium on New Research Techniques in Physics*, Rio de Janeiro, 1952; included in the meeting's proceedings; unpublished.

[JT17a] *'O ensino de física no curso secundário'* [*Teaching high-school physics*], in: *Ciência e Cultura* 5, 45, 1953 (with J. Leite Lopes).

[JT18] *'Non-relativistic equation for particles with spin 1'*, in: *Anais da Academia Brasileira de Ciências* 26, 327, 1954 (with J.J. Giambiagi).

[JT19] *'Invariance of field theory under time inversion'*, in: *Notas de Física*, No. 16, 1954 (from his thesis; unpublished).

[JT20]*§ *'Mass Reversal and the Universal Interaction'*, in: *Il Nuovo Cimento* 1, 226, 1955.

[JT21]* *'A causal Interpretation of the Pauli equation'*, in: *Il Nuovo Cimento* 1, 18, 1955 (with D. Bohm and R. Schiller).

[JT22] *'Relativistic theory of spinning point particles'*, in: *Anais da Academia Brasileira de Ciências* 27, 259, 1955.

[JT23]* *'Diagrams for processes involving hyperons'*, in: the *Physical Review* 103, 1589, 1956.

[JT24] *'Polarization of spin-one particles by nuclear scattering'*, in: *Anais da Academia Brasileira de Ciências*, 28, 157, 1956 (with S.W. MacDowell).

[JT24a] *'Histórico e realizações do Departamento de Física Teórica do Centro Brasileiro de Pesquisas Físicas'*, [*The history and achievements of the Department of Theoretical Physics at the Brazilian Center for Physics Research*], in: *Ciência e Cultura* 8, 8, 1956.

[JT25]*§ *'On the theory of hyperons and K mesons'*, in: *Il Nuovo Cimento* 6, 69, 1957.

[JT26]* *'Note on the gamma decay of neutral pi mesons'*, in: *Il Nuovo Cimento* 6, 255, 1957.

[JT27] *'Baryon and Meson Interactions'*, in: *Proceedings of the 1957 International Conference on High Energy Nuclear Physics*, Rochester, IX-22, Interscience Publishers, New York, 1957.

[JT28]*§ *'Non-conservation of parity and the Universal Fermi Interaction'*, in: *Il Nuovo Cimento* 6, 912, 1957.

[JT28a] *'O Ensino da física nas universidades brasileiras'* [*The teaching of physics in Brazilian universities*], in: *Ciência e Cultura*, 9, 1957 (with J. Leite Lopes).

[JT28b] *'Discurso de recebimento do Prêmio Moinho Santista de Ciências Exatas'* [*Lecture delivered on receiving the Moinho Santista Prize for the Exact Sciences*), in: *Ciência e Cultura*, 9, 241, 1957.

[JT29]* '*On the masses of elementary particles*', in: *Nuclear Physics* 9, 585, 1958 (with A. Salam).

[JT30] '*–electron decay and the Universal Interaction*', in: *Anais da Academia Brasileira de Ciências* 30, 455, 1958 (with C.G. Oliveira).

[JT31] '*Some Reflections in Quantum Field Theory*', communication to the *Academia Brasileira de Ciências*, 1959 (with S. Kamefuchi).

[JT32] '*The failure of the space reflection principle*', in: *Proceedings of the Mathematical Society of the University of Southampton* 21, 9, 1959.

1960s _____ (17 items).

[JT33]§ '*On the K' meson*', in: *Proceedings of the 1960 International Conference on High-Energy Physics,* Rochester, NY, Interscience Publishers, 1960, pp. 466 and 513.

[JT34]* '*Isotopic Spin Relations in Hyperon Production*', in: *Il Nuovo Cimento* 22 1287, 1961 (with A. Salam).

[JT35]*§ '*Possible existence of a new (K') Meson*', in: *Physical Review Letters* 6, 120, 1961 (with N. Zagury and A.L.L. Videira).

[JT36]* '*Representations of Dirac Equations in General Relativity*', in: *Il Nuovo Cimento* 24, 672, 1962 (with C.G. Oliveira).

[JT37] '*On S =—Q*', University of Wisconsin, 1962 (lecture, with Abdus Salam); unpublished.

[JT37a] '*Inter-American Cooperation in Higher Education*', in: Proceedings of the *18th Annual Meeting of the American Conference of Academic Deans*, Ohio, 1962; unpublished.

[JT38] '*Octonions and Super-global Symmetry*', in: *Theoretical Physics,* International Atomic Energy Agency, Vienna, 1963; p. 251.

[JT38a] '*The International Center for Research in Theoretical Physics*' (ICTP, Trieste). Report Gov/Inf/98, May 1963, International Atomic Energy Agency, Vienna (with R.E. Marshak and L. Van Hove): Report of the Expert Commission on founding the ICTP; unpublished.

[JT38b] '*O Instituto de Física Pura e Aplicada da Universidade de Brasília*' [*The Institute of Pure and Applied Physics of the University of Brasília*], (with J. Leite Lopes, G. Beck, G.E.A. Fialho, R.A. Palmeira and R.A.Salmeron), 1963: Recommendations of the External Advisory Commission on the founding of the Physics Institute at the UnB.

[JT38c] '*Eletromagnetismo I*' [*Electromagnetism I*], Course syllabus, in: *Coleção Monografias de Física* X, CBPF, 1963; unpublished.

[JT38d] '*Eletromagnetismo II*' [*Electromagnetism II*], Course syllabus, in: *Coleção Monografias de Física* XI, CBPF, 1963; unpublished.

[JT39] Book: *Why Teach Physics?* eds. S.C. Brown, N. Clarke and J. Tiomno, M.I.T. Press, Boston, 1964.

[JT40] *'Science Education in the Contemporary World'*: chapter in *Why Teach Physics?*, M.I.T. Press, Boston, 1964, p. 7.

[JT40a] *'Eletromagnetismo III'* [Electromagnetism III], Course syllabus, in: *Coleção Monografias de Física* XVI, CBPF, 1964; unpublished.

[JT41] *'Contribuições à Física das Partículas Elementares'* [*Contributions to the physics of elementary particles*]: Thesis submitted to the Competition for the Chair of Advanced Physics at the FFCL/USP, São Paulo, July 1966; unpublished.

[JT42]* *'On the Covariance of Equal-time Commutators and Sum Rules'*, in: *Il Nuovo Cimento* 51A, 717, 1967 (with C. Bollini and J.J. Giambiagi).

[JT42a] *'Perspectivas da física no Brasil e reforma universitária'* [*Perspectives of physics in Brazil and the university reform*], in: *Ciência e Cultura* 20, 4, 702, 1968.

1970s _____ (21 items).

[JT43]* *'A Linear Theory of Gravitation'*, in: *Lettere al Il Nuovo Cimento* 3, 65, 1970 (with C.G. Bollini and J.J. Giambiagi).

[JT44]* *'Equivalence of Lorentz Transformations and Foldy-Wouthuysen Transformation for Free Spinor Fields'*, in: *Physica* 53, 581–601, 1971.

[JT45]* *'Electromagnetic Field of a Particle Moving in a Spherically Symmetric Black-Hole Background'*, in: *Lettere al Il Nuovo Cimento* 3, 211, 1972 (with R. Ruffini and C.V. Vishveshwara).

[JT46]*§ *'Can Synchrotron Gravitational Radiation Exist?'*, in: *Physical Review Letters*, 28, 1352, 1972 (with Marc Davis, R. Ruffini and F. Zerilli).

[JT47]* *'Pulsars and Pair Production in Electric Fields'*, in: *Nature Physical Science*, 238, 57, 1972 (with L. Parker).

[JT48]*§ *'Pulses of Gravitational Radiation of a Particle Falling Radially into a Schwarzschild Black Hole'*, in: the *Physical Review* D5, 2932, 1972 (with Marc Davis and R. Ruffini).

[JT49]* *'Maxwell Equations in a Spherically Symmetric Black Hole Background'*, in: *Lettere al Nuovo Cimento* 5, 851–855, 1972.

[JT50] *'Pair-Producing Electric Fields and Pulsars'*, in: the *Astrophysical Journal*, 178, 809, 1972 (with L. Parker).

[JT51]* *'Polarization of Gravitational Synchrotron Radiation'*, in: *Lettere al Nuovo Cimento* 4, 857, 1972 (with R.A. Breuer and C.V. Vishveshwara).

[JT52]* *'Balancing of Electromagnetic and Gravitational Forces and Torque Between Spinning Particles at Rest'*, in: the *Physical Review* D7, 356, 1973.

[JT53]* *'Gyromagnetic Ratio of a Massive Body'*, in: the *Physical Review* D7, 998, 1973 (with J. M. Cohen and R.M. Wald).

[JT54]* *'Electromagnetic Field of Rotating Charged Bodies'*, in: the *Physical Review* D7, 992, 1973.

[JT55]*§ *'Vector and Tensor Radiation from Schwarzschild Relativistic Circular Geodesics'*, in: the *Physical Review* D7, 1002, 1973 (with R.A. Breuer, R. Ruffini and C.V. Vishveshwara).

[JT56]* *'On Gravitation-Induced Electromagnetic Fields'*, in: *Acta Physica Austriaca* 38, 206, 1973.

[JT57]* *'Polarization of Gravitational Geodesic Synchrotron Radiation'*, in: *Il Nuovo Cimento* 25B, 851, 1975 (with R.A. Breuer and C.V. Vishveshwara).

[JT58] *'A Cartesian Operator Algebra for Expansion of Tensor Quantities and Equations in a Spherically Symmetric Background'*, in: *Revista Brasileira de Física* 8, 304, 1978 (with A.L.L. Videira).

[JT59] *'Charged point particles with magnetic moments in General Relativity'*, in: *Revista Brasileira de Física* 8, 350, 1978 (with Ricardo M. Amorim).

[JT60] *'On the relation between fields and potentials in non-abelian gauge fields'*, in: *Revista Brasileira de Física* 9, 1, 1979 (with J.J. Giambiagi and C.G. Bollini).

[JT61]* *'Singular potentials and analytic regularizations in classical Yang-Mills Equations'*, in: *Journal of Math. Phys.* 20, 1967, 1979 (with C.G. Bollini and J.J. Giambiagi).

[JT62]* *'Gauge field copies'*, in: *Phys. Letters* 83B, 185, 1979 (with J.J. Giambiagi and C.G. Bollini).

[JT62a] *'Contribuição de Albert Einstein à teoria da gravitação e cosmologia'* [*Albert Einstein's contributions to Gravitation Theory and Cosmology*], in: *Ciência e Cultura*, 31 (12), December 1979 (commemorative edition for the centenary of Albert Einstein's birth).

1980s _____ (37 items).

[JT63]* *'Hidden Singularities in non-abelian gauge fields'*, in: *Il Nuovo Cimento* 55A, 91, 1980 (with C.G. Bollini).

[JT64]* *'Wilson Loops and Related Strings for the instanton and its variational derivatives'*, in: *Il Nuovo Cimento* 59, 412, 1980 (with C.G. Bollini and J.J. Giambiagi).

[JT65]* *'Wilson Loops in Kerr Gravitation'*, in: *Letters to Il Nuovo Cimento* 81, 13, 1981 (with C.G. Bollini and J.J. Giambiagi).

[JT66] *'On the Proposed Kolen-Torr Experiment'*, in: *Proceedings of the International Conference on Space-time Absoluteness*, Graz, Austria, p. 147, 1982 (with W. Rodrigues).

[JT67]*§ *'Geodesic Motion and Confinement in Gödel's Universe'*, in: the *Physical Review* D27, 779, 1983 (with M. Novello and I.D. Soares).

[JT68] *'The early period of the Universal Fermi Interaction'*, in: *Proceedings of the Racine Conference on 50 Years of Weak Interactions*, 1984 [published as

Conference Proceedings, eds. D. Cline and G. Riedasch, University of Wisconsin, Madison/WI (1984)].

[JT69] *'Gödel-type metric in Einstein-Cartan Spaces'*, in: *Contributed Papers to the X. International Conference on General Relativity and Gravitation*, p. 507, 1983 (with A. Teixeira and J. Duarte).

[JT70]* *'Homogeneity of Riemannian Space-times of Gödel type'*, in: the *Physical Review* D28, 1251, 1983 (with M. Rebouças).

[JT71]*§ *'Pseudoscalar Mesons and Scalar diquarks—decay constants'*, in: *Il Nuovo Cimento* 81A, 485, 1984 (with I. Bediaga, E. Predazzi, A.F.S. Santoro and M.H.G. Souza).

[JT72] *'Gravitational coupling of scalar and fermionic fields to matter vorticity: Microscopic asymmetries'*, in: *Revista Brasileira de Física*, Suppl. 14, 372, 1984 (with I.D. Soares).

[JT73] *'Einstein's special relativity* vs. *Lorentz ether theory'*, in: *Revista Brasileira de Física*, Suppl. 14, 450, 1984 (with W. Rodrigues).

[JT74]* *'Lifetimes in a quark-diquark system'*, in: *Lettere al Nuovo Cimento* 42, 92–96, 1985 (with A.F.S. Santoro, I. Bediaga, M.G.H. Souza and E. Predazzi).

[JT75] *'Ciência, Universidade e Desenvolvimento'* [*Science, the University and Development*], in: *"Ciência e Sociedade"*, CBPF CS–011/85; Inaugural Lecture at USP, March 1968; unpublished.

[JT76]*§ *'On Experiments to detect possible failures of Relativity Theory'*, in: *Foundations of Physics* 15, No. 9, pp. 945–961, 1985 (with W. Rodrigues).

[JT77] *'Possíveis Violações da Teoria da Relatividade'* [*Possible Violations of the Theory of Relativity*], in: *Revista Brasileiro de Física*, Suppl. 15, 1985. A review of previous works on SRT, presented at the Symposium in honor of the 70th birthday of Mario Schenberg, August 1984.

[JT78]* *'Vector meson and axial-vector diquark decay constants'*, in: *Lettere al Nuovo Cimento* 42, 54, 1985 (with A.F.S. Santoro, I. Bediaga, E. Predazzi and M.G.H. Souza).

[JT79]* *'Gluon and qq mixing: [1440] system'*, in: *Zeitschrift f. Physik* C30, 493, 1985 (with A.C.B. Antunes, F. Caruso and E. Predazzi).

[JT80]*§ *'Experiments to Detect Possible Weak Violations of Special Relativity'*, in: *Physical Review Letters* 55, 143, 1985 (with A.K.A. Maciel).

[JT81]* *'A class of Inhomogeneous Gödel-type models'*, in: *Il Nuovo Cimento* 90B, 204, 1985 (with M. Rebouças).

[JT82]* *'Stefan Marinov and "Friends" '*, in: *Nature* 317, 772, 1985.

[JT83] *'Experiências de uma vida na Universidade Brasileira'* [*Experiences of a lifetime in Brazilian Universities*], in: *Ciência e Sociedade*, CBPF–CS–005, 1986; unpublished (Semester inaugural lecture at the *Universidade de Pará*).

[JT84]*§ '*A New Scheme For Nonleptonic Decays: Predictions of F Meson*', in: *Physics Letters* 181B, 395, 1986 (with I. Bediaga and E. Predazzi).

[JT85]* '*Homogeneous Cosmos of Weyssenhoff fluid in Einstein-Cartan Space*', in: the *Physical Review* D 34, 1986 (with J. Duarte de Oliveira and A.F. da F. Teixeira).

[JT86] '*Experimental Evidence Against a Lorentz Aether Theory (LAT)*', in: Proceedings of the *Fourth Marcel Grossmann Meeting on Relativistic Astrophysics*, ed. R. Ruffini, Elsevier, 1986, p. 1347.

[JT87]*§ '*A new Hadronization Scheme: The case of explicit Charm Decay*', in: *Nuclear Physics* B294, 1071, 1987 (with J.L. Basdevant, I. Bediaga and E. Predazzi).

[JT88]* '*Inhomogeneous two-fluid Cosmologies*', in: *General Relativity and Gravitation* 20, 1019, 1988 (with J.A.S. Lima).

[JT89] '*Antipodal Universes in the Topology $S^3 X R$ and $H^3 X R$*', in: *Notas de Física*-004–CBPF, 1988 (with I.D. Soares and F.D. Sasse); unpublished.

[JT90] '*Neutrinos in Antipodal Universes: Parity Transformation and Asymmetries*', in: *Notas de Física*-005–CBPF, 1988 (with I.D. Soares and F.D. Sasse); unpublished.

[JT91] '*A class of Inhomonogeneous Cosmologies with Heat Flow*', in: *Leite Lopes Festschrift*, eds. N. Fleury, S. Joffily, J.A.M. Simões and A. Troper, World Scientific, Singapore, 1988, p. 303 (with J.A.S. Lima).

[JT92] '*Origens do Centro Internacional de Física Teórica*' (ICTP–Trieste) [*Origins of the International Centre for Theoretical Physics* (ICTP Trieste)], in: *Ciência e Sociedade*, CBPF–CS–007/88, 1988; unpublished.

[JT93]* '*On the Detection of Matter Vorticity and Spacetime Torsion*', in: *Physics Letters* A137, 99, 1989 (with B. Figueiredo and I.D. Soares).

[JT94]* '*Analysis of Absolute Space-time Theories*', in: *Foundations of Physics* 19, 505, 1989 (with A.K.A. Maciel).

[JT95]* '*Experimental Analysis of Absolute Space-time Theories*', in: *Foundations of Physics* 19, 521, 1989 (with A.K.A. Maciel).

[JT96]*§ '*On the Thermodynamics of one-fluid Szekeres-like Cosmologies*', in: *Classical Quantum Gravitation* 6, L93, 1989 (with J.A.S. Lima).

[JT97] '*Thermodynamical Analysis of Cosmological Models*', in: *Encontro de Física Teórica do Rio de Janeiro*, Ed. UFRJ, 1989, pp. 95–110 (with J.A.S. Lima).

[JT98] '*Collaboration in Physics in Latin America (Round Table)*, in: *Proceedings of the 3rd Symposium on Pan-American Collaboration in Experimental Physics*, eds. R. Rubinstein and A. Santoro, World Scientific, Singapore, 1989, p. 266.

[JT99]* '*Reply to 'Comment on Recent Interpretation of Lorentz Ether Theory*'', in: *Foundations of Physics Letters* 2, No. 6, 601, 1989 (with A.K.A. Maciel).

1990s _____ (5 items).

[JT100]*§ '*Geodesics in Gödel-Type Space-times*', in: *General Relativity and Gravitation* 22, 683, 1990 (with I.D. Soares and M.O. Calvão).

[JT101]* '*Gravitational Coupling of Klein-Gordon and Dirac Particles to matter vorticity and space-time torsion*', in: *Classical Quantum Gravity* 2, 1593, 1992 (with B. Figueiredo and I.D. Soares).

[JT102] '*The Early Period of the Universal Fermi Interaction*', pages 99–109 in *Discovery of Weak Neutral Currents: 'The weak interaction before and after*', 1993; published in *AIP Proceedings* 300, eds. Alfred K. Mann and David B. Cline, Santa Monica, CA/Woodbury, NY, 1994. (Later re-publication of an article from the Proceedings of the Racine Conference, 1984. They were issued less formally immediately after that conference; cf. [JT68]).

[JT103]* '*The Physics of the Sagnac-Mashhoon Effects*', in: the *Physical Review* D 54, 2808, 1996 (with I.D. Soares).

[JT104] '*The Physics of Spin-rotation Coupling*', in: *Award Lectures of the Third World Academy of Sciences* (TWAS), 1996; unpublished.

2000s _____ (4 items).

[JT105] '*The Physics of Rotating Experiments*', in: '*Roberto Salmeron Festschrift*', eds. A. Santoro, R. Aldrovandi and J.M. Gago, Rio de Janeiro, p. 327, 2003 (with I.D. Soares).

[JT106] '*Trabalho Duro*' [*Hard Work*], in *Algumas razões para ser cientista* [Some reasons for being a scientist] (interview in the book by Carolina Cronemberger), Ed. CBPF, Rio de Janeiro, 2005, p. 96.

[JT107] '*A física de partículas: o início dos trabalhos de física de partículas no Brasil*' [The Physics of Particles: Beginnings of work on particle physics in Brazil], in: MAST Colloquia—*Memória da Física*, Vol. II, MAST, Rio de Janeiro, 2005.

[[JT108] '*On the exact relativistic dynamics of the Sagnac effect*', in: *XXVIII Encontro Nacional de Física de Partículas e Campos*, September 24th–28th, 2007. See the abstract at:

http://www.sbf1.sbfisica.org.br/eventos/enfpc/xxviii/sys/resumos/R0050-1.pdf.

Total _____ 124 items.

Literature

Interviews

Henrique Lins de Barros.
Mario Novello.
Nicim Zagury.
Silvia Tiomno Tolmasquim.
Sonia Frota-Pessôa and Roberto Frota-Pessôa.

Articles, Books and Papers

'*Roberto Salmeron Festschrift: A Master and a Friend*', edited by Ruben Aldrovandi, A.F.S. Santoro, and José Mariano Gago. Editora AIAFEX, Rio de Janeiro (2003); ISBN 85-85806-02-8. This book is a collection of essays and articles in honor of Salmeron's 80th birthday.

A. Ali, C. Isham, T. Kibble, and Riazuddin, eds., '*Selected papers of Abdus Salam*', World Scientific, Singapore (1994); ISBN 978-981-02-1662-7. See pp. 170–174, '*On Fermi Interactions*'; online at https://www.worldscientific.com/doi/abs/10.1142/9789812795915_0022.

Margaret Alston, Luis W. Alvarez, Philippe Eberhard, Myron L. Good, William Graziano, Harold K. Ticho, and Stanley G. Wojcicki, '*Resonance in the K–π System*', in *Physical Review Letters*, Vol. 6, No. 6, pp. 300–302 (March 15th, 1961).

Ana Maria Ribeiro de Andrade. *Físicos, mésons e política: a dinâmica da ciência na sociedade.* Rio de Janeiro, Hucitec; MAST, 1998.

© Springer Nature Switzerland AG 2020
W. D. Brewer and A. T. Tolmasquim, *Jayme Tiomno*, Springer Biographies,
https://doi.org/10.1007/978-3-030-41011-7

Kétévi Adiklè Assamagan, PhD Thesis on 'A Precise Determination of the Pion Beta-Decay Rate: Design and Calibration' (University of Virginia, 1995). See in particular the Introduction, Part 3, 'Chiral Invariance and V–A'. [Note that the minus sign is apparently exchanged in Eqns. (1.34)/(1.35). Uploaded by Bernward Krause, January 15th, 1996, to the website http://pibeta.phys.virginia.edu/docs/publications/ketevi_diss/node6.html].

'*Science and Ultimate Reality – Quantum Theory, Cosmology, and Complexity*', eds. John D. Barrow, Paul C.W. Davies, and Charles L. Harper, Jr. Cambridge University Press/UK (2004); ISBN 0-521-83113-X. This is the *Festschrift* resulting from the Symposium held at Princeton in March, 2002 to honor John Wheeler on his 90th birthday. See also https://www.aps.org/publications/apsnews/200205/princeton.cfm.

José Maria Filardo Bassalo, '*Jayme Tiomno, os Mésons e a Física Paraense*', CBPF: *Ciência e Sociedade*, Document CBPF-CS-005/87 (1987).

José Maria Filardo Bassalo, '*Mortes de Universidades Brasileiras: Militar (1965) e Civil (1998)*'. CBPF: *Ciência e Sociedade*, Document CBPF-CS-024/98 (1998).

José Maria Filardo Bassalo and Olival Freire Jr., '*Wheeler, Tiomno, e a Física Brasileira*', in *Revista Brasileira de Ensino de Física*, Vol. 25, No. 4, p. 426 (2003).

José Maria Filardo Bassalo, '*O Instituto Central de Física Pura e Aplicada da Universidade de Brasília, em 1965*'. CBPF: *Ciência e Sociedade*, Document CBPF-CS-010/12 (June 2012).

Ignácio Bediaga and Francisco Caruso, '*Enrico Predazzi: 25 Anos Colaborando com o Brasil*', in *Revista Brasileira de Ensino de Física*, Vol. 18, No. 1, pp. 24–29 (March 1996).

Hans Bethe and Robert Marshak, in the *Physical Review*, Vol. 72, p. 506 (1947); 'The Two-Meson paper'.

Bhowmik, B.; Evans, D.; Falla, D.; Hassan, F.; Kamnal, A.A.; Nagpaul, K.K.; Prowse, D.J.; Rend, M.; Alexander, G.; Johnston, R.H.W.; O'Ceallaigh, C,; Keefe, D.; Burhop, EH.S.; Davis, D.H.; Kumor, R.C.; Lasich, W.B.; Shaukat, M.A.; Stannard, F.R.; Bacchella, G.; Bonetti, A.; Dilworth, C.; Occhialini, G.; Scarsi, L.; Grilli, M.; Guerriero, L.; von Lindern, L.; Merlin, M. and Salandin, A. [K^-*collaboration*]: 'The interaction and decay of K⁻-mesons in photographic emulsions, Part I. General characteristics of K⁻ interactions and analysis of events in which a charged π-meson is emitted', in: *Il Nuovo Cimento*, Vol. 13, No. 4 (1959), pp. 690–729. See also Bhowmik, B.; Evans, D.; Falla, D.; Hassan, F.; Kamal, A.A.; Nagpaul, K.K.; Prowse, D.J.J.; Rend, M.; Alexander, G.; Johnston, R.H.W.; O'Ceallaigh, C.; Keefe, D,; Burhop, E.H.S.; Davis, D.H.; Kumar, R.C.; Lasich, W.D.; Shaukat, M.A.; Stannard, F.R.; Bacchella, M.; Bonetti, A.; Dilworth, C.; Occhialini, G.; Scarsi, L.; Grilli, M.; Guerriero, L.; von Lindern, L.; Merlin, M. and Salandin, A. [K^-*collaboration*]: 'The Interaction of K⁻-Mesons with Photographic Emulsion Nuclei, Part II. The Emission of Hyperons from K⁻ interactions at Rest', in: *Il Nuovo Cimento*, Vol. 14, No. 2 (1959), pp. 315–364.

Wolfgang Bietenholz and Lillian Prado, '*Revolutionary Physics in Reactionary Argentina*', in *Physics Today*, Vol. 67, No. 2, p. 38 (2014); see https://physicstoday.scitation.org/doi/full/10.1063/PT.3.2277.

O.H. Blackwood, W.B. Herron and W.C Kelly, *Física na Escola Secundária, MEC*, Rio de Janeiro (1962). Translated by J. Leite Lopes and J. Tiomno.

Interview with David Bohm by Maurice Wilkins on Oct. 3rd, 1986: Niels Bohr Library & Archives, American Institute of Physics, College Park, MD/USA; see www.aip.org/history-programs/niels-bohr-library/oral-histories/32977-5.

Stephen Boughn, '*Electromagnetic radiation induced by a gravitational wave*', in the *Physical Review* D, Vol. 11, pp. 248–252 (1975).

'*Why Teach Physics?*', eds. Sanborn C. Brown, Norman Clarke, Jayme Tiomno. Published for the IUPAP by MIT Press, Cambridge, Mass./USA (2003); ISBN: 02-625-23760, 97-802-625-23769. Original edition 1964. See also [JT39].

'*The Birth of Particle Physics*', eds. Laurie Mark Brown and Lillian Hoddeson; based on a Fermilab Symposium on the *History of Particle Physics*, May 1980. Cambridge University Press, Cambridge/UK, New York/USA (1983); ISBN 0-521-24005-0.

'*From Pions to Quarks. Physics in the 1950's*', eds. Laurie Mark Brown, Max Dresden, and Lillian Hoddeson; based on a Fermilab Symposium, May 1985. Cambridge University Press, Cambridge/UK, New York/USA (1989); ISBN 0-521-30984-0.

'*Wrong for the Right Reasons*', eds. Jed Z. Buchwald and Allan Franklin, Springer *Archimedes* book series, Vol. 11, Springer Dordrecht (2005); ISBN 978-1-4020-3047-5.

'*Alberto Santoro, A Life of achievements*', eds. F. Caruso, E. Christoph, V. Oguri and R. Rubinstein, Editora AIAFEX, Rio de Janeiro (2011); ISBN 978-85-85806-03-3. This is a *Festschrift* for Santoro's 70th birthday and contains some information on his career and his collaborators.

Francisco Caruso, '*José Maria Filardo Bassalo, aos 80 anos*', in *Caderno Brasileiro de Ensino de Física*, Vol. 32, No. 2, pp. 297–298 (August 2015), and articles therein. This is an issue dedicated to the 80th birthday of José Maria Filardo Bassalo. Available online.

Idleu de Castro Moreira, '*Feynman e suas conferências sobre o ensino de física no Brasil*', in: *Revista Brasileira de Ensino de Física*, Vol. 40, No. 4, e4203 (2018); available at www.scielo.br/rbef. See also: Ildeu de Castro Moreira and Matheus Costa Paiva, '*Feynman e o ensino de ciências no Brasil*', in: *Física na Escola*, Vol. 14, No. 1 (2016).

Terry M. Christensen, '*John Archibald Wheeler: A Study of Mentoring in Modern Physics*' (2009). Available as a pdf at https://ir.library.oregonstate.edu/concern/graduate_thesis_or_dissertations/3r074x13x.

Ignazio Ciufolini, '*Wheeler, John Archibald. Gravitation and Inertia*', Princeton University Press, Princeton/NJ (1995); ISBN 0-691-03323-4.

David Cline and G. Riedasch, eds., '50 Years of Weak Interactions. Proceedings, Wingspread Conference', Racine/WI, USA, May 29th–June 6th, 1984.

M. Conversi, E. Pancini, and O. Piccioni, '*On the Disintegration of Negative Mesons*', in the *Physical Review*, Vol. 71, p. 209 (1947); The 'CPP Experiment'.

Joaquim da Costa Ribeiro, '*O Fenômeno Termo-Dielétrico*' (electric currents associated with changes of phase), *Universidade do Brasil*, Rio de Janeiro, 1945. This is Costa Ribeiro's thesis for the *concurso* (competition) for an open position there. See the file on Joaquim da Costa Ribeiro, Archives of the Museum of Astronomy and Related Sciences, Rio de Janeiro (MAST); [CR].

'*Algumas razões para ser um cientista*', eds. Carolina Cronemberger, Ricardo Magnus Osório Galvão, Ronald Cintra Shellard, Márcia Reis, and Mario Bag. CBPF, Rio de Janeiro, and ICTP, Trieste (2005). Based on a similar book published by the ICTP for its 40th anniversary in 2004. The book is available as a pdf at: http://estatico.cnpq.br/portal/premios/2018/pjc/assets/pdf/webaulas/web-01/algumas-razoes-para-ser-um-cientista.pdf.

'*ORIGENS da Fundação Getulio Vargas*', in: *FUNDAÇÃO Getulio Vargas: concretização de um ideal / Organizadora* Maria Celina D'Araújo. Rio de Janeiro, *Fundação Getulio Vargas* Publications (1999), 334 pp.

Max Dresden and Chen Ning Yang, '*Phase shift in a rotating neutron or optical interferometer*', in: the *Physical Review* D, Vol. 20, p. 1846 (15 October 1979).

Maria de Lourdes de Albuquerque Fávero, *Faculdade Nacional de Filosofia*, Vol.5—interview. Rio de Janeiro, *Universidade Federal do Rio de Janeiro*, 1992.

Maria de Lourdes de Albuquerque Fávero, '*A Faculdade Nacional de Filosofia: origens, construção e extinção*'. *Serie Estudos – Periódico do Mestrado em Educação da UCDB*, No. 16, July/Dec. 2003, pp.107–131.

T. Fazzini, G. Fidecaro, A.W. Merrison, H. Paul, and A.V. Tollestrup, '*Electron Decay of the Pion*', in: *Physical Review Letters*, Vol. 1, p. 247 (1958); see also Giuseppe Fidecaro, '*The Discoveries of Rare Pion Decays at the CERN Synchrocyclotron*', pdf from '*60 Years of CERN Experiments and Discoveries*', available at www.worldscientific.com.

(a) Enrico Fermi, '*Tentativo di una teoria dei raggi β*' in *Il Nuovo Cimento*, Vol. 11, No. 1, pp. 1–19 (1934); and (b) '*Versuch einer Theorie der Beta-Strahlen. I*', in *Zeitschrift für Physik*, Vol. 88, p. 161 (1934).

Marcio Luis Ferreira Nascimento, '*On the "Missing Letter" to Lattes and the Nobel Prize in Physics*', in the CBPF series *Ciência e Sociedade*, CBPF-CS-003/15, August 2015. Rio de Janeiro; ISSN 0101-9228.

Richard P. Feynman and Murray Gell-Mann, '*Theory of the Fermi Interaction*' in the *Physical Review*, Vol. 109, p. 193 (1958).

R.P. Feynman, '*The Problem of teaching Physics in Latin America*', in *Engineering and Science*, Vol. 27, p. 21 (1963).

R.P Feynman, R.B. Leighton, and Matthew Sands, '*The Feynman Lectures on Physics*', three volumes, Addison-Wesley, Boston (1964, 1966). New edition 2006; ISBN 0-8053-9045-6. Note that all three volumes can be read online (but not downloaded or copied) at the site http://www.feynmanlectures.caltech.edu/.

R.J. Finkelstein, '*My Century of Physics*', 2016. This is a memoir written for Finkelstein's centennial celebration at UCLA, where he had been professor of physics since 1948; it is available as a pdf at https://conferences.pa.ucla.edu/finkelstein-centennial/biography.pdf.

L.L. Foldy and S.A. Wouthuysen, in: the *Physical Review*, Vol. 78, p. 29 (1950). Introducing the 'Foldy-Wouthuysen transformation' (F-W).

Karin S.F. Fornazier and Antonio A.P. Videira, '*Os anos de formacão de um físico teórico brasileiro: Jayme Tiomno entre 1942 e 1950*', in *Ciência e Sociedade*, CBPF, Vol. 5, No. 1, pp. 1–12 (2018).

Fraser, Gordon, '*Cosmic Anger: Abdus Salam — The First Muslim Nobel Scientist*', Oxford University Press, Oxford/UK (2008); ISBN 978-0-19-920846-3.

For a discussion of Bohm's years in Brazil, as well has his scientific career, see Olival Freire, Jr., '*Science and Exile: David Bohm, the hot times of the Cold War, and his struggle for a new interpretation of quantum mechanics*' (2005); available at https://arxiv.org/pdf/physics/0508184. [This is a preprint of the paper published in *Historical Studies in the Physical and Biological Sciences*, Vol. 36, No. 1, September 2005; (pp. 1–34) https://doi.org/10.1525/hsps.2005.36.1].

Olival Freire Jr. and José Maria Filardo Bassalo, '*John Archibald Wheeler e a física brasileira*', in *Física na Escola*, Vol. 9, No. 1, pp. 44–47 (2008).

Olival Freire Jr., Antonio Augusto Passos Videira, and Aurino Ribeiro Filho, '*Ciência e política durante o regime militar (1964–1984): a percpção dos físicos brasileiros*', in *Boletim do Museu Paraense Emílio Goeldi. Ciências Humanas.* Belém, Vol. 4, No. 3, pp. 479–485, Sept.–Dec. 2009.

Cord Friebe, Meinard Kuhlmann, Holger Lyre, Paul M. Näger, Oliver Passon, and Manfred Stöckler, '*The Philosophy of Quantum Physics*', Springer Frontier Series, Springer Nature, Cham, Switzerland (2018); ISBN 978-3-391-78354-3.

J. I. Friedman and V. L. Telegdi, '*Nuclear Emulsion Evidence for Parity Non-conservation in the Decay Chain $\pi^+ \to \mu^+ \to e^+$*', in: the *Physical Review*, Vol. 105, p. 1681 (1957). See also the *Physical Review*, Vol. 106, p. 1290 (1957).

Elisa Frota-Pessôa and Neusa Margem, '*Sobre a desintegração do méson pesado positivo*', in: *Anais da Academia Brasileira de Ciências*, Vol. 22, pp. 371–383 (1950).

Elisa Frota-Pessoa, Interview by Ana Elisa Gerbasi da Silva and Lizete Castro Pereira Nunes, held in Rio de Janeiro for the *Projeto de Estudos de Educação e Sociedade* at UFRJ, on March 29th, 1990; supplemented by an interview in June, 2012 for *Cosmos & Contexto*. Available at: https://cosmosecontexto.org.br/elisa-frota-pessoa-suas-pesquisas-com-emulsoes-nucleares-e-a-fisica-no-brasil/.

Roberto Frota-Pessôa, '*Elisa Frota-Pessôa (1921–2018): à frente de seu tempo*'; cf. CBPF pages, at http://portal.cbpf.br/pt-br/ultimas-noticias/filho-de-pesquisadora-emerita-escreve-sobre-a-vida-e-a-obra-da-mae.

Ludmila Gama Pereira, Masters candidate at the UFF, Níterói: '*A construção do saber histórico e projeto social: Os historiadores da UFRJ na época da Ditadura Militar no Brasil (1964–1985)*'. (pdf available at: http://encontro2008.rj.anpuh.org/resources/content/anais/1212962424_ARQUIVO_ludmila_Gama_Pereira-_anpuh_rio.pdf).

E. Gardner and C.M.G. Lattes: '*Production of mesons by the 184-inch Berkeley cyclotron*', in *Science*, Vol. 107, pp. 270–271 (1948).

R. L. Garwin, L. M. Lederman, and M. Weinrich, '*Observations of the Failure of Conservation of Parity and Charge Conjugation in Meson Decays: the Magnetic Moment of the Free Muon*', in the *Physical Review*, Vol. 105, p. 1415 (1957).

Riccardo Giacconi, '*Report of the ICRANet Scientific Committee Meeting*', December, 2012. Online at https://www.icranet.org/documents/signedSCMeeting2012.pdf.

Myriam Segre de Giambiagi, '*O CBPF que eu conheci*', CBPF Series *Ciência e Sociedade*, (Document CBPF-CS-008/01), 2001.

Louisa Gilder, 'The Age of Entanglement. When Quantum Physics was Reborn', Alfred A. Knopf, New York (2008); ISBN: 978-1-4000-4417-7.

Sheldon Glashow, '*The renormalizability of vector meson interactions*', in *Nuclear Physics*, Vol. 10, p. 107 (1959).

James Gleick, '*Genius: The Life and Science of Richard Feynman*', Vintage Books, New York (1992); ISBN13: 978-0-67-974704-8.

Kurt Gödel, '*An example of a new type of cosmological solutions of Einstein's field equations of gravitation*', in: *Reviews of Modern Physics*, Vol. 21, p. 447 (1949).

José Luiz Goldfarb, '*Voar também e com os homens: o pensamento de Mário Schenberg*', São Paulo: *Editora da Universidade de São Paulo*, 1994.

Claudia Graf-Grossmann, '*Marcel Grossmann—For the Love of Mathematics*', Springer Biographies, Cham/CH (2018); ISBN 978-3-319-90076-6.

Daniel M. Greenberger and A.W. Overhauser, '*Coherence effects in neutron diffraction and gravity experiments*', in: *Reviews of Modern Physics*, Vol. 51, p. 43 (1 January 1979).

John Gribbin and Mary Gribbin, '*Richard Feynman: A Life in Science*', Penguin Publishing Group, New York (1997); ISBN13: 978-0-52-594124-8.

Amit Hagar, '*Squaring the Circle: Gleb Wataghin and the Prehistory of Quantum Gravity*', in *Studies in the History and the Philosophy of Modern Physics*, Vol 46, No. 2, pp. 217–227 (2014).

Amélia I. Hamburger, '*Um Experimental no Mundo das Interações*', in *Jornal da Unicamp*, Vol. 281, p. 14 (March/April 2005). Available online.

Amélia I. Hamburger, '*César Lattes, físico brasileiro*', Revista USP, Vol. 66, pp.132–38 (2005).

Stephanie Hanel, '*Lise Meitner—Fame without a Nobel Prize*', on the website of the Lindau Nobel Laureate Meetings (5.11.2015), at https://www.lindau-nobel.org/lise-meitner-fame-without-a-nobel-prize/.

Magdolna Hargittal, '*Credit where credit's due?*', in: *Physics World*, Sept. 13th, 2012. Online at https://physicsworld.com/a/credit-where-credits-due/. Discussion of the Nobel prize question for C.S. Wu.

Franz Hasselbach and Marc Nicklaus, '*Sagnac experiment with electrons: Observation of the rotational phase shift of electron waves in vacuum*', in: the *Physical Review* A, Vol. 48, No. 1, pp. 143–151 (July 1993).

Leopold Infeld and Alfred Schild, '*On the Motion of Test Particles in General Relativity*', in *Reviews of Modern Physics*, Vol. 21, p. 408 (1949).

S. Kamefuchi, L. O'Raifeartaigh, and Abdus Salam, '*Change of variables and equivalence theorems in quantum field theories*', in *Nuclear Physics*, Vol. 28, No. 4, pp. 529–549 (1961).

Roy P. Kerr, '*Gravitational Field of a Spinning Mass as an Example of Algebraically Special Metrics*', in: *Physical Review Letters*11, 237 (1963).

Nicola N. Khuri, in: the *Physical Review*, Vol. 107, p. 1148 (1957).

Kinoshita, Dina Lida: '*Mario Schenberg: o cientista e o político*'; *Fundação Astrogildo Pereira*, Brasília/DF (2014); ISBN 978-8-58-921647-0.

John R. Klauder: '*Magic without magic. John Archibald Wheeler, a collection of essays in honor of his sixtieth birthday*'. Freeman, San Francisco (1972); ISBN 978-0-71-670337-2.

Oskar Klein, '*Mesons and Nucleons*', in *Nature*, Vol. 161, pp. 897–899 (1948).

'*50 Anos da SBF, 1966–2016*', eds. Marcelo Knobel and Cássio Leite Viera. Published by the *Socidade Brasileiro de Física* (SBF). Online at http://www.sbfisica.org.br/arquivos/SBF-50-anos.pdf. See p. 20.

C.M.G. Lattes, H. Muirhead, G.P.S. Occhialini, and C.F. Powell: '*Processes involving charged mesons*', in: *Nature*, Vol. 159, pp. 694–697 (1947).

C.M.G. Lattes, G.P.S. Occhialini, and C.F. Powell: '*A determination of the ratio of the masses of pi-meson and mu-meson by the method of grain-counting*', in: *Proceedings of the Physical Society*, Vol. 61, No. 2, pp. 173–183 (1948).

Interview with César Lattes in 1995, quoted by Ana Maria Ribeiro de Andrade in '*Físicos, mésons e política: a dinâmica da ciência na sociedade*', São Paulo, HUCITEC (1998); archived at MAST, Rio de Janeiro.

César Lattes, Interview by Micheline Nussenzveig, Cássio Leite Vieria, and Fernando de Souza Barros (August 1995). See *Canal Ciência, Portal de Divulgacão Científica e Technológica*, IBICT (published 2011).

Max von Laue, '*Zur Dynamik der Relativitätstheorie*', in: *Annalen der Physik*, Vol. 340, No. 8, pp. 524–542 (1911). Online at http://gallica.bnf.fr/ark:/12148/bpt6k15338w/f535.item. An English translation is available at https://www.scribd.com/document/131731005/Max-Von-Laue-1911-Artigo-On-the-Dynamics-of-the-Theory-of-Relativity-Wikisource-The-Free-Online-Library.

G. F. Leal Ferreira, '*Há 50 Anos: O Efeito Costa Ribeiro*', in *Revista Brasileira de Ensino de Física*, Vol. 22, No. 3, September 2000; pp. 434–443.

T.D. Lee, M. Rosenbluth, and C.N. Yang, '*Interactions of Mesons with Nucleons and Light Particles*', in: the *Physical Review*, Vol. 75, p. 905 (1949).

T.D. Lee and C.N. Yang, '*Question of Parity Conservation in Weak Interactions*', in: the *Physical Review*, Vol.104, pp. 254–258 (1956).

Ralph Leighton and Richard P. Feynman, '*Surely you're joking, Mr. Feynman*', W.W. Norton, New York (1985); ISBN 0-393-01921-7.

Ralph Leighton and Richard P. Feynman, '*What do you care what other people think*', W.W. Norton, New York (1988); ISBN 0-393-02659-0.

J. Leite Lopes and J. Tiomno, '*O ensino de fisica no curso secundário*' [*Teaching high-school physics*] in: *Ciência e Cultura*, Vol. 5, p. 45 (1953). See also [JT17a].]

José Leite Lopes: Interview by Tjerk Franken, Strassbourg, France (July, 1977). Available at CPDOC, *Fundação Getúlio Vargas* (published 2010).

José Leite Lopes: '*Richard Feynman in Brazil: Recollections*' (in English), in: CBPF: *Ciência e Sociedade*, ISSN 0101-9228, Document CBPF-CS-013/88, available online.

José Leite Lopes, '*Weak Interaction Physics: From its Origin to the Electroweak Model*', *Notas de Física*, CBPF (April 1996), available as a pdf at https://inis.iaea.org/collection/NCLCollectionStore/_Public/27/073/27073436.pdf.

'*Uma História da Física no Brasil*', Livraria da Física, São Paulo (2004); ISBN 978-858-832519-7.

Maria Lucia de Camargo Linhares and Henrique César da Silva, '*Cientista e Mulher: Entre Discursos e Representações de Elisa Frota-Pessôa*', in: *Proceedings of the 13th Womens Worlds Congress*, Florianopolis, BR (2017); ISSN: 2179-510X. Available online in the Electronic Proceedings.

Maria Lucia de Camargo Linhares, '*Elisa Frota-Pessoa : a textualização de suas (auto)representações e questões de gênero nas ciências*', MS Dissertation, Universidade de Santa Catarina, Florianópolis, 2018. 170 pp.

'*Frontier Physics —Essays in Honour of Jayme Tiomno*', eds. S.W. MacDowell, H.M. Nussenzveig, and R.A. Salmeron, World Scientific, Singapore (1991); ISBN 978-98-145-3907-4. See the summary and contents at https://www.worldscientific.com/worldscibooks/10.1142/1371.

Erwin Madelung, '*Quantentheorie in hydrodynamischer Form*', in *Zeitschrift für Physik*, Vol. 40, No. 3–4, pp. 322–326 (March 1927).

H.M. Mahmoud and E.J. Konopinski, '*The Evidence of the once-forbidden Spectra for the Law of β decay*', in: the *Physical Review*, Vol. 88, p. 1266 (1952).

'*Discovery of Weak Neutral Currents*: '*The weak interaction before and after*, 1993*', AIP Conference Proceedings, Vol. 300 (1994), Eds. Alfred K. Mann and David B. Cline, Santa Monica/CA-Woodbury/NY.

Eduardo C. Marino, '*Jorge André Swieca: Uma figura ímpar na física brasileira*', in *Revista Brasileira de Ensino de Física*, Vol. 37, No. 3 (2015). Available online.

Alfredo Marques, '*CBPF: da Descoberta do Méson pi aos dez primeiros anos*', in CBPF: *Ciência e Sociedade* (Document CBPF—CS–031/97); unpublished. Available online.

Alfredo Marques, '*Em Memória de César Lattes*' (An obituary for Lattes written shortly after his death in March, 2005 and published by the CBPF: *Ciência e Sociedade* (Document CBPF-CS-004/05; and in abbreviated form in *Ciência Hoje*). Available online.

Alfredo Marques, '*Neusa Amato dos Vinte aos Oitenta*' ('Neusa Amato from the Twenties to the Eighties'), in CBPF: *Ciência e Sociedade* (Document CBPF - CS - 0001/07). Available online.

Alfredo Marques, '*Jayme Tiomno*', in CBPF: *Ciência e Sociedade* (Document CBPF-CS-003/11), (January 2011). An obituary for Tiomno, available online.

Alfredo Marques, '*César Lattes * 1924–† 2005*', in *Ciência e Sociedade* (Document CBPF-CS-001/13). Available online.

Robert E. Marshak, 'The Pain and Joy of a Major Scientific Discovery', in Zeitschrift für Naturforschung, Vol. 52a, pp. 3–8 (1997). This is a talk given at a banquet in September, 1997, to celebrate the 60th birthday of his former student, George Sudarshan. See also the essay by E.C.G. Sudarshan and R.E. Marshak in [Mac-Dowell et al. 1991].

Bahram Mashhoon, 'Neutron interferometry in a rotating frame of reference', in: Physical Review Letters, Vol. 61, p. 2639 (1988). The 'Mashhoon effect'.

Luisa Massarani, Nara Azevedo, 'Carlos Chagas Filho: o "cientista-elétrico",' Rio de Janeiro: Museu da Vida/Casa de Oswaldo Cruz/Fiocruz, 2011.

Jagdish Mehra, 'The Beat of a Different Drum', Oxford University Press, New York (1994); ISBN 0-19-853948-7.

Charles W. Misner, Kip S. Thorne, and John Archibald Wheeler, 'Gravitation', W.H. Freeman and Co.: San Francisco (1973); ISBN 0-7167-0334-3.

Charles W. Misner, Kip S. Thorne and Wojciech H. Zurek, 'John Wheeler, relativity, and quantum information' in Physics Today, April 2009. See https://www.its.caltech.edu/~kip/PubScans/VI-50.pdf.

Charles W. Misner, pp. 9-28 in 'General Relativity and John Archibald Wheeler', eds. Ignazio Ciufolini and Richard A. Matzner, Springer: Dordrecht, Heidelberg, London, New York (2010); ISBN: 978-90-481-3734-3. Available online from springer.com.

Marieta de Moraes Ferreira, 'Ditadura militar, universidade e ensino de história: da Universidade do Brasil à UFRJ', in: Ciência e Cultura, Vol. 66, No. 4, São Paulo (Oct./Dec., 2014). Available as a pdf at: http://cienciaecultura.bvs.br/scielo.php?script=sci_arttext&pid=S0009-67252014000400012.

Rodrigo Patto Sá Motta, 'As universidades e o regime militar', Zahar, Rio de Janeiro (2014).

Jeffrey S. Nico, 'Neutron Beta Decay', in: Journal of Physics G: Nuclear & Particle Physics, Vol. 36, Article 104001 (36 pp.) (2009).

Bruno Nobre and Antonío Augusto Passos Videira, '"Mas seja tudo pelo bem da física" ["May it all be for the sake of physics"]: notes on the scientific trajectory of Francisco Xavier Roser, SJ (1904–1967)', in: Revista Brasileira de Ensino de Física, Vol. 40, No. 2 (2018); available online.

Salvador Nogueira, article in the 'Science' section of the Folha de São Paulo, issue of Monday, March 9th, 2005, entitled 'CBPF publica tese"inédita" do cientista' ['CBPF publishes the "unpublished" thesis of a scientist']. Available online.

Proceedings MG-X: 'The Tenth Marcel Grossmann Meeting on Recent Developments in Theoretical and Experimental General Relativity, Gravitation and Relativistic Field Theories', Eds. Mario Novello, Santiago Perez Bergliaffa, and Remo Ruffini. World Scientific, Singapore (2006); ISBN 978-981-256-667-6. See the summary and contents at https://www.worldscientific.com/worldscibooks/10.1142/6033.

Mario Novello, 'Os cientistas da minha formação', Editora Livraria da Física, São Paulo (2016); ISBN 978-857-861-391-4.

Moysés Nussenzveig, '*Algumas reminiscências pessoais sobre a ótica no Brasil*', in: *Memória da Física*, org. Heloisa Maria Bertol Domingues, MAST, Rio de Janeiro, pp.83–100, 2005.

J.R. Oppenheimer and G.M. Volkoff, '*On Massive Neutron Cores*', in: the *Physical Review*, Vol. 55, pp. 374–381 (January 1939).

J.R. Oppenheimer and H. Snyder, '*On Continued Gravitational Contraction*', in: the *Physical Review*, Vol. 56, pp. 455–459 (September 1939).

Tadashi Ouchi, Kei Senba, and Minoru Yonezawa, '*Theory of Mass Reversal in the Quantized Field Theory*', in *Progress of Theoretical Physics*, Vol. 15, No. 5, pp. 431–444 (1956).

Antonio Paim, '*A UDF e a Ideia de Universidade*', *Biblioteca Tempo Brasileiro*, Rio de Janeiro (1981).

Abraham Pais, interview conducted by Lillian Hoddson for AIP, August 27th, 1974. Online at htpps://www.aip.org/history-programs/niels-bohr-library/oral-histories/5047.

F. David Peat, '*Infinite Potential: The Life and Times of David Bohm*', Addison Wesley, Reading/MA (1997); ISBN 0-201-32820-8.

José N. Pecina-Cruz, '*Time Reversal Induces Negative Mass and Charge Conjugation*' (2005). Preprint available at: https://arxiv.org/pdf/hep-ph/0505188.

Daniel Piza, '*Jayme Tiomno, o físico brasileiro que viu o Prêmio Nobel passar*', in: *O Estado de São Paulo*, 19/11/2006, Vida&, p. A30. Online at: http://www2.senado.leg.br/bdsf/handle/id/323455.

Daniel Piza, '*Uma lagrima e um escandalo*', opinion/blog published in the *Estadão*, January 18th, 2011. Online at: https://www.estadao.com.br/blogs/daniel-piza/uma-lagrima-e-um-escandalo/. See also the blog http://profs.if.uff.br/tjpp/blog/entradas/jayme-tiomno-e-o-silencio-da-midia.

Bruno Pontecorvo, '*Nuclear capture of mesons and the meson decay*', in: the *Physical Review*, Vol. 72, p. 246 (1947). [Manuscript received 21st June, 1947].

Hans Postma, W.J. Huiskamp, A.R. Miedema, M.J. Steenland, H.A. Tolhoek and C.J. Gorter, '*Asymmetry of the positon emission by polarized ^{58} Co nuclei*', in: *Physica*, Vol. 23, Issue 1, pp. 259–260 (Jan. 1957).

Enrico Predazzi, '*In memory of Jayme Tiomno*', in: *Norte Ciência*, Vol. 2, No. 1, pp. 104–107 (2011).Online at: http://aparaciencias.org/vol-2.1/10_In_memory_of_Jayme_Tiomno.pdf.

Giampietro Puppi, '*Sui mesoni dei raggi cosmici*', in: *Il Nuovo Cimento*, Vol. 5, No. 6, pp. 587–588 (1948).

G. Rajasekaran, '*Fermi and the Theory of Weak Interactions*'; preprint: arXiv:1403.3309v1 [physics.hist-ph] (February 25th, 2014), published in: *Resonance*, Vol. 19, Issue 1, pp. 18–44 (January 2014).

'*Black Holes, Gravitational Waves, and Cosmology: An Introduction to Current Research*', eds. Martin Rees, Remo Ruffini and John A. Wheeler. Gordon and Breach, New York (1974); ISBN 978-06-770-4580-1.

Darcy Ribeiro, '*UnB - Invenção e Descaminho*', Editora Avenir, Rio de Janeiro (1978).

Wolfgang Rindler, '*Gödel, Einstein, Mach, Gamow and Lanczos: Gödel's remarkable excursion into cosmology*', in: *American Journal of Physics*, Vol. 77, p. 498 (2009). Online at https://doi.org/10.1119/1.3086933.

João Augusto de Lima Rocha, '*Desmontada a versão da ditadura de 1964 sobre a morte de Anísio Teixeira*'. *Jornal GGN*, April 22nd, 2018. Online at: https://jornalggn.com.br/ditadura/desmontada-a-versao-da-ditadura-de-1964-sobre-a-morte-de-anisio-teixeira-por-joao-augusto-de-lima-rocha/.

G.D. Rochester and C.C. Butler, '*Evidence for the existence of new unstable elementary particles*', in *Nature*, Vol. 160, pp. 855–857 (1947).

Juan G. Roederer, '*The Constant yet Ever-Changing Abdus Salam International Centre for Theoretical Physics*', in *Physics Today*, Vol. 54, No.9, p.31 (2001); cf. https://physicstoday.scitation.org/doi/full/10.1063/1.1420509.

M. Rudermann and R.J. Finkelstein, '*Note on the Decay of the π-Meson*', in: the *Physical Review*, Vol. 76, p. 1458 (1949).

Remo Ruffini and John A. Wheeler, '*Introducing the black hole*', in *Physics Today* (January 1971), pp. 30–41.

M.G. Sagnac, in: *Comptes Rendus*, Vol. 157, p. 208 (1913), and references therein. The 'Sagnac effect'.

Jun John Sakurai, '*Mass Reversal And Weak Interactions*', in: *Il Nuovo Cimento*, Vol. 7, pp. 649–660 (1958).

Jun John Sakurai, '*Invariance principles and Elementary Particles*', Princeton University Press, Princeton NJ (2015); ISBN 978-1-4008-7787-4.

Abdus Salam and J.C. Ward, '*Weak and electromagnetic interactions*', in *Il Nuovo Cimento*, Vol. 11, No. 4, pp. 568–577 (1959).

Roberto A. Salmeron, '*Depoimento do Professor Roberto Aureliano Salmeron na Commisão Parla-mentar de Inquerito sobre a Universidade de Brasília*', Brasília (Nov. 10th, 1965).

Roberto A. Salmeron, '*A universidade interrompida: Brasília 1964–1965*', Editora UnB,
Brasília/DF (2012); ISBN 978-85-230-0966-3. Earlier editions 1999 and 2007.

'*Entrevista com o Professor Roberto Aureliano Salmeron*', in *Passages de Paris*, No. 2, 2005; ISSN 773-0341. This was a magazine published by and for the Brazilian expat community in France. Available online at http://www.apebfr.org/passagesdeparis/edition2/p6-salmeron.pdf.

Roberto A. Salmeron, '*Origens da Universidade de Brasília*', lecture at UnB in 2005, reprinted in the *e-bulletin of physics* of the Institute of Physics, UnB (November 2013).

Interview with Mario Schenberg, by Oliveiros S. Ferreira, Frederico Branco, and Lourenço Dantas Mota (1978); published in the newspaper '*O Estado de São Paulo*' on December 10th, 1978. Available online at: http://www.sbfisica.org.br/rbef/pdf/vol01a18.pdf. See also the interview for '*o Globo*' on 12.08.1968, in the Schenberg archives, USP; [MS].

Mário Schenberg (interview, 1978). Rio de Janeiro, CPDOC, 2010; 93pp.

Winston Gomes Schmiedecke, '*Costa Ribeiro: Ensino, pesquisa e desenvolvimento da física no Brasil*', in: *Revista Brasileira de Ensino de Física*, Vol. 37, No. 1 (2015).

Bert Schroer, '*Jorge A. Swieca's contributions to quantum field theory in the 60's and 70's and their relevance in present research*', arXiv:0712.0371v3 [physics.hist-ph], (May 2010). Published in *The European Physical Journal* H, Vol. 35, No. 1, pp. 53–88 (2010). Available online.

Wanderley Vítorino da Silva Filho, '*W.V. Costa Ribeiro: Ensino, pesquisa e desenvolvimento da física no Brasil*', Campina Grande: EDUEPB, Livraria da Física, São Paulo (2013).

Carlos Alberto da Silva Lima, Sérgio Joffily, and Roberto A. Salmeron, '*Homenagem à Professora Elisa Frota-Pessôa*', in the *Brazilian Journal of Physics*, Vol. 34, No. 4A (December 2004).

Ruth Lewin Sime, '*Lise Meitner—A Life in Physics*', University of California Press, Berkeley and Los Angeles (1996); ISBN-13: 978-0520208605.

Ruth Lewin Sime, '*Marietta Blau in the history of cosmic rays*', in *Physics Today*, Vol. 65, Issue 10, p. 8 (2012).

E.C.G. Sudarshan and R.E. Marshak, '*Chirality Invariance and the Universal Fermi Interaction*' in the *Physical Review*, Vol. 109, p. 1860 (1958). See also E.C.G. Sudarshan and R.E. Marshak, *Proc. of Padua-Venice Conference on 'Mesons and Newly Discovered Particles*', 1957, p. V-14.

J.A. Swieca and A.L.L. Videira, '*O Tempo de Plínio*', 2013, in CBPF: *Ciência e Sociedade* (Document CBPF-CS-002/13) (2013) (original manuscript: November 1972). Available online.

Odilon A.P. Tavares, '*CBPF, 60 anos de Física Nuclear*', in: CBPF Series *Ciência e Sociedade* (Document CBPF-CS-002/09), July 2009.

Heráclio Duarte Tavares, '*O Centro Brasileiro de Pesquisas Físicase o Instituto de Física Teórica sob a ótica militar*', in: *Contemporanea: Historia y problemas del siglo XX*, Año 6, Vol. 6, 2015, pp.67–82; ISSN: 1688-7638.

Jayme Tiomno, Lecture on receiving the *Moinho Santista* Prize. Published in *Ciência e Cultura*, Vol. 9, No. 4 (Dec. 1957); cf. also [JT75].

Jayme Tiomno, '*Memorial Relativo às Atividades Profissionais de Jayme Tiomno*'; Tiomno's resumé submitted to the FFCL/USP, in 1966 as part of a *concurso* (competition) for an open position there. Archived in [JT].

'*Contribuições à Física das Partículas Elementares*', ('Contributions to the Physics of Elementary Particles'): Thesis submitted to the Competition for the Chair of Advanced Physics at the FFCL/USP, São Paulo, July 1966; [JT]. See also [JT41].

Jayme Tiomno, Interview by Carla Costa, Márcia Bandeira de Mello Leite Arriela, and Ricardo Guedes Pinto, in Rio de Janeiro, 1977. Available at CPDOC, *Fundação Getúlio Vargas* (published 2010).

Jayme Tiomno, '*The early period of the universal Fermi application*', in AIP Conference Proceedings, Vol 300, p. 99 (1994) ([Mann & Cline 1994]); preprint available from the CBPF, *Notas de Física*. Originally published in [Cline & Riedasch 1984]; cf. also [JT68 and 102].

Jayme Tiomno, list of his own 'Papers of Major Importance', supporting his nomination for the TWAS Physics Prize, 1995; [JT].

Jayme Tiomno, '*A física de partículas: o início dos trabalhos de física de partículas no Brasil*' [The Physics of Particles: Beginnings of work on particle physics in Brazil], in: MAST Colloquia – *Memória da Física*, Vol. II, MAST, Rio de Janeiro, 2005. This is Tiomno's last autobiographical memoir, recounting his role in particle-physics research in Brazil in the early period, 1946–1961; he contributed it to a scientific–historical series at MAST, complementing his personal archive [JT] there. Also listed in Appendix B as [JT107].

Silvia Tiomno Tolmasquim, '*Histórias de invernos e verões*', *Verve*, Rio de Janeiro (2014); ISBN 978-85-66031-77-5.

Bruno Touschek, '*Excitation of Nuclei by Electrons*', in: *Nature*, Vol. 160, p. 500 (Oct. 1947).

José Carlos Valladão de Mattos, '*Memórias, Sonhos e Outros Incômodos*', Biblioteca24horas, São Paulo, 2015; ISBN 978-85-4160-905-0.

Antonio Luciano Leite Videira, '*Quarenta anos de Física de Jayme Tiomno*', Talk given at the session in honor of Tiomno's 60th birthday at the *II. Encontro Nacional de Partículas e Teoria de Campos*, Cambuquira, MG (Sept. 1980). The manuscript of this talk is available as a pdf from the CBPF (Document CBPF-CS-001/85, ISSN 0-101-9338).

Antonio Luciano Leite Videira, '*Da Relatividade às partículas (ida e volta): quarenta anos de física de Jayme Tiomno*', Rio de Janeiro: Centro Brasileiro de Pesquisas Físicas, Doc. CBPF-CS001/85, 1985.

Antonio Luciano Leite Videira, '*Leibniz e a Cadeira de Cinema*' (1997), in CBPF: *Ciência e Sociedade* (Document CBPF-CS-018/97), 1997. See also '*Plínio Süssekind Rocha: um mestre"excentrico"*' (2017) (https://doi.org/10.12957/emconstrucao.2017.31842), by the same author. Both available online.

Antonio Augusto Passos Videira, '*Um vienense nos trópicos: A vida e a obra de Guido Beck entre 1943 e 1988*', in: Videira, Antonio Augusto Passos, Bibiloni, Anibal C. (orgs.), '*Encontro de história da ciência: análises comparativas das relações científica no Século XX entre os países do Mercosul no campo da física*', Rio de Janeiro: CBPF, 2001; pp.146-181.

Cássio Leite Vieira, '*Dia triste para a História da Ciência*', in *Ciência Hoje* (January 2011).

Cássio Leite Vieira and Antônio Augusto Passos Videira, '*Carried by History: César Lattes, Nuclear Emulsions, and the Discovery of the Pi-meson*', in: *Physics in Perspective* (ISSN 1422-6944), Vol. 16, No. 1 (2014). Available online.

Gleb Wataghin, Interview by Cylon Eudxio Silva, held in Rio de Janeiro, 1975. Available at CPDOC, *Fundação Getúlio Vargas* (published 2010).

Steven Weinberg, '*A Model of Leptons*', in: *Physical Review Letters*, Vol. 19, pp. 1264–66 (1967).

S.A. Werner, J.L. Staudenmann, and R. Colella, '*Effect of Earth's Rotation on the Quantum Mechanical Phase of the Neutron*', in: *Physical Review Letters*, Vol. 42, p. 1103 (23 April 1979).

John A. Wheeler's autobiography (with Kenneth Ford): '*Geons, Black Holes, and Quantum Foam*', W.W. Norton and Co., Inc., New York (1998); ISBN 0-393-31991-1.

Eugene Paul Wigner and Andrew Szanton: '*The Recollections of Eugene P. Wigner*'. Springer, Heldelberg, New York (1992); ISBN 978-0-306-44326-8.

C.S. Wu, E. Ambler, R.W. Hayward, D.D. Hoppes, and R.P. Hudson, '*Experimental Test of Parity Conservation in Beta Decay*', in: the *Physical Review*, Vol. 105, pp. 1413–1415 (1957).

C.S. Wu, '*The Universal Fermi Interaction and the Conserved Vector Current in Beta Decay*', in: *Reviews of Modern Physics*, Vol. 36, p. 618 (1964). This is a rather complete review of the various approaches to the Universal V–A theory, written by a pioneer in the field.

Chen Ning Yang, '*Selected Papers, 1945-1950, with commentary*', World Scientific Publishing Co., Singapore (2005); ISBN 981-256-367-9.

Yi-Bo Yang, Jian Liang, Yu-Jiang Bi, Ying Chen, Terrence Draper, Keh-Fei Liu and Zhaofeng Liu, '*Proton Mass Decomposition from the QCD Energy Momentum Tensor*', in: *Physical Review Letters*, Vol. 121, Paper 212001 (November 2018).

Hideki Yukawa, *Models and Methods in the Meson Theory, Reviews of Modern Physics*, Vol. 21, No. 3, pp. 474–479 (July 1949).

'*Subnuclear Phenomena*', Part A, edited by A. Zichichi, Academic Press, New York (1970); ISBN 978-0-12-780580-1.

About the Authors

William Dean Brewer was born in Boise, ID/USA in 1943. After a BA in Chemistry/Mathematics at the University of Oregon, he obtained his Ph.D. in Physical Chemistry ('Weak-Interactions Studies by Nuclear Orientation') at UC Berkeley, working in the group of David A. Shirley. Following post-doctoral work at the *Freie Universität Berlin*/Germany and the *Laboratoire de Physique des Solides* in Orsay/France, he completed his *Habilitation* at the FU Berlin in 1975 and was University Professor there from 1977 until retirement in 2008. He has worked on nuclear and solid-state physics via low-temperature nuclear orientation, magnetism on surfaces and in thin films, valence determinations and magnetism by XAS and XMCD with synchrotron radiation. Visiting Scholar at Stanford and IBM Yorktown Heights/USA, KUL Leuven/Belgium, USP and CBPF/ Brazil. He has translated over 20 scientific textbooks and monographs (German to English), and translates for the Einstein Papers Project (Pasadena/Jerusalem/Princeton).

© Springer Nature Switzerland AG 2020
W. D. Brewer and A. T. Tolmasquim, *Jayme Tiomno*, Springer Biographies,
https://doi.org/10.1007/978-3-030-41011-7

Alfredo Tiomno Tolmasquim born in Rio de Janeiro in 1960, studied Chemical Engineering, but redirected his field of interest to the History of Science and Science Communication. He took his Ph.D. in Communication and Culture at UFRJ in Brazil, and did post-doctoral work at the Edelstein Center for the History and Philosophy of Science, Technology and Medicine of the Hebrew University, Jerusalem. He was Visiting Scholar at the Institute for Advanced Studies of the Hebrew University and at the Max Planck Institute for the History of Science in Berlin. He was Researcher at the *Museum of Astronomy and Related Sciences* (MAST) in Rio de Janeiro and served as its Director from 2003 to 2011. He was also Head of Research and Teaching at the *Brazilian Institute of Information on Science and Technology*, and is currently Director of Scientific Development at the *Museum of Tomorrow*, also in Rio de Janeiro. Among his publications, his book *Einstein: the traveller of Relativity in South America*, published by Vieira & Lent Eds., occupies a distingished position. He is the nephew of Jayme Tiomno.

Index

Printed in the United States
by Baker & Taylor Publisher Services